Springer-Verlag
Science Press

Hua Loo Keng Wang Yuan

# Applications of Number Theory to Numerical Analysis

Springer-Verlag
Berlin Heidelberg New York
Science Press, Beijing
1981

Hua Loo Keng Wang Yuan
Institute of Mathematics, Academia Sinica
Beijing
The People's Republic of China

*Revised edition of the original Chinese edition published by Science Press Beijing 1978 as the first volume in the Academia Sinica's Series in Pure and Applied Mathematics.*

Distribution rights throughout the world, excluding The People's Republic of China, granted to Springer-Verlag Berlin Heidelberg New York

AMS Subject Classification (1980): 10-XX, 12Axx, 65-XX

ISBN-13: 978-3-642-67831-8 e-ISBN-13: 978-3-642-67829-5
DOI: 10.1007/978-3-642-67829-5

Library of Congress Cataloging in Publication Data. Hua, Loo Keng, 1911–. Applications of number theory to numerical analysis. Bibliography: p. 1. Numerical analysis. 2. Numbers, Theory of. I. Wang Yuan, joint author. II. Title. QA297.H83. 511. 80-22434


Typesetting: Science Press, Beijing, The People's Republic of China

Softcover reprint of the hardcover 1st edition 1981

2141/3140-543210

# *Preface*

Owing to the developments and applications of computer science, mathematicians began to take a serious interest in the applications of number theory to numerical analysis about twenty years ago. The progress achieved has been both important practically as well as satisfactory from the theoretical view point. For example, from the seventeenth century till now, a great deal of effort was made in developing methods for approximating single integrals and there were only a few works on multiple quadrature until the 1950's. But in the past twenty years, a number of new methods have been devised of which the number theoretic method is an effective one.

The number theoretic method may be described as follows. We use number theory to construct a sequence of uniformly distributed sets in the $s$-dimensional unit cube $G_s$, where $s \geqslant 2$. Then we use the sequence to reduce a difficult analytic problem to an arithmetic problem which may be calculated by computer. For example, we may use the arithmetic mean of the values of integrand in a given uniformly distributed set of $G_s$ to approximate the definite integral over $G_s$ such that the principal order of the error term is shown to be of the best possible kind, if the integrand satisfies certain conditions. It is worth mentioning that the principal order of the error term of the Cartesian product formula for a classical single quadrature formula depends on and increases very rapidly with the dimension $s$. And though the error term in the Monte Carlo method is independent of $s$, it is in the sense of probability there, not in the usual sense of error. The number theoretic method may also be used to construct an approximate polynomial for the periodic function of $s$ variables and to treat the problems of the approximate solutions to integral equations and partial differential equations of certain types.

Many important methods and results in number theory, especially those concerning the estimation of trigonometrical sums and simultaneous Diophantine approximations as well as those of classical algebraic number theory, may be used to construct the uniformly distributed sequence in $G_s$. The fundamental concepts in the number theoretic method were advanced in 1957—1962. N. M. Korobov (1957) introduced the $p$ set with the aid of the estimation of a complete exponential sum. Using the Sun Zi theorem (Chinese remainder theorem), J. H. Halton (1960) generalized the J. G. Van der Corput sequence. N. S. Bahvalov (1959) and C. B. Haselgrove (1962) introduced independently the *gp* (good point) set and N. M. Korobov (1959) and E. Hlawka (1962) proposed independently the *glp* (good lattice point) set. It was suggested by us in 1960 to define the uniformly distributed sets in $G_s$ by means of a set of independent units of the cyclotomic field by which an effective algorithm for obtaining a sequence of sets of rational numbers with the same denominators that approximate a basis of the field simultaneously was obtained, where the principal order of the error term is of the best possible kind. Perhaps, it is worth mentioning that the classical methods of best simultaneous rational approximations are ineffective from the view point of numerical analysis. In 1974, we proposed also a method for defining the uniformly distributed sequence by the recurrence formula defined by a PV (Pisot-Vijayaraghavan) number. In this book, we first illuminate these methods and give the estimates of the discrepancies of the sets so defined. Then we shall give various applications of them to numerical analysis and a table of *glp* sets as an appendix.

Aside from a knowledge of elementary number theory (see Hua Loo Keng [2]), we shall need several deeper theorems in number theory for which the references are given. Concerning the more extended methods and problems in the theory of uniform distribution and the theory of multiple quadrature, we refer the reader to monographs of S. Haber [1], Hsu Li Zhi and Zhou Yun Shi [1], L. Kuiper and H. Niederreiter [1], H. Niederreiter [5] and A. H. Stroud [1].

It is with great pleasure and gratitude that we acknowledge conversation and correspondence with professors Feng Kang, He Zuo Xiu, Hsu Li Zhi, Wang Guang Yin and Xu Zhong Ji and assistant professors Wan Qing Xuan, Wei Gong Yi and Xu Feng. We are indebted to professor B. J. Birch for his suggestion to refine the concept of effectiveness and to make a distinction between theoretical effectiveness and the effectiveness which can be attained by a computer. Finally, we are grateful to Science Press (Beijing) and Springer-Verlag for all their help and patience during the course of publication.

*January*, 1980

Hua Loo Keng
Wang Yuan

# Contents

## *Chapter* 4. **Estimation of Discrepancy**



## *Chapter* 5. **Uniform Distribution and Numerical Integration**



## *Chapter* 6. **Periodic Functions**



## *Chapter* 7. **Numerical Integration of Periodic Functions**



## *Chapter* 8. **Numerical Error for Quadrature Formula**

*Chapter 1*

# Algebraic Number Fields and Rational Approximation

## § 1.1. The units of algebraic number fields

Let $Q$ denote the rational number field and $\alpha$ be an algebraic number of degree $s$. Then the algebraic number field $\mathscr{F}_s = Q(\alpha)$ is the field given by the polynomials in $\alpha$ of degree $< s$ with rational coefficients.

Let $s = r_1 + 2r_2$. Let $\alpha^{(1)}(=\alpha), \alpha^{(2)}, \cdots, \alpha^{(s)}$ be the conjugates of $\alpha$, where $\alpha^{(1)}, \cdots, \alpha^{(r_1)}$ are real numbers, $\alpha^{(r_1+1)}, \cdots, \alpha^{(r_1+2r_2)}$ are complex numbers and $\alpha^{(r_1+r_2+1)} = \overline{\alpha^{(r_1+1)}}, \cdots, \alpha^{(r_1+2r_2)} = \overline{\alpha^{(r_1+r_2)}}$. For any $\xi \in \mathscr{F}_s$, we have

$$\xi^{(r_1+r_2+j)} = \overline{\xi^{(r_1+j)}}, \quad 1 \leqslant j \leqslant r_2.$$

In other words, there are at most $r_1 + r_2$ different absolute values

$$|\xi^{(1)}|, \cdots, |\xi^{(r_1)}|, |\xi^{(r_1+1)}|, \cdots, |\xi^{(r_1+r_2)}|$$

among the conjugates of $\xi$.

Suppose that $\omega_1, \cdots, \omega_s$ is an integral basis of $\mathscr{F}_s$. Form the matrix

$$\Omega = (\omega_j^{(i)}), \quad 1 \leqslant i, j \leqslant s,$$

the matrix

$$S = \Omega'\Omega = \left(\sum_{k=1}^{s} \omega_i^{(k)}\omega_j^{(k)}\right), \quad 1 \leqslant i, j \leqslant s$$

is called the fundamental matrix of $\mathscr{F}_s$. Clearly, it is a symmetric matrix with rational integer entries. The invariants of a fundamental matrix under the modular group are characteristic properties of the algebraic number field. The determinant $\det S$ of $S$ is called the discriminant of the field.

Let $r = r_1 + r_2$. Let $\varepsilon_1, \cdots, \varepsilon_{r-1}$ be a set of units of $\mathscr{F}_s$. If

$$\det(\ln|\varepsilon_j^{(i)}|) \neq 0, \quad 2 \leqslant i \leqslant r, \quad 1 \leqslant j \leqslant r-1,$$

then the set of units $\varepsilon_1, \cdots, \varepsilon_{r-1}$ is called a set of independent units of $\mathscr{F}_s$. It follows by Dirichlet's unit theorem that there always exists a set of independent units in any algebraic number field. (Cf. E. Landau [1].)

Hereafter, we use $c(f, \cdots, g)$ to denote a positive constant depending on $f, \cdots, g$ only, but not always with the same value. We use $a, c, C, \cdots$ to denote absolute positive constants.

**Theorem 1.1.** *Let $\gamma_1, \cdots, \gamma_r$ be a given set of real numbers satisfying*

$$\sum_{j=1}^{r_1} \gamma_j + 2 \sum_{j=r_1+1}^{r} \gamma_j = 0. \tag{1.1}$$

*Then there exists a unit $\eta \in \mathscr{F}_s$ such that*

$$c^{-1} e^{\gamma_i} \leqslant |\eta^{(i)}| \leqslant c e^{\gamma_i}, \quad 1 \leqslant i \leqslant r, \tag{1.2}$$

*where $c = c(\mathscr{F}_s)$.*

*Proof.* Let $\varepsilon_1, \cdots, \varepsilon_{r-1}$ be a set of independent units of $\mathscr{F}_s$. Let

$$\xi^{(i)} = \varepsilon_1^{(i) a_1} \cdots \varepsilon_{r-1}^{(i) a_{r-1}}, \quad 1 \leqslant i \leqslant r.$$

Further let $c = e^a$, where

$$a = 2^{-1} \max_{1 \leqslant i \leqslant r} \left( \sum_{j=1}^{r-1} |\ln |\varepsilon_j^{(i)}|| \right). \tag{1.3}$$

Then

$$\prod_{i=1}^{r_1} |\xi^{(i)}| \prod_{j=1}^{r_2} |\xi^{(r_1+j)}|^2 = \prod_{k=1}^{r-1} \left( \prod_{i=1}^{r_1} |\varepsilon_k^{(i)}| \prod_{j=1}^{r_2} |\varepsilon_k^{(r_1+j)}|^2 \right)^{a_k} = 1. \tag{1.4}$$

Consider the system of linear equations

$$a_1 \ln |\varepsilon_1^{(i)}| + \cdots + a_{r-1} \ln |\varepsilon_{r-1}^{(i)}| = \gamma_i, \quad 1 \leqslant i \leqslant r. \tag{1.5}$$

Since

$$\det (\ln |\varepsilon_j^{(i)}|) \neq 0, \quad 2 \leqslant i \leqslant r, \quad 1 \leqslant j \leqslant r-1,$$

(1.5), except for the equation corresponding to $i = 1$, has a unique solution and it follows by (1.1) and (1.4) that this solution satisfies the equation for $i = 1$ in (1.5).

Let $b_k (1 \leqslant k \leqslant r-1)$ be the integers such that

$$|b_k - a_k| \leqslant \frac{1}{2}, \quad 1 \leqslant k \leqslant r-1.$$

Then we may define a unit $\eta$ of $\mathscr{F}_s$ by

$$\eta(=\eta^{(1)}) = \varepsilon_1^{b_1} \cdots \varepsilon_{r-1}^{b_{r-1}}.$$

From (1.3) and (1.5), we have

$$|\ln|\eta^{(i)}| - \ln|\xi^{(i)}|| = |\ln|\eta^{(i)}| - \gamma_i|$$

$$\leqslant \sum_{k=1}^{r-1} |b_k - a_k| |\ln|\varepsilon_k^{(i)}|| \leqslant a.$$

Hence we have (1.2). The theorem is proved.

For a real algebraic number field $\mathscr{F}_s$, set $\gamma_1 = \gamma$ and $\gamma_2 = \cdots = \gamma_r = \gamma'$. Then the condition (1.2) becomes

$$\gamma + (s-1)\gamma' = 0 \quad \text{or} \quad \gamma' = -\frac{\gamma}{s-1}.$$

Hence we have

**Theorem 1.2.** *Let $\mathscr{F}_s$ be a real algebraic number field of degree $s$. Then for any given real number $\gamma$, there exists a unit $\eta \in \mathscr{F}_s$ such that*

$$c^{-1}e^{\gamma} \leqslant \eta \leqslant ce^{\gamma}$$

*and*

$$c^{-1}e^{-\frac{\gamma}{s-1}} \leqslant |\eta^{(i)}| \leqslant ce^{-\frac{\gamma}{s-1}}, \quad 2 \leqslant i \leqslant s,$$

*where $c = c(\mathscr{F}_s)$.*

*Remark.* We may assume that $\eta > 0$ in Theorem 1.2, otherwise we may use $-\eta$ instead of $\eta$.

## § 1.2. The simultaneous Diophantine approximation of an integral basis

Let $\mathscr{F}_s$ be a real algebraic number field of degree $s$. Let $\omega_1, \cdots, \omega_s$ be an integral basis of $\mathscr{F}_s$. Take $\gamma = 1, 2, \cdots$ in Theorem 1.2. Then we may obtain a sequence of units $\eta_l (l = 1, 2 \cdots)$ such that

$$\eta_l > l, \quad |\eta_l^{(i)}| \leqslant c(\mathscr{F}_s)\eta_l^{-\frac{1}{s-1}}, \quad 2 \leqslant i \leqslant s. \tag{1.6}$$

Put

$$n_l = \sum_{i=1}^{s} \eta_l^{(i)} \tag{1.7}$$

and

$$h_{lj} = \sum_{i=1}^{s} \eta_l^{(i)} \omega_j^{(i)}. \tag{1.8}$$

Then $n_l$ and $h_{lj}(1 \leqslant j \leqslant s)$ are rational integers and we have

**Theorem 1.3.**

$$\left| \frac{h_{lj}}{n_l} - \omega_j \right| \leqslant c(\mathscr{F}_s) n_l^{-1-\frac{1}{s-1}}, \quad 1 \leqslant j \leqslant s. \tag{1.9}$$

*Proof.* For simplicity, we omit the index $l$. By (1.6), (1.7) and (1.8), we have

$$n = \eta + O(\eta^{-\frac{1}{s-1}}) = \eta + O(n^{-\frac{1}{s-1}}) = \eta(1 + O(n^{-\frac{1}{s-1}}))$$

and

$$h_j = \eta\omega_j + O(\eta^{-\frac{1}{s-1}}) = \eta\omega_j(1 + O(n^{-1-\frac{1}{s-1}})).$$

Hence

$$\begin{aligned} \frac{h_j}{n} &= \omega_j(1 + O(n^{-1-\frac{1}{s-1}}))(1 + O(n^{-1-\frac{1}{s-1}}))^{-1} \\ &= \omega_j + O(n^{-1-\frac{1}{s-1}}), \quad 1 \leqslant j \leqslant s, \end{aligned}$$

where the constants implied by the symbol "$O$" depend on $\mathscr{F}_s$ only. The theorem is proved.

Now we shall give the expressions of $n$ and $h_j(1 \leqslant j \leqslant s)$. Let

$$\eta = \sum_{j=1}^{s} k_j \omega_j, \tag{1.10}$$

where $k_j(1 \leqslant j \leqslant s)$ are rational integers. Then we have

$$(\eta^{(1)}, \cdots, \eta^{(s)}) = (k_1, \cdots, k_s)\Omega'$$

and

$$(\eta^{(1)}, \cdots, \eta^{(s)})\Omega = (k_1, \cdots, k_s)S = (h_1, \cdots, h_s). \tag{1.11}$$

Let

$$\sum_{j=1}^{s} a_j \omega_j = 1, \tag{1.12}$$

where $a_j (1 \leqslant j \leqslant s)$ are rational integers. Then

$$n = \sum_{i=1}^{s} \eta^{(i)} \sum_{j=1}^{s} a_j \omega_j^{(i)} = \sum_{j=1}^{s} a_j \sum_{i=1}^{s} \eta^{(i)} \omega_j^{(i)} = \sum_{j=1}^{s} a_j h_j. \qquad (1.13)$$

Hence we obtain the set of integers $(n, h_1, \cdots, h_s)$ corresponding to $\eta$ by (1.10), (1.11), (1.12) and (1.13).

*Remarks.* 1. Theorem 1.3 is also true if the set $\omega_i (1 \leqslant i \leqslant s)$ is a basis of $\mathscr{F}_s$ and $\eta$ can be represented as a linear combination of $\omega_i' s$ with rational integer coefficients.

2. By Schmidt's theorem on simultaneous Diophantine approximation of algebraic numbers (Cf. W. M. Schmidt [2, 4]). We know that the estimate $c(\mathscr{F}_s) n_l^{-1-\frac{1}{s-1}}$ given in (1.9) is the best possible and it cannot be replaced even by $c(\mathscr{F}_s, \varepsilon) n_l^{-1-\frac{1}{s-1}-\varepsilon}$. Hereafter we use $\varepsilon$ to denote any pre-assigned positive number. But we have not yet considered here the best constant $c(\mathscr{F}_s)$ of (1.9). By the argument in the proof of Theorem 1.3, we know that it depends not only on the choice of $\eta_l$, but also on that of the integral basis. Let

$$\overline{|\omega|} = \max_{1 \leqslant i,\, j \leqslant s} |\omega_j^{(i)}|.$$

Then the right hand side of (1.9) may be written as

$$\overline{|\omega|}\, c(\varepsilon_1, \cdots, \varepsilon_{r-1}) n^{-1-\frac{1}{s-1}}.$$

3. Theorem 1.3 is not new. Our purpose here is to suggest a computational method for obtaining $(n, h_1, \cdots, h_s)$. For $s = 2$, we can use continued fractions to treat the present problem, In the case of $s > 2$, the situation is entirely different. The classical methods can only prove the existence of infinitely many sets of $(n, h_1, \cdots, h_s)$ satisfying (1.9). But they do not suggest any effective way (in the sense of numerical analysis) for finding $(n, h_1, \cdots, h_s)$. It is shown here that the problem of finding the sets of integers $(n, h_1, \cdots, h_s)$ is equivalent to the problem for finding a set of independent units in $\mathscr{F}_s$ and this requires only $c(\mathscr{F}_s) \ln n$ elementary operations for obtaining the set $(n, h_1, \cdots, h_s)$. Though Dirichlet's unit theorem is an existence theorem too, there are however many real algebraic number fields for which sets of independent units are known.

## § 1.3. The real cyclotomic field

Let $m$ be an integer $\geqslant 5$ and $s = \frac{\varphi(m)}{2}$. The real cyclotomic field

$$\mathscr{R}_s = Q\left(\cos\frac{2\pi}{m}\right)$$

is an algebraic number field of degree $s$. The field has an integral basis

$$\omega_1 = 1, \quad \omega_l = 2\cos\frac{2\pi(l-1)}{m}, \quad 2 \leqslant l \leqslant s$$

(Cf. J. J. Liang [1]). Let $h_1(=1), h_2, \cdots, h_s$ be the integers satisfying $1 \leqslant h < m/2$ and $(h, m) = 1$. The transformation

$$\sigma_j\colon\ \omega_1 \to \omega_1, \quad \omega_l \to 2\cos\frac{2\pi(l-1)h_j}{m}, \quad 2 \leqslant l \leqslant s$$

is an automorphism of the cyclotomic field $\mathscr{R}_s$ and the $s$ automorphisms

$$\sigma_1(=1), \sigma_2, \cdots, \sigma_s$$

form the group of automorphism of the field. A number $\xi$ of $\mathscr{R}_s$ has $s$ conjugates under these automorphisms. Form the matrix

$$\Omega = (\sigma_i \omega_j), \quad 1 \leqslant i, j \leqslant s.$$

Then

$$S = \Omega'\Omega = (a_{ij}),$$

where $a_{11} = s$, $a_{1j} = a_{j1} = C_m(j-1)(2 \leqslant j \leqslant s)$ and $a_{ij} = C_m(i+j-2) + C_m(i-j)(2 \leqslant i, j \leqslant s)$ in which $C_m(k)$ denotes the trigonometric sum (Ramanujan sum)

$$C_m(k) = \sum_{(a, m)=1} e^{2\pi i a k/m},$$

which can be evaluated by the following lemma.

**Lemma 1.1.** *$C_m(k)$ is a multiplicative function of $m$, i.e.,*

$$C_{m_1}(k) C_{m_2}(k) = C_{m_1 m_2}(k),$$

*if* $(m_1, m_2) = 1$. *And*

$$C_{p^l}(k) = \begin{cases} p^l - p^{l-1}, & \text{if } p^l \mid k, \\ -p^{l-1}, & \text{if } p^{l-1} \| k, \\ 0, & \text{if } p^{l-1} \nmid k. \end{cases}$$

*Hereafter we always use* $p$ *to denote the prime number and* $p^l \| b$ *to denote* $p^a \mid b$ *but* $p^{a+1} \nmid b$ (Cf. Hua Loo Keng [2], Chap. 7).

In particular, for the case $m = p$, since

$$\sum_{l=1}^{s} 2\cos\frac{2\pi l}{p} = -1,$$

we have an integral basis

$$\omega_l = 2\cos\frac{2\pi}{p}g^l, \quad 1 \leqslant l \leqslant s,$$

of $\mathscr{R}_s = Q\left(\cos\frac{2\pi}{p}\right)$, where $g$ denotes a primitive root mod $p$. Hence

$$S = pI - M,$$

where $I$ is the $s \times s$ identity matrix and $M = (m_{i_j})$ is the matrix, where

$$m_{i_j} = 1 \quad (1 \leqslant i, j \leqslant s).$$

Suppose that $\varepsilon_1, \cdots, \varepsilon_{s-1}$ is the set of independent units of $Q\left(\cos\frac{2\pi}{m}\right)$ which will be given in next section. With the aid of $\varepsilon_i's$, we may con struct by the method of §§ 1.1—1.2, a sequence of sets of integers $(n_l, h_{l1}, \cdots, h_{ls})$ $(l = 1, 2, \cdots)$ such that

$$\left|\frac{h_{lj}}{n_l} - 2\cos\frac{2\pi(j-1)}{m}\right| \leqslant c(\mathscr{R}_s)\, n_l^{-1-\frac{1}{s-1}}, \quad 2 \leqslant j \leqslant s, \qquad (1.14)$$

where $h_{l1} = n_l$.

Hereafter we use the notations

$$c_{l1} = 1 \quad \text{and} \quad c_{lj} = h_{lj} \quad (2 \leqslant j \leqslant s).$$

*Remark.* Since $Q\left(\cos\frac{2\pi}{5}\right) = Q(\sqrt{5})$, the sequence of sets $(n_l, c_{l1}, \cdots, c_{ls})$ $(l = 1, 2, \cdots)$ which give the best simultaneous Diophantine approximation of the integral basis of $\mathscr{R}_s$ may be recognized as a generalization of the

Fibonacci sequence and the integral basis of $Q\left(\cos\frac{2\pi}{m}\right)$ may be regarded as a generalization of the golden ratio in higher dimensional space.

## § 1.4. The units of a cyclotomic field

We always use $p_1, p_2, \cdots$ to denote different prime numbers. Let $Z_m$ denote the multiplicative group of reduced residue classes modulo $m$. If the elements $h$ and $-h$ of $Z_m$ are identified, then we have the quotient group of $Z_m$ by $\{\pm 1\}$

$$Z_m/\{\pm 1\} = \{h_1, \cdots, h_s\}.$$

The $s$ characters of the group $Z_m/\{\pm 1\}$ are induced by the characters of the group $Z_m$ with $\chi(-1)=1$ and they are denoted by

$$\chi_1, \cdots, \chi_s,$$

where $\chi_s$ denotes the principal character.

Set $h_{s+j} = h_j$. Let

$$\varepsilon(h) = \prod_{\substack{n|m,\ n>1 \\ p^l \| n \Rightarrow p^l \| m}} 2\sin\frac{\pi h}{n}, \quad (h, m) = 1$$

and

$$\eta_j = \varepsilon(h_{j+1})/\varepsilon(h_j), \quad 1 \leqslant j \leqslant s.$$

where $A \Rightarrow B$ means that $A$ implies $B$. The $\eta_j (1 \leqslant j \leqslant s)$ are all units of $\mathscr{R}_s$, since

$$\prod_{j=1}^{s} \eta_j = 1.$$

**Theorem 1.4.** *$\eta_1, \cdots, \eta_{s-1}$ form a set of independent units of $\mathscr{R}_s$.*

To prove the theorem, we shall need the following lemmas.

**Lemma 1.2.**

$$\sum_{j=1}^{s} \chi(h_j) = \begin{cases} s, & \text{if } \chi = \chi_s, \\ 0, & \text{otherwise.} \end{cases}$$

*Proof.* Since

$$\sum_{(h,m)=1} \chi(h) = \begin{cases} 2s, & \text{if } \chi = \chi_s, \\ 0, & \text{otherwise,} \end{cases}$$

(Cf. Hua Loo Keng [2], Chap. 7), hence

$$\sum_{j=1}^{s} \chi(h_j) = \frac{1}{2} \sum_{(h,m)=1} \chi(h) = \begin{cases} s, & \text{if } \chi = \chi_s \\ 0, & \text{otherwise.} \end{cases}$$

The lemma is proved.

Let

$$\eta_j^{(i)} = \frac{\varepsilon(h_i h_{j+1})}{\varepsilon(h_i h_j)}, \quad 1 \leqslant i \leqslant s.$$

Then $\eta_j^{(1)}(=\eta_j), \eta_j^{(2)}, \cdots, \eta_j^{(s)}$ are all the conjugates of $\eta_j$. Let

$$C = (c_{ij}), \quad 1 \leqslant i, j \leqslant s-1,$$

where $c_{ij} = \ln |\eta_j^{(i)}| (1 \leqslant i, j \leqslant s-1)$.

**Lemma 1.3.**

$$|\det C| = \prod_{j=1}^{s-1} \left| \sum_{i=1}^{s} \chi_j(h_i) \ln |\varepsilon(h_i)| \right|.$$

*Proof.* Clearly

$$|\det C| = |\det C^*|,$$

where $C^* = (c_{ij}^*), 1 \leqslant i, j \leqslant s$, in which

$$c_{ij}^* = \begin{cases} \ln |\varepsilon(h_i h_j)|, & \text{if } 1 \leqslant i \leqslant s-1, 1 \leqslant j \leqslant s, \\ 1, & \text{otherwise.} \end{cases}$$

Let

$$P = (\chi_j(h_i)), \quad 1 \leqslant i, j \leqslant s.$$

Then it follows by Lemma 1.2 that

$$|\det C \cdot \det P| = |\det C^* P|$$

$$= s \prod_{j=1}^{s-1} \left| \sum_{i=1}^{s} \chi_j(h_i) \ln |\varepsilon(h_i)| \right| |\det D|, \tag{1.15}$$

where

$$D = (\bar{\chi}_j(h_i)), \quad 1 \leqslant i, j \leqslant s-1.$$

Since

$$s|\det D| = |\det D^*| = |\det \bar{P}|, \tag{1.16}$$

where

$$D^* = \begin{pmatrix} D & \mathbf{I} \\ \mathbf{0} & s \end{pmatrix},$$

in which $\mathbf{0}$ and $\mathbf{I}$ denote the zero row vector and identity column vector respectively, and

$$|\det P| = |\det \bar{P}| = |\det \bar{P}'P|^{1/2} = s^{s/2}. \tag{1.17}$$

The lemma follows by substituting (1.16) and (1.17) into (1.15).

Since

$$2\sum_{j=1}^{s} \chi(h_j) \ln |\varepsilon(h_j)| = \sum_{(a,\, m)=1} \chi(a) \ln|\varepsilon(a)|,$$

hence by the definition of the set of independent units, we have

**Lemma 1.4.** *A necessary and sufficient condition that* $\eta_1, \cdots, \eta_{s-1}$ *form a set of independent units is that*

$$R_\chi = \sum_{(a,\, m)=1} \chi(a) \ln |\varepsilon(a)| \neq 0$$

*holds for any non-principal character* $\chi$ mod $m$ *with* $\chi(-1) = 1$.

**Lemma 1.5.**

$$\prod_{h=0}^{m-1} 2\sin\left(\frac{h\pi}{m} + \theta\right) = 2\sin m\theta.$$

*Proof.*

$$\begin{aligned}
\prod_{h=0}^{m-1} 2\sin\left(\frac{h\pi}{m} + \theta\right) &= (-i)^m \prod_{h=0}^{m-1} \left(e^{\left(\frac{h\pi}{m}+\theta\right)i} - e^{-\left(\frac{h\pi}{m}+\theta\right)i}\right) \\
&= (-i)^m e^{\frac{\pi i}{m}\sum_{h=0}^{m-1} h - m\theta i} \prod_{h=0}^{m-1} \left(e^{2\theta i} - e^{-\frac{2\pi h i}{m}}\right) \\
&= (-i)^m e^{\frac{\pi i(m-1)}{2} - m\theta i} \left(e^{2m\theta i} - 1\right) \\
&= -i\left(e^{m\theta i} - e^{-m\theta i}\right) = 2\sin m\theta.
\end{aligned}$$

The lemma is proved.

**Lemma 1.6.** *Suppose that* $0<\theta<1$. *Then*

$$\sum_{n=1}^{\infty}\frac{e^{2\pi in\theta}}{n}=-\ln(2\sin\pi\theta)+\left(\frac{\pi}{2}-\pi\theta\right)i.$$

*Proof.* Since

$$\sum_{n=1}^{\infty}\frac{r^n e^{2\pi in\theta}}{n}=-\ln\left(1-re^{2\pi i\theta}\right)$$

for $0<r<1$ and the series $\sum_{n=1}^{\infty}\frac{e^{2\pi in\theta}}{n}$ is convergent, hence, by Abel's theorem, we have

$$\begin{aligned}\sum_{n=1}^{\infty}\frac{e^{2\pi in\theta}}{n}&=-\ln\left(1-e^{2\pi i\theta}\right)\\&=-\ln\left(e^{\pi i\theta}(e^{-\pi i\theta}-e^{\pi i\theta})\right)\\&=-\ln\left((2\sin\pi\theta)e^{\left(\pi\theta-\frac{\pi}{2}\right)i}\right)\\&=-\ln\left(2\sin\pi\theta\right)+\left(\frac{\pi}{2}-\pi\theta\right)i.\end{aligned}$$

The lemma is proved.

**Lemma 1.7.** *Let* $\chi$ *be a primitive character* mod $d$ *with* $\chi(-1)=1$. *Then*

$$\sum_{(a,d)=1}\chi(a)\ln\left|2\sin\frac{\pi a}{d}\right|=-\tau(\chi)L(1,\bar{\chi}),$$

*where*

$$\tau(\chi)=\sum_{(r,d)=1}\chi(r)e^{2\pi ir/d},\quad L(1,\chi)=\sum_{n=1}^{\infty}\frac{\chi(n)}{n}.$$

*Proof.* Let

$$S(n,\chi)=\sum_{(r,d)=1}\chi(r)e^{2\pi inr/d}.$$

Then

$$\bar{\chi}(n)\tau(\chi)=S(n,\chi)$$

(Cf. Hua Loo Keng [2], Chap. 7) and so

$$\tau(\chi)L(1,\bar{\chi})=\sum_{n=1}^{\infty}\frac{\bar{\chi}(n)\tau(\chi)}{n}=\sum_{n=1}^{\infty}\frac{1}{n}\sum_{(r,d)=1}\chi(r)e^{2\pi inr/d}$$

$$= \sum_{(r,d)=1} \chi(r) \sum_{n=1}^{\infty} \frac{e^{2\pi i n r/d}}{n}$$

$$= \sum_{(r,d)=1} \chi(r) \left( -\ln\left|2\sin\frac{\pi r}{d}\right| + \left(\frac{\pi}{2} - \frac{\pi r}{d}\right) i \right)$$

$$= -\sum_{(r,d)=1} \chi(r)\ln\left|2\sin\frac{\pi r}{d}\right| - \frac{\pi i}{d}\sum_{(r,d)=1}\chi(r)r \tag{1.18}$$

by Lemma 1.6. Since $\chi(-1) = 1$, we have

$$\sum_{(r,d)=1} \chi(r)r = \sum_{(r,d)=1} \chi(d-r)(d-r) = -\sum_{(r,d)=1} \chi(r)r$$

and so

$$\sum_{(r,d)=1} \chi(r)r = 0.$$

Substituting into (1.18), the lemma follows.

**Lemma 1.8.** *Let* $m = p_1^{l_1}\cdots p_r^{l_r}$ *and* $d = p_1^{l'_1}\cdots p_r^{l'_r}$, *where* $l_i \geqslant 1$ *and* $0 \leqslant l'_i \leqslant l_i$ $(1 \leqslant i \leqslant r)$. *Let* $\chi$ *be a primitive character* mod $d$ *and also a character* mod $m$ *with* $\chi(-1) = 1$. *Then*

$$\sum_{(a,m)=1} \chi(a)\ln\left|2\sin\frac{\pi a}{m}\right| = -F(m_1)\tau(\chi)L(1,\bar{\chi}),$$

*where*

$$m_1 = \prod_{\substack{p \mid m \\ p \nmid d}} p, \quad F(m_1) = \prod_{p \mid m_1} (1 - \chi(p)).$$

*Proof.* First, suppose that $m = dd'$ and $l'_i > 0 (1 \leqslant i \leqslant r)$. Then

$$r_\chi = \sum_{(a,m)=1} \chi(a)\ln\left|2\sin\frac{\pi a}{m}\right|$$

$$= \sum_{a_1=1}^{d'-1} \sum_{(a_2,d)=1} \chi(a_1 d + a_2)\ln\left|2\sin\frac{\pi(a_1 d + a_2)}{m}\right|$$

$$= \sum_{(a_2,d)=1} \chi(a_2) \sum_{a_1=0}^{d'-1} \ln\left|2\sin\left(\frac{\pi a_1}{d'} + \frac{\pi a_2}{m}\right)\right|$$

$$= \sum_{(a_2,d)=1} \chi(a_2)\ln\left|2\sin\frac{\pi a_2 d'}{m}\right|$$

$$= \sum_{(a,d)=1} \chi(a)\ln\left|2\sin\frac{\pi a}{d}\right| \tag{1.19}$$

by Lemma 1.5.

Next, suppose that $m=m_1m_2$, $m_1=p_1^{l_1}\cdots p_j^{l_j}$, $m_2=p_{j+1}^{l_{j+1}}\cdots p_r^{l_r}$ and $d\mid m_2$, where $1\leqslant j<r$. Hence $\chi$ is also a character mod $m_2$ and

$$\begin{aligned}
r_\chi &= \sum_{(a,m)=1} \chi(a)\ln\left|2\sin\frac{\pi a}{m}\right| \\
&= \sum_{(a_1,m_1)=1}\sum_{(a_2,m_2)=1} \chi(a_1m_2+a_2m_1)\ln\left|2\sin\frac{\pi(a_1m_2+a_2m_1)}{m}\right| \\
&= \sum_{(a_2,m_2)=1} \chi(a_2m_1)\sum_{a_1=1}^{m_1}\ln\left|2\sin\left(\frac{\pi a_1}{m_1}+\frac{\pi a_2}{m_2}\right)\right|\sum_{k\mid(a_1,m_1)}\mu(k) \\
&= \sum_{(a_2,m_2)=1} \chi(a_2m_1)\sum_{k\mid m_1}\mu(k)\sum_{l=1}^{\frac{m_1}{k}}\ln\left|2\sin\left(\frac{\pi lk}{m_1}+\frac{\pi a_2}{m_2}\right)\right|.
\end{aligned}$$

It follows by Lemma 1.5 and (1.19) that

$$\begin{aligned}
r_\chi &= \sum_{(a_2,m_2)=1} \chi(a_2m_1)\sum_{k\mid m_1}\mu(k)\ln\left|2\sin\frac{\pi a_2m_1}{km_2}\right| \\
&= \sum_{k\mid m_1}\mu(k)\chi(k)\sum_{(a_2,m_2)=1}\bar{\chi}(k)\chi(a_2m_1)\ln\left|2\sin\frac{\pi a_2m_1}{km_2}\right| \\
&= F(m_1)\sum_{(a,m_2)=1}\chi(a)\ln\left|2\sin\frac{\pi a}{m_2}\right| \\
&= F(m_1)\sum_{(a,d)=1}\chi(a)\ln\left|2\sin\frac{\pi a}{d}\right|.
\end{aligned}$$

The lemma follows by Lemma 1.7.

**Lemma 1.9.** *Suppose that* $m=m_1m_2$, *where* $(m_1,m_2)=1$. *Let* $\chi$ *be a character* mod $m$. *Then*

$$\sum_{(a,m)=1}\chi(a)\ln\left|2\sin\frac{\pi a}{m_2}\right| = \begin{cases} \varphi(m_1)\displaystyle\sum_{(a,m_2)=1}\chi(a)\ln\left|2\sin\frac{\pi a}{m_2}\right|, & \\ \qquad \text{if } \chi \text{ is a character mod } m_2, & \\ 0, \quad \text{otherwise.} & \end{cases}$$

*Proof.* Suppose that $\chi$ is not a character mod $m_2$. Then there exists an integer $t$ such that

$$(1+tm_2,m)=1,\quad \chi(1+tm_2)\neq 1.$$

Hence

$$\chi(1+tm_2)\sum_{(a_1,m_1)=1}\chi(a_2m_1+a_1m_2)$$

$$=\sum_{(a_1,m_1)=1}\chi(a_2m_1+a_1m_2(1+tm_2))$$

$$=\sum_{(a,m_1)=1}\chi(a_2m_1+am_2)$$

and so

$$\sum_{(a_1,m_1)=1}\chi(a_2m_1+a_1m_2)=0. \tag{1.20}$$

Suppose that $\chi$ is a character mod $m_2$. Then

$$\sum_{(a_1,m_1)=1}\chi(a_2m_1+a_1m_2)=\varphi(m_1)\chi(a_2m_1). \tag{1.21}$$

Since

$$\sum_{(a,m)=1}\chi(a)\ln\left|2\sin\frac{\pi a}{m_2}\right|$$

$$=\sum_{(a_1,m_1)=1}\sum_{(a_2,m_2)=1}\chi(a_2m_1+a_1m_2)\ln\left|2\sin\frac{\pi a_2m_1}{m_2}\right|$$

$$=\sum_{(a_2,m_2)=1}\ln\left|2\sin\frac{\pi a_2m_1}{m_2}\right|\sum_{(a_1,m_1)=1}\chi(a_2m_1+a_1m_2). \tag{1.22}$$

The lemma follows by substituting (1.20) and (1.21) into (1.22).

The proof of Theorem 1.4. Let $\chi$ be any non-principal character mod $m$ with $\chi(-1)=1$. Then $\chi$ is a primitive character mod $d$, where $d|m$ and $d\geqslant 3$. Hence it follows by Lemmas 1.8 and 1.9 that

$$R_\chi=\sum_{(a,m)=1}\chi(a)\ln|\varepsilon(a)|$$

$$=\sum_{\substack{n|m,n>1\\ p^l\|n\Rightarrow p^l\|m}}\sum_{(a,m)=1}\chi(a)\ln\left|2\sin\frac{\pi a}{n}\right|$$

$$=\sum_{\substack{n|m,n>1\\ p^l\|n\Rightarrow p^l\|m}}\sum_{\chi(\mathrm{mod}\, n)}\varphi\left(\frac{m}{n}\right)\sum_{(a,n)=1}\chi(a)\ln\left|2\sin\frac{\pi a}{n}\right|$$

$$=-\tau(\chi)L(1,\bar{\chi})\sum_{\substack{n|m,n>1\\ p^l\|n\Rightarrow p^l\|m\\ d|n}}\varphi\left(\frac{m}{n}\right)F(n_1) \tag{1.23}$$

where $n_1 = \prod_{\substack{p^l \| n \\ p \nmid d}} p^l$.

Let $d' = \prod_{\substack{p^l \| n \\ p \mid d}} p^l$. Then $n = d'n_1$ and $(d', n_1) = 1$. Hence

$$\sum_{\substack{n \mid m, n>1 \\ p^l \| n \Rightarrow p^l \| m \\ d \mid n}} \varphi\left(\frac{m}{n}\right) F(n_1) = \varphi\left(\frac{m}{d'}\right) \sum_{\substack{n_1 \mid \frac{m}{d'} \\ p^l \| n_1 \Rightarrow p^l \| \frac{m}{d'}}} \frac{F(n_1)}{\varphi(n_1)}$$

$$= \varphi\left(\frac{m}{d'}\right) \prod_{p^l \| \frac{m}{d'}} \left(1 + \frac{1 - \chi(p)}{\varphi(p^l)}\right) \neq 0. \tag{1.24}$$

Since $\tau(\chi) \neq 0$ and $L(1, \bar{\chi}) \neq 0$ (Cf. Hua Loo Keng [2], Chap. 7 and Chap. 9), therefore $R_\chi \neq 0$ by (1.23). and (1.24). Hence the theorem follows by Lemma 1.4.

*Remark.* In general, $\eta_1, \cdots, \eta_{s-1}$ is not a set of fundamental units (Cf. J. M. Masley and H. L. Montgomery [1] and W. Sinnott [1]).

## §1.5. Continuation

**Theorem 1.5.** *Suppose that* $m = p_1^{l_1} \cdots p_r^{l_r}$, *where* $r \geqslant 2$ *and* $l_i \geqslant 1$ $(1 \leqslant i \leqslant r)$. *Then a necessary and sufficient condition that*

$$2 \sin \frac{\pi h}{m}, \quad (h, m) = 1, \quad 1 < h \leqslant \frac{m}{2} \tag{1.25}$$

*form a set of independent units is that the group* $Z_{m/p_i^{l_i}}$ *is generated by* $-1$ *and* $p_i$ *for any* $i$ *with* $1 \leqslant i \leqslant r$, *this is denoted by*

$$Z_{m/p_i^{l_i}} = \langle -1, p_i \rangle, \quad 1 \leqslant i \leqslant r. \tag{1.26}$$

To prove the theorem, we shall need the following lemmas.

**Lemma 1.10.** *Suppose that* $n$ *is an integer* $> 1$. *Then*

$$\prod_{\substack{h=0 \\ (h,n)=1}}^{n-1} 2 \sin \frac{\pi h}{n} = \begin{cases} p, & \text{if } n = p^l, \\ 1, & \text{otherwise.} \end{cases}$$

*Proof.* By Lemma 1.5,

$$\ln \prod_{\substack{h=0\\(h,n)=1}}^{n-1} 2\sin\left(\frac{h\pi}{n}+\theta\right) = \sum_{h=0}^{n-1} \ln\left(2\sin\left(\frac{h\pi}{n}+\theta\right)\right) \sum_{d\mid(h,n)} \mu(d)$$

$$= \sum_{d\mid n} \mu(d) \sum_{k=0}^{\frac{n}{d}-1} \ln\left(2\sin\left(\frac{kd\pi}{n}+\theta\right)\right)$$

$$= \sum_{d\mid n} \mu(d) \ln\left(2\sin\frac{n\theta}{d}\right) = \ln \prod_{d\mid n} (\sin d\theta)^{\mu\left(\frac{n}{d}\right)}.$$

Let $\theta \to 0$. Then the lemma follows.

We shall suppose hereafter that $m$ has at least two different prime factors. Let

$$H = (h_{ij}) = \left(\ln\left|2\sin\frac{\pi h_i h_j^*}{m}\right|\right), \quad 1 \leqslant i, j \leqslant s,$$

where $h_j^*$ denotes the inverse element of $h_j$ in $Z_m/\{\pm 1\}$. Then we have by Lemma 1.2 that

$$P'HP'^{-1} = (k_{ij}), \quad 1 \leqslant i, j \leqslant s,$$

where

$$k_{ij} = \frac{1}{s} \sum_{r,t=1}^{s} \chi_i(h_r) \ln\left|2\sin\frac{\pi h_r h_t^*}{m}\right| \bar{\chi}_j(h_t)$$

$$= \frac{1}{s} \sum_{r=1}^{s} \chi_i(h_r)\bar{\chi}_j(h_r) \sum_{t=1}^{s} \bar{\chi}_j(h_t h_r^*) \ln\left|2\sin\frac{\pi h_r h_t^*}{m}\right|$$

$$= \delta_{ij} \sum_{l=1}^{s} \chi_j(h_l) \ln\left|2\sin\frac{\pi h_l}{m}\right|$$

$$= \frac{1}{2}\delta_{ij} \sum_{(a,m)=1} \chi_j(a) \ln\left|2\sin\frac{\pi a}{m}\right|,$$

in which $\delta_{ij}$ denotes the Kronecker symbol, i.e., $\delta_{ij} = 1$, if $i = j$ and $\delta_{ij} = 0$ otherwise. This means that the $s$ eigenvalues of the matrix $H$ are

$$\lambda_i = \frac{1}{2} \sum_{(a,m)=1} \chi_i(a) \ln\left|2\sin\frac{\pi a}{m}\right|, \quad 1 \leqslant i \leqslant s.$$

Let $H_{ij}$ denote the cofactor for $h_{ij}$ in $H$. Then all the $H_{ij}$'s are equal to each other, since the sums for every row and every column of $H$ are equal to zero. In particular, we have

$$H_{11} = H_{22} = \cdots = H_{ss}.$$

Denote the characteristic polynomial of $H$ by

$$\det(xI - H) = x^s + k_1 x^{s-1} + \cdots + k_{s-1}x + k_s.$$

Since

$$\lambda_s = \frac{1}{2} \sum_{(a,m)=1} \ln \left| 2 \sin \frac{\pi a}{m} \right| = 0$$

by Lemma 1.10, hence

$$k_{s-1} = (-1)^{s-1}(H_{11} + \cdots + H_{ss}) = (-1)^{s-1} s H_{11}$$

and

$$k_s = (-1)^s \lambda_1 \cdots \lambda_s = 0.$$

Since $H_{11} \neq 0$ means that (1.25) is a set of independent units, we have

**Lemma 1.11.** *A necessary and sufficient condition that* (1.25) *is a set of independent units is that* 0 *is a simple root of the equation* $\det(xI - H) = 0$, *i.e.,*

$$\lambda_\chi = \sum_{(a,m)=1} \chi(a) \ln \left| 2 \sin \frac{\pi a}{m} \right| \neq 0$$

*holds for any non-principal character* mod $m$ *with* $\chi(-1) = 1$.

The proof of Theorem 1.5. First, suppose that (1.26) holds. Let $\chi$ be a non-principal character mod $m$ and also a primitive character mod $d$ with $\chi(-1) = 1$. Then $d = p_1^{l_1'} \cdots p_r^{l_r'}$, where $0 \leqslant l_i' \leqslant l_i (1 \leqslant i \leqslant r)$. If $l_i' > 0$ $(1 \leqslant i \leqslant r)$, then

$$\lambda_\chi = -\tau(\chi) L(1, \bar{\chi}) \neq 0$$

by Lemma 1.8. If $l_1' = \cdots = l_j' = 0$ and $l_k' > 0 (j+1 \leqslant k \leqslant r)$, where $1 \leqslant j < r$, then by Lemma 1.8,

$$\lambda_\chi = -\prod_{i=1}^{j} (1 - \chi(p_i)) \tau(\chi) L(1, \bar{\chi}).$$

If there exists $p_i$ such that $\chi(p_i) = 1$, where $1 \leqslant i \leqslant j$, then it follows from $\chi(-1) = 1$ and (1.26) that $\chi(n) = 1$ for any $n \in Z_{m/p_i^{l_i}}$. Hence $\chi$ is a principal character mod $d$, since $d \left| \frac{m}{p_i^{l_i}} \right.$. This leads to a contradiction and so $\lambda_\chi \neq 0$. By Lemma 1.11, we know that (1.25) is a set of independent units.

Next, suppose that there exists an $i$ with $1 \leqslant i \leqslant s$ such that

$$Z_{m/p_i^{l_i}} \neq \langle -1, p_i \rangle.$$

Let $d = \frac{m}{p_i^{l_i}}$. Then there exists a primitive character mod $d^*$ such that $\chi(-1) = \chi(p_i) = 1$, where $d^* | d$. Hence $\lambda_\chi = 0$ by Lemma 1.8 and so (1.25) is not a set of independent units by Lemma 1.11. The theorem is proved.

From Theorem 1.5, we derive

**Theorem 1.6.** *Suppose that* $m = p_1^{l_1} \cdots p_r^{l_r}$ *where* $r \geqslant 4$ *and* $l_i \geqslant 1$ $(1 \leqslant i \leqslant r)$ *or* $r \geqslant 3, p_1 = 2, l_1 \geqslant 3$ *and* $l_i \geqslant 1 (2 \leqslant i \leqslant r)$. *Then* (1.25) *is not a set of independent units.*

For example, (1.25) is independent for $m = 21$ and dependent for $m = 35$.

Now, we shall study the units of $\mathscr{R}_s$ for the case $m = p$. It follows by Theorem 1.4 that

$$\rho_j = \frac{\sin \frac{\pi}{p} g^{j+1}}{\sin \frac{\pi}{p} g^j}, \quad 1 \leqslant j \leqslant s-1$$

form a set of independent units of $\mathscr{R}_s$, where $g$ denotes a primitive root mod $p$. Clearly

$$\rho_j = \rho_{j+s}, \quad \rho_1 \cdots \rho_s = \pm 1.$$

We shall give the linear expression of $\rho_l$ by using the basis

$$\omega_l = 2 \cos \frac{2\pi}{p} g^l, \quad 1 \leqslant l \leqslant s$$

of $\mathscr{R}_s$ as follows.

$$\rho_l = ((e^{\frac{\pi i g^l}{p}})^g - (e^{-\frac{\pi i g^l}{p}})^g)(e^{\frac{\pi i g^l}{p}} - e^{-\frac{\pi i g^l}{p}})^{-1}$$

$$= \sum_{m=0}^{g-1} (e^{\frac{\pi i g^l}{p}})^{g-1-m} (e^{-\frac{\pi i g^l}{p}})^m$$

$$= \sum_{m=0}^{g-1} e^{\frac{\pi i g^l(g-1-2m)}{p}} = \sum_{m=0}^{g-1} \cos \frac{\pi g^l(g-1-2m)}{p}.$$

First, suppose that $2|g$. Then

$$\rho_l = \sum_{m=0}^{\frac{1}{2}g-1} 2\cos \frac{\pi g^l(g-1-2m)}{p}.$$

Let

$$g-1-2m \equiv 2g^{e_m} (\operatorname{mod} p), \quad m = 0, 1, \cdots, \frac{1}{2}g-1.$$

Then

$$\rho_l = \sum_{m=0}^{\frac{1}{2}g-1} \omega_{l+e_m}.$$

Since $g^s \equiv -1 \pmod p$, $l + e_m$ may be replaced by $l + e_m - ks$, where $ks < l + e_m \leqslant (k+1)s$. Hence

$$(\rho_1, \cdots, \rho_s) = (\omega_1, \cdots, \omega_s)M,$$

where $M$ is a circular matrix

$$M = \begin{pmatrix} a_1 & a_s & \cdots & a_2 \\ & & \cdots & \\ a_s & a_{s-1} & \cdots & a_1 \end{pmatrix},$$

in which $a_t = 1$, if $t$ equals $1 + e_m - s$ or $1 + e_m$ and $a_t = 0$ otherwise.

Next, suppose that $2 \nmid g$. Then

$$\rho_l = 1 + \sum_{m=0}^{\frac{1}{2}(g-3)} 2\cos \frac{\pi g^l(g-1-2m)}{p}.$$

Let

$$\frac{g-1}{2} - m \equiv g^{e_m} (\operatorname{mod} p), \quad m = 0, 1, \cdots, \frac{1}{2}(g-3).$$

Then

$$\rho_l = 1 + \sum_{m=0}^{\frac{1}{2}(g-3)} \omega_{l+e_m}.$$

Since

$$-1=\sum_{l=1}^{s}\omega_l,$$

hence

$$(\rho_1,\cdots,\rho_s)=(\omega_1,\cdots,\omega_s)N,$$

where $N$ is also a circular matrix and its elements are 0 or $-1$.

By this way, any unit $\eta=\rho_1^{l_1}\cdots\rho_{s-1}^{l_{s-1}}$ can easily be expanded as a linear combination of $\omega_i$'s.

Let

$$x+x^{-1}=y \quad \text{or} \quad x=\frac{y\pm\sqrt{y^2-4}}{2}.$$

Then from the equation

$$x^{p-1}+x^{p-2}+\cdots+x+1=0$$

we have the minimal equation of $\omega_l(1\leqslant l\leqslant s)$

$$1+\sum_{\nu=1}^{s}\left(\left(\frac{y+\sqrt{y^2-4}}{2}\right)^{\nu}+\left(\frac{y-\sqrt{y^2-4}}{2}\right)^{\nu}\right)=0.$$

Hence

$$(-1)^s\prod_{l=1}^{s}\omega_l=1+\sum_{l=1}^{[\frac{s}{2}]}\frac{(-4)^l}{2^{2l-1}}=1+2\sum_{l=1}^{[\frac{s}{2}]}(-1)^l=(-1)^{[\frac{s}{2}]},$$

$$\prod_{l=1}^{s}\omega_l=(-1)^{[\frac{1}{2}(s+1)]}=(-1)^{\frac{p^2-1}{8}}$$

and so $\omega_l(1\leqslant l\leqslant s)$ are all the units of $\mathscr{R}_s$.

**Theorem 1.7.** *A necessary and sufficient condition that $\omega_1,\cdots,\omega_{s-1}$ form a set of independent units is that 2 is a primitive root* mod $p$ *or 2 belongs to the exponent $s$ and $p\equiv 7 \pmod 8$.*

*Proof.* If 2 is a primitive root mod $p$, then by taking $g=2$, we have

$$\rho_l=2\cos\frac{\pi 2^l}{p}=\omega_{l-1}, \quad 1<l\leqslant s-1$$

and

$$\rho_1=2\cos\frac{2\pi}{p}=2\cos\frac{2^{s+1}\pi}{p}=\omega_s.$$

Now suppose that 2 is not a primitive root mod $p$. Let

$$2 \equiv g^l \pmod p.$$

Then $(l, p-1) > 1$ and

$$\omega_\mu = 2\cos\frac{2\pi}{p} g^\mu = \pm 2\cos\frac{\pi}{p} g^{\mu+l} = \frac{\sin\frac{\pi}{p} g^{\mu+2l}}{\sin\frac{\pi}{p} g^{\mu+l}}$$

$$= \rho_{l+\mu}\rho_{l+\mu+1}\cdots\rho_{2l+\mu-1}$$

and so

$$\omega_1\omega_{1+l}\cdots\omega_{1+(t-1)l} = (\rho_{l+1}\cdots\rho_{2l})(\rho_{2l+1}\cdots\rho_{3l})\cdots(\rho_{lt+1}\cdots\rho_{(l+1)t}).$$

The number of $\omega_i$'s in the product of the left hand side is $t$ and the number of $\rho_i$'s in the right hand side is $lt$.

Take $t = \dfrac{p-1}{(l, p-1)}$. If $(l, p-1) > 2$, then $t < s$ and $\dfrac{l(p-1)}{(l, p-1)}$ is a multiple of $p-1$ and so the right hand side is equal to $\pm 1$. Hence $\omega_1, \cdots, \omega_{s-1}$ is not a set of independent units. Suppose that $(l, p-1) = 2$. Then the exponent of 2 is $s$ and 2 is a quadratic residue mod $p$, i.e.,

$$\left(\frac{2}{p}\right) = (-1)^{\frac{p^2-1}{8}} = 1.$$

Hence $p \equiv \pm 1 \pmod 8$ (Cf. Hua Loo Keng [2], Chap. 3).

First, suppose that $p = 1 \pmod 8$. Then $l = 2m$, $2 \nmid m$ and $(m, p-1) = 1$, hence $g^m$ is a primitive root mod $p$ too. Without loss of generality, we may suppose that

$$2 \equiv g^2 \pmod p. \tag{1.27}$$

Hence there is an identity between $\dfrac{p-1}{4}$ of $\omega_i$'s

$$\omega_1\omega_3\cdots\omega_{2\frac{p-1}{4}-1} = \rho_3\rho_4\rho_5\rho_6\cdots\rho_1\rho_2 = \pm 1,$$

i.e., $\omega_1, \cdots, \omega_{s-1}$ is not a set of independent units.

Next, suppose that $p \equiv 7 \pmod 8$. Since $2 \| (p-1)$, we may suppose that $2 \| l$, otherwise $2 \| (l + p - 1)$. Hence we may assume that (1.27) holds also. If there exist $l_1, \cdots, l_s$ such that

$$\omega_1^{l_1}\omega_2^{l_2}\cdots\omega_s^{l_s} = \pm 1,$$

then

$$\pm 1 = (\rho_3\rho_4)^{l_1}(\rho_5\rho_6)^{l_2}\cdots(\rho_2\rho_3)^{l_s} = \rho_1^{l_{s-2}+l_{s-1}}\rho_2^{l_{s-1}+l_s}\cdots\rho_s^{l_{s-3}+l_{s-2}}.$$

Since the set $\rho_1, \cdots, \rho_{s-1}$ is independent and

$$\rho_1\cdots\rho_s = \pm 1,$$

we have

$$l_{s-2} + l_{s-1} = l_{s-1} + l_s = \cdots = l_{s-3} + l_{s-2}.$$

Since $2\nmid s$, hence

$$l_1 = l_2 = \cdots = l_s.$$

If $l_s = 0$, then $l_1 = \cdots = l_{s-1} = 0$. Hence $\omega_1, \cdots, \omega_{s-1}$ form a set of independent units. The lemma is proved.

## §1.6. The Dirichlet field

Let $s = 2^t$. The field $\mathscr{D}_s = Q(\sqrt{p_1}, \cdots, \sqrt{p_t})$ is a real algebraic number field of degree $s$ which is called the Dirichlet field.

Consider the solution of Pell's equation

$$x^2 - p_{i_1}\cdots p_{i_k}y^2 = \pm 4, \quad x \geqslant 0, y \geqslant 0$$

with smallest $\frac{x}{2} + \frac{\sqrt{p_{i_1}\cdots p_{i_k}}\,y}{2}$, where $k \geqslant 1$ and $1 \leqslant i_1 < \cdots < i_k \leqslant t$ is any choice of $1, \cdots, t$. We order the $d_i = p_{i_1}\cdots p_{i_k}$ such that $d_i \equiv 1 \pmod 4$ for $1 \leqslant i \leqslant m$. Set

$$\varepsilon_i = \begin{cases} \dfrac{x_i}{2} + \dfrac{\sqrt{d_i}}{2}y_i, & x_i \equiv y_i \equiv 1 (\mathrm{mod}\, 2), \quad \text{if} \quad 1 \leqslant i \leqslant m, \\ x_i + \sqrt{d_i}\,y_i, & \text{if} \quad m+1 \leqslant i \leqslant s-1. \end{cases}$$

Then $\varepsilon_1, \cdots, \varepsilon_{s-1}$ form a set of independent units of $\mathscr{D}_s$. We have also a basis of $\mathscr{D}_m$

$$\omega_1 = 1, \omega_2 = \varepsilon_1, \cdots, \quad \omega_{m+1} = \varepsilon_m, \omega_{m+2} = \sqrt{d_{m+1}}, \cdots, \quad \omega_s = \sqrt{d_{s-1}}.$$

The transformation

$$\sigma_{i_1\cdots i_k}: \sqrt{p_\nu} \to \begin{cases} -\sqrt{p_\nu}, & \text{if} \quad \nu = i_j (1 \leqslant j \leqslant k), \\ \sqrt{p_\nu}, & \text{otherwise} \end{cases}$$

is an automorphism of the field $\mathscr{D}_s$. $s-1$ transformations $\sigma_{i_1\cdots i_k}(k \geqslant 1, 1 \leqslant i_1 < \cdots < i_k \leqslant t)$ and the identity transformation $\sigma_0$ form the group of automorphism of the field. A number $\xi$ of $\mathscr{D}_s$ has its $s$ conjugates under these automorphisms.

**Lemma 1.12.** *For any given integer $l \geqslant 1$, there exists a unit $\eta_l$ of $\mathscr{D}_l$ such that*

$$\eta_l \geqslant \varepsilon_1^{(s-1)l} \tag{1.28}$$

*and*

$$|\eta_l^{(i)}| \leqslant c\eta_l^{-\frac{1}{s-1}}, \quad 2 \leqslant i \leqslant s, \tag{1.29}$$

*where* $c = c(\mathscr{D}_s)$.

*Proof.* Let $\Sigma_{i_1\cdots i_k}$ denote the number of $\varepsilon_\nu$'s which change into $\pm\varepsilon_\nu^{-1}$ under the transformation $\sigma_{i_1\cdots i_k}$. Suppose that $\varepsilon_\nu = X_\nu + \sqrt{d_\nu}Y_\nu$. Then

$$\sigma_{i_1\cdots i_k}\varepsilon_\nu = \begin{cases} \pm\varepsilon_\nu^{-1}, & \text{if } d_\nu \text{ is divisible by odd number of } p_{i_j}\ (1 \leqslant j \leqslant k), \\ \pm\varepsilon_\nu, & \text{otherwise.} \end{cases}$$

Since the number of $d_\nu$'s which possess $2r+1$ prime factors of $p_{i_j}(1 \leqslant j \leqslant k)$ and $h$ other prime factors is

$$\binom{k}{2r+1}\binom{t-k}{h},$$

therefore the number of $d_\nu$'s which have $2r+1$ prime factors of $p_{i_j}(1 \leqslant j \leqslant k)$ is equal to

$$\binom{k}{2r+1}\left(1 + \binom{t-k}{1} + \cdots + \binom{t-k}{t-k}\right) = 2^{t-k}\binom{k}{2r+1}.$$

Hence

$$\Sigma_{i_1\cdots i_k} = \sum_{2r+1 \leqslant k} 2^{t-k}\binom{t}{2r+1}.$$

Since

$$\sum_{i=0}^{k}(-1)^i\binom{k}{i} = (1-1)^k = 0$$

and

$$\sum_{i=0}^{k}\binom{k}{i} = (1+1)^k = 2^k,$$

hence

$$\sum_{2r\leqslant k}\binom{k}{2r}=\sum_{2r+1\leqslant k}\binom{k}{2r+1}=2^{k-1}$$

and so

$$\Sigma_{i_1\cdots i_k}=2^{t-1}. \tag{1.30}$$

Take $c=\varepsilon_1^{l_0 2^{t-1}}$, where $\varepsilon_1^{l_0}=\max\limits_{2\leqslant j\leqslant s}\varepsilon_j$. For any given positive integer $l$, we may define the integers $l_j(2\leqslant j\leqslant s)$ by

$$\varepsilon_1^l\leqslant\varepsilon_j^{l_j}<\varepsilon_1^{l+l_0},\quad 2\leqslant j\leqslant s-1. \tag{1.31}$$

Let

$$\eta_l=\varepsilon_1^l\varepsilon_2^{l_2}\cdots\varepsilon_{s-1}^{l_{s-1}}.$$

Then we have (1.28) by (1.31). (1.29) may be derived by (1.30) and (1.31) as follows.

$$\begin{aligned}|\eta_l^{(i)}|&<\varepsilon_1^{(l+l_0)(s-1-2^{t-1})-l2^{t-1}}=\varepsilon_1^{-l+l_0 2^{t-1}-l_0}\\&=c\varepsilon_1^{-l-l_0}\leqslant c(\varepsilon_1^{-l}\varepsilon_2^{-l_2}\cdots\varepsilon_{s-1}^{-l_{s-1}})^{\frac{1}{s-1}}\\&=c\eta_l^{-\frac{1}{s-1}}(2\leqslant i\leqslant s).\end{aligned}$$

The lemma is proved.

Form the matrix

$$\Omega=(\omega_j^{(i)}),\quad 1\leqslant i,j\leqslant s.$$

Then

$$S=\Omega'\Omega=\begin{pmatrix}A&0\\0&B\end{pmatrix},$$

where $A=(a_{ij})(1\leqslant i,j\leqslant m+1)$ and $B=(b_{ij})\ (1\leqslant i,j\leqslant s-m-1)$ in which $a_{11}=2^t$, $a_{ii}=2^{t-2}(x_{i-1}^2+d_{i-1}y_{i-1}^2)(2\leqslant i\leqslant m+1)$, $a_{1j}=a_{j1}=2^{t-1}x_{j-1}\,(2\leqslant j\leqslant m+1)$, $a_{ij}=2^{t-2}x_{i-1}x_{j-1}(2\leqslant i,\ j\leqslant m+1,\ i\neq j)$, $b_{ii}=2^t d_{m+i}(1\leqslant i\leqslant s-m-1)$ and $b_{ij}=0(i\neq j)$.

Let $\eta_l(l=1,2,\cdots)$ be a sequence of units of $\mathscr{D}_s$ satisfying (1.28) and (1.29). Clearly $\eta_l$ may be expressed as a linear combination of $\omega_i$'s with rational integer coefficients

$$\eta_l=\sum_{i=1}^{s}k_{li}\omega_i.$$

Hence from

$$(h_{l1}, \cdots, h_{ls}) = (k_{l1}, \cdots, k_{ls})S,$$

we have

$$n_l = h_{l1} = 2^{t-1}\left(2k_{l1} + \sum_{i=1}^{m} x_i k_{l,i+1}\right),$$

$$h_{lj} = \begin{cases} 2^{t-2}\left(2x_{j-1}k_{l1} + d_{j-1}y_{j-1}^2 k_{lj} + \sum_{i=2}^{m+1} x_{i-1}x_{j-1}k_{li}\right), & \text{if} \quad 2 \leqslant j \leqslant m+1, \\ 2^t k_{lj} d_{j-1}, \quad \text{if} \quad m+2 \leqslant j \leqslant s \end{cases}$$

and the simultaneous Diophantine approximation of the basis of $\mathscr{D}_s$

$$\left|\frac{h_{lj}}{n_l} - \omega_j\right| < c(\mathscr{D}_s) n_l^{-1-\frac{1}{s-1}}, \quad 1 \leqslant i \leqslant s.$$

*Remark.* The proof of Lemma 1.12 gives a simple algorithm for the computation of the sequence of units $\eta_l (l = 1, 2, \cdots)$ satisfying (1.28) and (1.29) which does not involve the solution of the system of linear equation (1.5). But in practice the field $\mathscr{D}_s$ is not as convenient as $\mathscr{R}_s$, since the $\eta_l$ given by $\mathscr{D}_s$ increases too fast as $l$ increases and moreover $s$ is equal to $2^t$.

There are also many real algebraic number fields for which the sets of independent units are known (Cf. L. Bernstein, [1]).

## § 1.7. The cubic field

Let $Q(\alpha)$ be a real cubic algebraic number field and let $\alpha$ satisfy the equation

$$x^3 - a_2x^2 - a_1x - a_0 = 0, \tag{1.32}$$

where $a_2, a_3$ are rational integers and $a_0 = \pm 1$. Suppose that (1.32) has only one real root and $\alpha > 1$. The field $Q(\alpha)$ has a basis

$$1, \alpha, \alpha^2$$

and a sequence of units $\eta_l = \alpha^l (l = 1, 2, \cdots)$ such that

$$|\eta_l^{(i)}| \leqslant \eta_l^{-\frac{1}{2}}, \quad i = 2, 3.$$

Hence we have

$$n_l = \alpha^l + \alpha^{(2)l} + \alpha^{(3)l}, \quad l = 1, 2, \cdots,$$

$$h_{lj} = \alpha^{l+j} + \alpha^{(2)l+j} + \alpha^{(3)l+j} = n_{l+j}, \quad j = 1, 2$$

and the simultaneous Diophantine approximation of the basis

$$\left|\frac{n_{l+j}}{n_l} - \alpha^j\right| \leqslant c(\alpha)n_l^{-\frac{3}{2}}, \quad j = 1, 2.$$

Here $n_l$ and $h_{l_j}(j = 1, 2)$ are taken from the same sequence of integers. It follows by (1.32) that

$$n_0 = 3, \quad n_1 = a_2, \quad n_2 = a_2^2 + 2a_1$$

and

$$n_l = \sum_{i=1}^{3} \alpha^{(i)l} = \sum_{i=1}^{3} \alpha^{(i)l-3}\alpha^{(i)3}$$

$$= \sum_{i=1}^{3} \alpha^{(i)l-3}(a_2\alpha^{(i)2} + a_1\alpha^{(i)} + a_0)$$

$$= a_2 n_{l-1} + a_1 n_{l-2} + a_0 n_{l-3}$$

for $l \geqslant 3$. Hence $n_l$ satisfies a simple recurrence relation.

For example, suppose that the minimal polynomial of $\alpha$ is

$$x^3 - x - 1 = 0.$$

Then $\alpha$ satisfies $1 < \alpha < 2$. Since

$$x^3 - x - 1 = (x - \alpha)(x^2 + \alpha x + \alpha^{-1})$$

and

$$\alpha^2 - 4\alpha^{-1} = (\alpha^3 - 4)\alpha^{-1} = (\alpha - 3)\alpha^{-1} < 0,$$

therefore $\alpha^{(2)}$ and $\alpha^{(3)}$ are conjugate complex numbers. Let

$$\eta_l = \alpha^l, \quad l = 1, 2, \cdots.$$

Then $n_l(l = 1, 2, \cdots)$ satisfy the recurrent formula

$$n_0 = 3, \quad n_1 = 0, \quad n_2 = 2, \quad n_l = n_{l-2} + n_{l-3} \quad (l \geqslant 3).$$

For real quadratic field $Q(\alpha)$, where $\alpha$ is a unit $> 1$, the rational approximation of $\alpha$ thus obtained is in essence the continued fraction.

*Remark.* It is easily seen that we may also obtain the less precise result of simultaneous Diophantine approximation of the $\omega_i$'s, if $\eta_l(l = 1, 2, \cdots)$ is a sequence of algebraic integers and

$$\sum_{i=2}^{s} |\eta_l^{(i)}| = o(\eta_l).$$

In particular, if we take $\eta_l = \alpha^l (l = 1, 2, \cdots)$, then the rational approximation of $\alpha$ so obtained is in essence the Jacobi-Perron algorithm. Here we may take $\alpha > 1$ and the absolute values of its conjugates are all less than 1 which then $\alpha$ is called a PV number. We shall discuss them in the next chapter.

## Notes

Theorem 1.4: Cf. K. Ramachandra, [1].

The other results: Cf. Hua Loo Keng and Wang Yuan [1, 4, 5, 6, 7, 8] and Hua Loo Keng, Wang Yuan and Pei Ding Yi [1].

*Chapter 2*

# Recurrence Relations and Rational Approximation

## § 2.1. The recurrence formula for the elementary symmetric function

Let $\mathscr{F}_s = Q(\alpha)$ be a real algebraic number field of degree $s$. We shall give in this chapter an algorithm for the simultaneous Diophantine approximation obtained by $\eta_l = \alpha^l (l = 1, 2, \cdots)$ which is essentially the Jacobi-Perron algorithm (Cf. L. Bernstein [1]). It yields less precise results but the computations of $n_l$ and $h_{l_j}(1 \leqslant j \leqslant s)$ are comparatively simple.

Let $\alpha$ satisfy the irreducible equation

$$f(x) = x^s - a_{s-1}x^{s-1} - \cdots - a_1 x - a_0 = 0, \tag{2.1}$$

where $a_{s-1}, \cdots, a_1, a_0$ are rational integers. Let

$$\alpha(= \alpha^{(1)}) > 1, \quad |\alpha^{(2)}| \leqslant \cdots \leqslant |\alpha^{(s)}| < 1.$$

An algebraic number with this property is called a PV number. Let

$$\rho = -\frac{\ln|\alpha^{(s)}|}{\ln \alpha}.$$

Then

$$|\alpha^{(i)}| \leqslant \alpha^{-\rho}, \quad 2 \leqslant i \leqslant s \tag{2.2}$$

and

$$0 < \rho \leqslant \frac{1}{s-1} - \frac{\ln|a_0|}{(s-1)\ln \alpha}, \tag{2.3}$$

since

$$|\alpha^{(s)}| = \frac{|a_0|}{\alpha|\alpha^{(2)} \cdots \alpha^{(s-1)}|} \geqslant \frac{|a_0|}{\alpha|\alpha^{(s)}|^{s-2}}.$$

Let $S_l$ denote the elementary symmetric functions of the roots of $f(x)$

$$S_l = \alpha^l + \alpha^{(2)l} + \cdots + \alpha^{(s)l}, \quad l = 0, 1, \cdots.$$

Then $S_l$'s are all rational integers. Without loss of generality, we may assume that $S_l > 0$, since $\alpha$ is a PV number. By (2.2), we have

$$|S_l - \alpha^l| \leqslant (s-1)|\alpha^{(s)}|^l = (s-1)\alpha^{-\rho l} \leqslant (s-1)s^\rho S_l^{-\rho}, \tag{2.4}$$

since

$$S_l \leqslant \alpha^l + (s-1) < s\alpha^l.$$

**Theorem 2.1.**

$$\left|\frac{S_{n+k}}{S_n} - \alpha^k\right| \leqslant c(\alpha) S_n^{-1-\rho}, \quad 1 \leqslant k \leqslant s-1.$$

*Proof.* From (2.4), we have

$$\begin{aligned}\frac{S_{n+k}}{S_n} &= \frac{\alpha^{n+k} + O(S_{n+k}^{-\rho})}{\alpha^n + O(S_n^{-\rho})} \\ &= \alpha^k(1 + O(S_n^{-1-\rho}))(1 + O(S_n^{-1-\rho}))^{-1} \\ &= \alpha^k + O(S_n^{-1-\rho}).\end{aligned}$$

The theorem is proved.

It is well-known that $S_l$ can be evaluated by Newton's formula

$$S_0 = s, \quad S_1 = a_{s-1}, \cdots, \quad S_{s-1} = a_{s-1}S_{s-2} + \cdots + a_1 S_0$$

and

$$S_n = a_{s-1}S_{n-1} + a_{s-2}S_{n-2} + \cdots + a_1 S_{n-s+1} + a_0 S_{n-s} \quad (n \geqslant s).$$

*Remark.* It is easily seen from (2.3) that to take $\alpha$ to be a unit is more advantageous.

## § 2.2. The generalization of $S_n$

Let $\xi$ be a number of $Q(\alpha)$ and

$$Q_n = \sum_{i=1}^{s} \xi^{(i)} \alpha^{(i)n}. \tag{2.5}$$

Then $Q_n$ is a rational number which is called the generalization of $S_n$. It also satisfies a recurrence relation. By (2.1), we have

$$Q_n = \sum_{i=1}^{s} \xi^{(i)} \alpha^{(i)n-s} \alpha^{(i)s}$$

$$= \sum_{i=1}^{s} \xi^{(i)} \alpha^{(i)n-s} (a_{s-1} \alpha^{(i)s-1} + \cdots + a_1 \alpha^{(i)} + a_0)$$

$$= a_{s-1} \sum_{i=1}^{s} \xi^{(i)} \alpha^{(i)n-1} + \cdots + a_1 \sum_{i=1}^{s} \xi^{(i)} \alpha^{(i)n-s+1} + a_0 \sum_{i=1}^{s} \xi^{(i)} \alpha^{(i)n-s}$$

$$= a_{s-1} Q_{n-1} + \cdots + a_1 Q_{n-s+1} + a_0 Q_{n-s} \tag{2.6}$$

for $n \geqslant s$. Hence the sequences $(Q_n)$ and $(S_n)$ differ only in their initial values.

Now we shall define $\xi$ such that $(Q_0, \cdots, Q_{s-1})$ is the given initial value. Choose

$$\Omega = (\alpha^{(i)j}), \quad 1 \leqslant i \leqslant s, \quad 0 \leqslant j \leqslant s-1.$$

Then

$$S = \Omega' \Omega$$

is a non-singular matrix with integer coefficients. From

$$(Q_0, \cdots, Q_{s-1}) = (\xi^{(1)}, \cdots, \xi^{(s)}) \Omega,$$

we have

$$(Q_0, \cdots, Q_{s-1}) = (\xi^{(1)}, \cdots, \xi^{(s)}) \Omega'^{-1} \Omega' \Omega$$

and so

$$(\xi^{(1)}, \cdots, \xi^{(s)}) = (Q_0, \cdots, Q_{s-1}) S^{-1} \Omega'. \tag{2.7}$$

Hence $\xi \neq 0$, if $Q_0, \cdots, Q_{s-1}$ are not all equal to zero. And for any given integral initial vector $(Q_0, \cdots, Q_{s-1})$, we obtain a sequence of rational integers $Q_n (n = s, s+1, \cdots)$ by (2.6).

**Theorem 2.2** *Let* $\mathbf{Q} = (Q_0, \cdots, Q_{s-1})$ *be a non-zero integral vector. Then there exists a constant* $c_1(\mathbf{Q}, \alpha)$ *such that* $|Q_n| > 1$ *and that*

$$\left| \frac{Q_{n+k}}{Q_n} - \alpha^k \right| \leqslant c(\mathbf{Q}, \alpha) |Q_n|^{-1-\rho}, \quad 1 \leqslant k \leqslant s-1$$

*holds for* $n > c_1(\mathbf{Q}, \alpha)$.

*Proof.* By (2.5) and (2.7), we have $|Q_n| > 1$ and

$$Q_n = \xi \alpha^n + O(|\alpha^{(s)}|^n) = \xi \alpha^n + O(\alpha^{-\rho n}) = \xi \alpha^n + O(|Q_n|^{-\rho})$$

for $n > c_1(\mathbf{Q}, \alpha)$, where the constant implied by the symbol "$O$" depends only on $\mathbf{Q}$ and $\alpha$. The theorem follows.

Let $\omega_1(=1), \omega_2, \cdots, \omega_s$ be any given basis of $Q(\alpha)$. Then

$$\omega_j = \sum_{k=1}^{s} t_{jk}\alpha^{k-1}, \quad 2 \leqslant j \leqslant s,$$

where $t_{jk}(2 \leqslant j \leqslant s, 1 \leqslant k \leqslant s)$ are rational numbers. Let

$$Q_n(j) = \sum_{k=1}^{s} t_{jk}Q_{n+k-1}, \quad 2 \leqslant j \leqslant s.$$

Clearly, $Q_n(j)$ satisfies also the recurrence relation

$$Q_n(j) = a_{s-1}Q_{n-1}(j) + \cdots + a_1Q_{n-s+1}(j) + a_0Q_{n-s}(j) \quad (n \geqslant s)$$

for $2 \leqslant j \leqslant s$, where the initial values $Q_0(j), \cdots, Q_{s-1}(j)$ are determined by $Q_0, \cdots, Q_{2s-2}$ and $t_{jk}(1 \leqslant k \leqslant s)$. From Theorem 2.2, we can derive

**Theorem 2.3.** *Under the assumption of Theorem 2.2, the relation*

$$\left|\frac{Q_n(j)}{Q_n} - \omega_j\right| = O(|Q_n|^{-1-\rho}), \quad 2 \leqslant j \leqslant s$$

*holds for $n > c(\mathbf{Q}, \alpha)$, where the constant implied by the symbol "O" depends only on $Q, \alpha$ and $\omega_i$'s.*

We may also use the "Yang Hui triangular method" to prove Theorem 2.2. (Cf. Hua Loo Keng [3]). Suppose that $(Q_0, \cdots, Q_{s-1})$ is a given non-zero integral initial vector and that $Q_n(n \geqslant s)$ is a sequence of integers defined by (2.6). Then

$$\begin{aligned}(1 - a_{s-1}x - \cdots - a_1x)\sum_{n=0}^{\infty} Q_nx^n \\ &= Q_0 + (Q_1 - a_{s-1}Q_0)x + \cdots \\ &\quad + (Q_{s-1} - a_{s-1}Q_{s-2} - \cdots - a_1Q_0)x^{s-1} \\ &= P_{s-1}(x)(\text{say}).\end{aligned}$$

Hence

$$\sum_{n=0}^{\infty} Q_nx^n = \frac{P_{s-1}(x)}{\prod_{i=1}^{s}(1 - \alpha^{(i)}x)} = \sum_{i=1}^{s}\frac{A_i}{1 - \alpha^{(i)}x}, \tag{2.8}$$

where

$$A_i = \frac{\alpha^{(i)s-1}P_{s-1}(\alpha^{(i)-1})}{\prod_{j\neq i}(\alpha^{(i)} - \alpha^{(j)})}, \quad 1 \leqslant i \leqslant s.$$

Expand the right hand of (2.8) as a power series and then we have

$$Q_n = \sum_{i=1}^{s} A_i \alpha^{(i)n} \tag{2.9}$$

by comparing the coefficients of $x^n$ of (2.8). Since $(Q_0, \cdots, Q_{s-1})$ is not a zero vector and $1, \alpha, \cdots, \alpha^{s-1}$ is a basis of $Q(\alpha)$, we have $A_i \neq 0 (1 \leqslant i \leqslant s)$ and so Theorem 2.2 follows by (2.9).

*Remarks.* 1. In practical use, we often take the initial values $Q_0 = \cdots = Q_{s-2} = 0, Q_{s-1} = 1$ and the basis $1, \omega_2, \cdots, \omega_s$, where

$$\omega_j = \alpha^{j-1} - a_{s-1}\alpha^{j-2} - \cdots - a_{s-j+2}\alpha - a_{s-j+1}, \quad 2 \leqslant j \leqslant s.$$

(Cf. L. Bernstein [1]). Since

$$\omega_j = (a_{s-j}\alpha^{s-j} + \cdots + a_1\alpha + a_0)\alpha^{-s+j-1}, \quad 2 \leqslant j \leqslant s$$

by (2.1), we have

$$\overline{|\omega|} \leqslant s \max_{0 \leqslant j \leqslant s-1} |a_j|.$$

2. By (2.3), we have

$$0 < \rho \leqslant \frac{1}{s-1}.$$

Perron proved that the relation

$$\left|\frac{Q_{n+1}}{Q_n} - \alpha\right| < c\,(\mathbf{Q}, \alpha)\,|Q_n|^{-1-\frac{1}{s-1}}$$

holds only for $s = 2$ (the continued fraction) and for $s = 3$ and $\alpha^{(2)}, \alpha^{(3)}$ are conjugate complex numbers (Cf. O. Perron [1]). Hence the results given here are rougher than the corresponding results of chapter 1. However compared with $(n_l, h_{l1}, \cdots, h_{ls})$, the number of elementary operations required for obtaining $Q_n$ is decreased, since $Q_n$ satisfies a simple recurrence relation

## § 2.3. PV numbers

Let $\mathscr{F}_s$ be a real algebraic number field of degree $s$. We shall show that it requires only $c(\mathscr{F}_s)$ elementary operations for obtaining a PV number $\alpha$ of degree $s$ in $\mathscr{F}_s$. Hence when $\mathscr{F}_s = Q(\alpha)$ we may obtain the simultaneous Diophantine approximation of the basis of $\mathscr{F}_s$ by the method stated in § 2.2.

First, we mention two lemmas.

**Lemma 2.1.** *Let $s = r_1 + 2r_2$. Let $\xi_1, \cdots, \xi_s$ be $s$ linear forms of $x_1, \cdots, x_s$ with determinant $\Delta \neq 0$, where $\xi_1, \cdots, \xi_{r_1}$ are forms with real coefficients and the other forms have complex coefficients such that $\xi_{r_1+r_2+j} = \overline{\xi_{r_1+j}}\,(1 \leqslant j \leqslant r_2)$. Further let $\lambda_1, \cdots, \lambda_s$ be $s$ positive numbers satisfying $\lambda_{r_1+j} = \lambda_{r_1+r_2+j}\,(1 \leqslant j \leqslant r_2)$ and $\lambda_1 \cdots \lambda_s \geqslant \left(\frac{2}{\pi}\right)^{r_2} |\Delta|$. Then there exists a non-zero integer point such that*

$$|\xi_1| \leqslant \lambda_1, \cdots, |\xi_s| \leqslant \lambda_s$$

(Cf. Hua Loo Keng [2], Chap. 20).

**Lemma 2.2.** *Let $\alpha$ be a number of $\mathscr{F}_s$ satisfying the irreducible equation*

$$h(x) = 0, \quad \partial^0 h = l.$$

*Let*

$$g(x) = \prod_{i=1}^{s} (x - \alpha^{(i)}).$$

*Then $l \mid s$ and $g(x)$ is a polynomial with rational coefficients such that*

$$g(x) = c h(x)^{s/l},$$

*where $c$ is a rational number.* (Cf. Hua Loo Keng [2], Chap. 16).

Let $\omega_1, \cdots, \omega_s$ be a basis of $\mathscr{F}_s$, where the $\omega_i$'s are algebraic integers. Let $s = r_1 + 2r_2$ and

$$\alpha^{(i)} = \omega_1^{(i)} x_1 + \cdots + \omega_s^{(i)} x_s, \quad 1 \leqslant i \leqslant s, \tag{2.10}$$

where $\alpha^{(i)}(1 \leqslant i \leqslant r_1)$ have real coefficients and $\alpha^{(i)}(r_1 + 1 \leqslant i \leqslant s)$ have complex coefficients such that

$$\omega^{(r_1+r_2+j)} = \overline{\omega^{(r_1+j)}}, \quad 1 \leqslant j \leqslant r_2.$$

Let

$$\Omega = (\omega_j^{(i)}), \quad 1 \leqslant i, j \leqslant s.$$

Take

$$\lambda_1 = \left(\frac{2}{\pi}\right)^{r_2} |\Delta| (1 - \varepsilon)^{-(s-1)}, \quad \lambda_2 = \cdots = \lambda_s = 1 - \varepsilon,$$

where $0 < \varepsilon < 1$ and $\Delta = \det \Omega$. Then it follows by Lemma 2.1 that there exists a non-zero integer point $(x_1, \cdots, x_s)$ such that

$$|\alpha| \leqslant \lambda_1 \quad |\alpha^{(i)}| \leqslant 1 - \varepsilon \left(\leqslant \left(\frac{2}{\pi}\right)^{\frac{r_2}{s-1}} |\Delta|^{\frac{1}{s-1}} \alpha^{-\frac{1}{s-1}}\right),$$

$$2 \leqslant i \leqslant s. \tag{2.11}$$

Hence $\alpha$ is a non-zero algebraic integer and we may assume that $\alpha > 0$. By

$$|N(\alpha)| = \alpha |\alpha^{(2)} \cdots \alpha^{(s)}| \geqslant 1,$$

we have

$$\alpha \geqslant (1 - \varepsilon)^{-(s-1)} > 1 \tag{2.12}$$

and so $\alpha$ is a PV number of degree $s$ by (2.11), (2.12) and Lemma 2.2. From (2.10), we have

$$(x_1, \cdots, x_s) = (\alpha^{(1)}, \cdots, \alpha^{(s)}) \Omega'^{-1}$$

and so

$$|x_i| \leqslant \left(\frac{2}{\pi}\right)^{r_2} \frac{W}{(1-\varepsilon)^{s-1}} + (s-1)(1-\varepsilon) \frac{W}{|\Delta|}$$

$$= c(\mathscr{F}_s), \quad 1 \leqslant i \leqslant s, \tag{2.13}$$

where $W$ denotes the maximum of the absolute values of the cofactors of $\Delta$. Hence there is a non-zero integer point in the parallelpiped (2.13) satisfying (2.11) and so we obtain a PV number $\alpha$ of degree $s$ in $\mathscr{F}_s$.

## § 2.4. The roots of the equation $F(x) = 0$

Let $s \geqslant 2$. We denote the largest real root of the equation

$$F(x) = x^s - x^{s-1} - \cdots - x - 1 = 0$$

by $\eta (= \eta^{(1)})$ and its other roots by $\eta^{(2)}, \cdots, \eta^{(s)}$.

**Lemma 2.3.**

$$2 - 2^{-(s-1)} < \eta < 2 - 2^{-s} \tag{2.14}$$

*and*

$$|\eta^{(i)}| \leqslant \eta - 1, \quad 2 \leqslant i \leqslant s. \tag{2.15}$$

To prove Lemma 2.3, we shall need

**Lemma 2.4.** *If the coefficients of the polynomial*

$$g(x) = a_s x^s + a_{s-1}x^{s-1} + \cdots + a_1 x + a_0$$

*satisfy* $a_s \geqslant a_{s-1} \geqslant \cdots \geqslant a_1 \geqslant a_0 > 0$, *then no root of the equation* $g(x)=0$ *has modulus greater than* 1.

*Proof.* Since

$$\begin{aligned}|(1-x)g(x)| &\geqslant a_s|x|^{s+1} - ((a_s - a_{s-1})|x|^s \\ &\quad + (a_{s-1} - a_{s-2})|x|^{s-1} + \cdots + (a_1 - a_0)|x| + a_0) \\ &> a_s|x|^s(|x| - 1) > 0\end{aligned}$$

for $|x| > 1$. The lemma follows.

The proof of Lemma 2.3. Denote

$$Q(x) = (x-1)F(x) = x^{s+1} - 2x^s + 1.$$

Then

$$\begin{aligned}Q(2-2^{-s}) &= (2-2^{-s})^{s+1} - 2(2-2^{-s})^s + 1 \\ &= 1 - 2^{-s}(2-2^{-s})^s = 1 - (1-2^{-(s+1)})^s > 0\end{aligned}$$

and

$$\begin{aligned}Q(2-2^{-(s-1)}) &= (2-2^{-(s-1)})^{s+1} - 2(2-2^{-(s-1)})^s + 1 \\ &= 1 - 2^{-(s-1)}(2-2^{-(s-1)})^s = 1 - (2^{s^{-1}} - 2^{-s+s^{-1}})^s.\end{aligned}$$

Let

$$R(x) = 2^s - 1 - 2^{s-s^{-1}}.$$

Then

$$\begin{aligned}R'(s) &= 2^s \ln 2 - 2^{s-s^{-1}}(1+s^{-2})\ln 2 \\ &= 2^s(1 - 2^{-s^{-1}}(1+s^{-2}))\ln 2.\end{aligned}$$

Since

$$2^s \geqslant (1+s^{-2})^{s^2},$$

i.e., $R'(s) > 0$ for $s \geqslant 2$, therefore $R(s)$ increases, if $s \geqslant 2$. Hence

$$2^s - 1 - 2^{s-s^{-1}} > 0,$$

$$2^{s^{-1}} - 2^{-s+s^{-1}} > 1$$

and so $Q(2-2^{-(s-1)}) < 0$. (2.14) is thus proved.

Let

$$F(x)=(x-\eta)f(x)$$

and

$$f(x)=x^{s-1}+\beta_{s-2}x^{s-2}+\cdots+\beta_1x+\beta_0.$$

Then we have a system of linear equations

$$\begin{aligned}-\eta\beta_0&=-1,\\ \beta_0-\eta\beta_1&=-1,\\ \cdots\cdots&\\ \beta_{s-2}-\eta&=-1.\end{aligned}$$

Hence

$$\beta_0=\frac{1}{\eta},\quad \beta_1=\frac{\eta+1}{\eta^2},\cdots,\quad \beta_{s-3}=\frac{\eta^{s-3}+\eta^{s-4}+\cdots+\eta+1}{\eta^{s-2}},$$

$$\beta_{s-2}=\frac{\eta^{s-2}+\eta^{s-3}+\cdots+\eta+1}{\eta^{s-1}}=\eta-1.$$

Let

$$\gamma_j=\begin{cases}\beta_j/\beta_{j+1}, & \text{if}\quad 0\leqslant j\leqslant s-3,\\ \beta_j, & \text{if}\quad j=s-2.\end{cases}$$

Then

$$\gamma_{s-2}>\gamma_{s-3}>\cdots>\gamma_1>\gamma_0, \tag{2.16}$$

since $\dfrac{x}{x+1}$ is an increasing function of $x$ for $x\geqslant 0$ and

$$\gamma_j=\frac{\eta^{j+1}+\eta^j+\cdots+\eta}{\eta^{j+1}+\eta^j+\cdots+\eta+1}\qquad (0\leqslant j\leqslant s-2).$$

Let $x=\beta_{s-2}y$ and $g(y)=f(\beta_{s-2}y)$. Then

$$g(y)=\beta_{s-2}^{s-1}y^{s-1}+\beta_{s-2}^{s-1}y^{s-2}+\beta_{s-3}\beta_{s-2}^{s-3}y^{s-3}+\cdots+\beta_1\beta_{s-2}y+\beta_0.$$

From (2.16), we have

$$\beta_{s-2}^{s-1}>\beta_{s-3}\beta_{s-2}^{s-3}>\cdots>\beta_1\beta_{s-2}>\beta_0.$$

Hence it follows by Lemma 2.4 that the moduli of the roots of $g(y)=0$ are all $\leqslant 1$, i.e., the moduli of the roots of $f(x)=0$ are all $\leqslant\beta_{s-2}=$ $=\eta-1$. (2.15) and so the lemma is proved.

**Lemma 2.5.** *$|\eta^{(2)}|=\eta^{-1}$ for $s=2$ and $|\eta^{(2)}|=|\eta^{(3)}|=\eta^{-\frac{1}{2}}$ for $s=3$.*

*Proof.* Obviously $|\eta^{(2)}| = \eta^{-1}$ for $s=2$. For $s=3$, since

$$x^3 - x^2 - x - 1 = (x-\eta)(x^2 + (\eta-1)x + \eta^{-1}), \quad \eta > 1$$

and

$$(\eta-1)^2 - 4\eta^{-1} = \frac{\eta^3 - 2\eta^2 + \eta - 4}{\eta} = \frac{-\eta^2 + 2\eta - 3}{\eta}$$

$$= -\frac{(\eta-1)^2+2}{\eta} < 0,$$

therefore $\eta^{(2)}$ and $\eta^{(3)}$ are conjugate complex numbers. Hence $|\eta^{(2)}| = |\eta^{(3)}| = \eta^{-\frac{1}{2}}$. The lemma is proved.

## §2.5. The roots of the equation $G(x)=0$

Let $s \geqslant 2$. We denote the largest real root of the equation

$$G(x) = x^s - Lx^{s-1} - 1 = 0$$

by $\tau(=\tau^{(1)})$ and its other roots by $\tau^{(2)}, \cdots, \tau^{(s)}$, where $L$ is an integer $\geqslant 2$.

**Lemma 2.6.**

$$L < \tau < L + L^{-(s-1)} \tag{2.17}$$

*and*

$$(L + L^{-\frac{1}{s-1}})^{-\frac{1}{s-1}} < |\tau^{(i)}| < (L - (L-1)^{-\frac{1}{s-1}})^{-\frac{1}{s-1}},$$
$$2 \leqslant i \leqslant s. \tag{2.18}$$

*Proof.* (2.17) follows immediately by

$$G(L) = -1 < 0$$

and

$$G(L + L^{-(s-1)}) = L^{-(s-1)}(L + L^{-(s-1)})^{s-1} - 1 > 0.$$

Let $\chi$ be the real root of the equation

$$g(x) = x^s + Lx^{s-1} - 1 = 0$$

in the interval $(0,1)$. Since $g'(x) \neq 0$ in $(0,1)$, therefore $\chi$ is the only root of $g(x)=0$ in $(0,1)$. It follows from

$$g(0) = -1 < 0$$

and

$$g(1) = L > 0$$

that $g(X - \delta) < 0$ for any $\delta$ satisfying $0 < \delta < X$. Hence on the circle in the complex plane $|x| = X - \delta$, we have

$$1 > |x^s + Lx^{s-1}|.$$

It follows by Rouché's theorem that 1 and $G(x)$ have the same number of zero in the circle $|x| < X - \delta$, i.e., $G(x)$ has no zero in the circle $|x| < X - \delta$. Let $\delta \to 0$. Then the moduli of roots of $G(x) = 0$ are all $\geqslant X$. Since

$$\begin{aligned} g((L + L^{-\frac{1}{s-1}})^{-\frac{1}{s-1}}) &= ((L + L^{-\frac{1}{s-1}})^{-\frac{1}{s-1}} + L)(L + L^{-\frac{1}{s-1}})^{-1} - 1 \\ &< (L^{-\frac{1}{s-1}} + L)(L + L^{-\frac{1}{s-1}})^{-1} - 1 = 0, \end{aligned}$$

we have the left hand side of (2.18)

$$|\tau^{(i)}| \geqslant X > (L + L^{-\frac{1}{s-1}})^{-\frac{1}{s-1}}, \quad 2 \leqslant i \leqslant s.$$

Suppose that $L > 2$. Let $\Omega$ be the only real root of the equation

$$h(x) = x^s - Lx^{s-1} + 1 = 0$$

in the interval $(0, 1)$. Since

$$h(0) = 1 > 0$$

and

$$h(1) = -L + 2 < 0,$$

we have $h(\Omega + \delta) < 0$ for any $\delta$ satisfying $0 < \delta < 1 - \Omega$. Hence on the circle $|x| = \Omega + \delta$, we have

$$|Lx^{s-1}| > |x^s + 1|.$$

It follows by Rouché's theorem that $x^{s-1}$ and $G(x)$ have the same number of zeros in the circle $|x| < \Omega + \delta$, i.e., $G(x)$ has $s - 1$ zeros in the circle $|x| < \Omega + \delta$. Let $\delta \to 0$. Then $G(x) = 0$ has $s - 1$ roots in the circle $|x| \leqslant \Omega$. Since

$$\begin{aligned} &h((L - (L-1)^{-\frac{1}{s-1}})^{-\frac{1}{s-1}}) \\ &\quad = ((L - (L-1)^{-\frac{1}{s-1}})^{-\frac{1}{s-1}} - L)(L - (L-1)^{-\frac{1}{s-1}})^{-1} + 1 \\ &\quad = (L - (L-1)^{-\frac{1}{s-1}})^{-1}((L - (L-1)^{-\frac{1}{s-1}})^{-\frac{1}{s-1}} \\ &\qquad - (L-1)^{-\frac{1}{s-1}}) < 0, \end{aligned}$$

we have

$$|\tau^{(i)}| \leqslant \Omega < (L-(L-1)^{-\frac{1}{s-1}})^{-\frac{1}{s-1}}, \quad 2 \leqslant i \leqslant s.$$

Suppose that $L=2$. We proceed to prove that the equation

$$G(x) = x^s - 2x^{s-1} - 1 = 0$$

has $s-1$ roots with moduli $<1$. This is equivalent to proving that the equation

$$G^*(y) = y^s + 2y - 1 = 0$$

has $s-1$ roots with moduli $>1$. Let $\delta > 0$. Then it follows from Rouché's theorem that $G^*_\delta(y) = y^s + (2+\delta)y - 1$ and $y$ have the same number of zeros in the circle $|y| < 1$, i.e., $G^*_\delta(y) = 0$ has $s-1$ roots with moduli $\geqslant 1$. Let $\delta \to 0$. Then $G^*(y) = 0$ has also $s-1$ roots with moduli $\geqslant 1$. Since $G^*(y) = 0$ has no root satisfying $|y| = 1, G^*(y) = 0$ has $s-1$ roots with moduli $>1$. The right hand side of (2.18) and also the lemma is proved.

**Lemma 2.7.** $|\tau^{(2)}| = \tau^{-1}$ *for* $s=2$ *and* $|\tau^{(2)}| = |\tau^{(3)}| = \tau^{-\frac{1}{2}}$ *for* $s=3$.

*Proof.* Clearly $|\tau^{(2)}| = \tau^{-1}$ for $s=2$. For $s=3$, since

$$x^3 - Lx^2 - 1 = (x-\tau)(x^2 + (\tau - L)x + \tau^{-1}), \quad \tau > L$$

and

$$(\tau - L)^2 - 4\tau^{-1} = \frac{\tau^3 - 2L\tau^2 + L^2\tau - 4}{\tau} = \frac{-L\tau^2 + L^2\tau - 3}{\tau} < 0,$$

hence $\tau^{(2)}$ and $\tau^{(3)}$ are conjugate complex numbers and

$$|\tau^{(2)}| = |\tau^{(3)}| = \tau^{-\frac{1}{2}}.$$

The lemma is proved.

*Remark.* Minkowski proved that except when $Q(\alpha)$ is a real quadratic field or $Q(\alpha)$ is a cubic field and $\alpha^{(2)}, \alpha^{(3)}$ are conjugate complex numbers, the real algebraic number field $Q(\alpha)$ does not contain a unit $\eta$ such that

$$\eta > 1, \quad |\eta^{(2)}| = \cdots = |\eta^{(s)}|.$$

(Cf. H. Minhowshi [1], O. Perron [1]). However the above lemma shows that the absolute values of the conjugates of the unit $\tau$ are approximately equal if $L$ is sufficiently large.

## § 2.6. The roots of the equation $H(x)=0$

Let $s \geqslant 2$ and $A_1, \cdots, A_{s-1}$ be integers defined by the relation

$$A_1 = \binom{2s}{1}, \quad A_k = \binom{2s}{k} - A_1\binom{2s-2}{k-1} - \cdots - A_{k-1}\binom{2s-2k+2}{1},$$
$$s-1 \geqslant k > 1.$$

Let $r$ be the positive integer satisfying

$$s^2 > \frac{2}{r^{s-1}} + \frac{|A_1|}{r^{s-2}} + \cdots + \frac{|A_{s-2}|}{r}.$$

Further let $\omega(=\omega^{(1)})$ denote the largest real root of the equation

$$H(x) = x^s - s^2 r^{s-1} x^{s-1} + (-1)^{s-2} A_{s-2} r^{s-2} x^{s-2} + \cdots - A_1 r x - 1 = 0$$

and $\omega^{(2)}, \cdots, \omega^{(s)}$ its other roots.

**Lemma 2.8.**

$$\omega = s^2 r^{s-1} + O(r^{\frac{s}{2}-1}) \tag{2.19}$$

*and*

$$\omega^{(i)} = -\frac{1}{4\left(\sin\frac{\pi i}{s}\right)^2 r} + O(r^{-\frac{s}{2}-1}), \quad 1 \leqslant i \leqslant s-1, \tag{2.20}$$

*where the constant implied by the symbol "O" depends on s only.*

To prove Lemma 2.8, we shall need

**Lemma 2.9.**

$$\begin{aligned}(a+b)^{2s} &- A_1 ab(a+b)^{2s-2} - A_2 a^2 b^2 (a+b)^{2s-4} - \cdots \\ &- A_{s-2} a^{s-2} b^{s-2} (a+b)^4 + (-1)^{s-1} s^2 a^{s-1} b^{s-1} (a+b)^2 \\ &- (a^s + (-1)^{s-1} b^s)^2 = 0.\end{aligned} \tag{2.21}$$

*Proof.*
$$(a + b)^{2s} - A_1 ab(a + b)^{2s-2}$$

is a symmetric function of $a, b$ without the term $ab^{2s-1}$. $(a + b)^{2s} - A_1 ab(a + b)^{2s-2} - A_2 a^2 b^2 (a + b)^{2s-4}$ is also a symmetric function of $a, b$ without the terms $ab^{2s-1}$ and $a^2 b^{2s-2}$ and so on. Hence

$$\begin{aligned}(a + b)^{2s} - A_1 ab(a + b)^{2s-2} - A_2 a^2 b^2 (a + b)^{2s-4} - \cdots \\ - A_{s-1} a^{s-1} b^{s-1} (a + b)^2 - a^{2s} - b^{2s} - k a^s b^s = 0,\end{aligned}$$

where $k$ is a constant. Let $a = -b$. Then

$$2b^{2s} + (-1)^s b^{2s} k = 0$$

and so

$$k = (-1)^{s-1} 2.$$

Hence

$$\begin{aligned}(a + b)^{2s} - A_1 ab(a + b)^{2s-2} - \cdots - A_{s-1} a^{s-1} b^{s-1} (a + b)^2 \\ - (a^s + (-1)^{s-1} b^s)^2 = 0.\end{aligned}$$

Divide the above formula by $(a + b)^2$ and then put $a = -b$. This yields that

$$\begin{aligned}(-1)^s A_{s-1} b^{2s-2} - s^2 b^{2s-2} = 0, \\ A_{s-1} = (-1)^s s^2.\end{aligned}$$

The lemma is proved.

The proof of Lemma 2.8. Put $a + b = y, ab = -r$ and $a^s + (-1)^{s-1} b^s = -1$ in (2.21). Then we have the equation

$$\begin{aligned}y^{2s} + A_1 r y^{2s-2} - A_2 r^2 y^{2s-4} - \cdots + (-1)^{s-1} A_{s-2} r^{s-2} y^4 \\ + s^2 r^{s-1} y^2 - 1 = 0 \qquad (2.22)\end{aligned}$$

has $a + b$ as a solution, where $a, b$ satisfy

$$a = -\frac{r}{b}, \quad a^s + (-1)^{s-1} b^s = -1. \qquad (2.23)$$

By (2.23), we have

$$b^{2s} + (-1)^{s-1} b^s - r^s = 0.$$

Denote $\zeta = e^{\pi i/s}$. Then

$$\begin{aligned}b &= \left(\frac{(-1)^{s+l} + (1 + 4r^s)^{1/2}}{2}\right)^{1/s} \zeta^l \\ &= \sqrt{r}\left(\frac{(-1)^{s+l}}{2r^{s/2}} + \left(1 + \frac{1}{4r^s}\right)^{1/2}\right)^{1/s} \zeta^l\end{aligned}$$

$$= \sqrt{r}\left(1 + \frac{(-1)^{s+l}}{2r^{s/2}} + O(r^{-s})\right)^{1/s} \zeta^l$$

$$= \sqrt{r}\left(1 + \frac{(-1)^{s+l}}{2sr^{s/2}} + O(r^{-s})\right)^{1/s} \zeta^l,$$

$$1 \leqslant l \leqslant 2s.$$

Substituting into (2.23), we have

$$a = -\sqrt{r}\left(1 + \frac{(-1)^{s+l-1}}{2sr^{s/2}} + O(r^{-s})\right)\bar{\zeta}^l, \quad 1 \leqslant l \leqslant 2s$$

and so the roots of (2.22)

$$y = (\zeta^l - \bar{\zeta}^l)\sqrt{r} + \frac{(-1)^{s+l}(\zeta^l + \bar{\zeta}^l)}{2sr^{\frac{s-1}{2}}} + O(r^{-s+\frac{1}{2}}), \quad 1 \leqslant l \leqslant 2s.$$

Setting the variable $z = y^2$ in equation (2.22), we know that the roots of the equation

$$z^s + A_1 r z^{s-1} - A_2 r^2 z^{s-2} - \cdots + (-1)^{s-1} A_{s-2} r^{s-2} z^2 + s^2 r^{s-1} z - 1 = 0 \tag{2.24}$$

are

$$s^{-2} r^{-s+1} + O(r^{-\frac{3s}{2}+1}), \quad -4\left(\sin\frac{\pi i}{s}\right)^2 r + O(r^{-\frac{s}{2}+1}), \quad 1 \leqslant i \leqslant s-1.$$

Hence the roots of $H(x) = 0$ are (2.19) and (2.20) by substituting $x = y^{-1}$ in (2.24). The lemma is proved.

## § 2.7. The irreducibility of a polynomial

Let

$$g(x) = x^s + a_{s-1}x^{s-1} + \cdots + a_1 x + a_0,$$

where $a_i (0 \leqslant i \leqslant s-1)$ are rational integers and $a_0 \neq 0$.

**Theorem 2.4.** *If*

$$|a_1| > |a_0^{s-1}| + |a_{s-1}a_0^{s-2}| + \cdots + |a_2 a_0| + 1, \tag{2.25}$$

*then $g(x)$ is irreducible over $Q$.*

*Proof.* By (2.25), we have

$$|a_1 a_0| > |a_0^s| + |a_{s-1} a_0^{s-1}| + \cdots + |a_2 a_0^2| + |a_0|. \tag{2.26}$$

It follows by Rouché's theorem that $g(x)$ and $x$ have the same number of zeros in the circle $|x| < |a_0|$, i.e., $g(x)$ has only one zero $\vartheta$ in the circle $|x| < a_0$. By (2.26), the equation $g(x) = 0$ has no root with modulus $|a_0|$.

If $g(x) = u(x)v(x)$, where $u(x)$ and $v(x)$ are polynomials with integral coefficients and with degrees $\geqslant 1$, and if $u(\vartheta) = 0$, then the moduli of the roots of $v(x) = 0$ are all $> |a_0|$. Hence

$$|a_0| = |g(0)| = |u(0)v(0)| \geqslant |v(0)| > |a_0|$$

which leads to a contradiction. Thus we have the theorem.

**Theorem 2.5.** *If $g(x) = 0$ has only a root $\vartheta$ with modulus $\geqslant 1$, then $g(x)$ is irreducible over $Q$.*

*Proof.* If $g(x) = u(x)v(x)$, where $u(x)$ and $v(x)$ are polynomials with integral coefficients and with degrees $\geqslant 1$, and if $u(\vartheta) = 0$, then the moduli of the roots of $v(x)$ are all $< 1$. Hence $|v(0)| < 1$ which leads to a contradiction. The theorem follows.

**Theorem 2.6.** *If*

$$|a_{s-1}| > |a_{s-2}| + \cdots + |a_1| + |a_0| + 1, \tag{2.27}$$

*then $g(x)$ is irreducible over $Q$.*

*Proof.* It follows by (2.27) and Rouché's theorem that $x^{s-1}$ and $g(x)$ have the same number of zeros in the circle $|x| < 1$. Hence $g(x)$ has only one zero $\vartheta$ with modulus $\geqslant 1$ and thus the theorem follows by Theorem 2.5.

**Theorem 2.7.** *$\eta, \tau, \omega$ are all* PV *numbers of degree s.*

*Proof.* Since

$$\eta > 1, \quad |\eta^{(i)}| < 1, \quad 2 \leqslant i \leqslant s$$

by Lemma 2.3, $F(x)$ is irreducible over $Q$ by Theorem 2.5 and so $\eta$ is a

PV number of degree $s$. Similarly, we may prove that $\tau$ is a PV number of degree $s$ too. By Theorem 2.6 and its proof, we know that $H(x)$ is irreducible over $Q$ and $\omega$ is a PV number of degree $s$. The theorem is proved.

## § 2.8. The rational approximations of $\eta, \tau, \omega$

1. Let $F_n(=F_{s,n})$ $(n=0,1,\cdots)$ be a sequence of integers defined by the recurrence relation

$$F_0 = F_1 = \cdots = F_{s-2} = 0, \quad F_{s-1} = 1,$$

$$F_{n+s} = F_{n+s-1} + F_{n+s-2} + \cdots + F_{n+1} + F_n, \quad n \geqslant 0.$$

As usual, $(F_n)$ is called the generalized Fibonacci sequence of dimension $s$. Let

$$\rho = -\frac{\ln|\eta^{(s)}|}{\ln \eta}.$$

Then

$$\rho \geqslant -\frac{\ln(\eta-1)}{\ln \eta} \geqslant -\frac{\ln(1-2^{-s})}{\ln 2} \geqslant \frac{1}{2^s \ln 2} + \frac{1}{2^{2s+1}}.$$

Take a basis of $Q(\eta)$

$$\omega_1 = 1, \quad \omega_2 = \eta - 1, \cdots, \quad \omega_s = \eta^{s-1} - \eta^{s-2} - \cdots - \eta - 1.$$

Set

$$F_n(j) = F_{n+j-1} - F_{n+j-2} - \cdots - F_{n+1} - F_n, \quad 2 \leqslant j \leqslant s$$

Then we can derive from Theorem 2.3 the following

**Theorem 2.8.** *For $n \geqslant s$, we have*

$$\left|\frac{F_n(j)}{F_n} - \omega_j\right| \leqslant c(\eta) F_n^{-1-\frac{1}{2^s \ln 2}-\frac{1}{2^{2s+1}}}, \quad 2 \leqslant j \leqslant s. \tag{2.28}$$

For the cases $s=2$ and $s=3$, if we use Lemma 2.5 to replace Lemma 2.3, then we have

**Theorem 2.9.** *The right hand side of (2.28) may be replaced by $c(\eta)F_n^{-2}$ and $c(\eta)F_n^{-\frac{3}{2}}$ for the cases $s=2$ and $s=3$ respectively.*

2. Let $G_n(=G_{s,n})(n=0,1,\cdots)$ be the sequence of integers defined by the recurrence relation

$$G_0=G_1=\cdots=G_{s-2}=0, \quad G_{s-1}=1,$$

$$G_{n+s}=LG_{n+s-1}+G_n, \quad n\geqslant 0.$$

Let

$$\rho=-\frac{\ln|\tau^{(s)}|}{\ln\tau}.$$

Since

$$\begin{aligned}\ln\left(L-(L-1)^{-\frac{1}{s-1}}\right)&=\ln L+\ln\left(1-L^{-1}(L-1)^{-\frac{1}{s-1}}\right)\\&\geqslant \ln L-L^{-1}(L-1)^{-\frac{1}{s-1}}-L^{-2}(L-1)^{-\frac{2}{s-1}}\\&\geqslant \ln L-L^{-1}-L^{-2}\end{aligned}$$

and

$$\ln\left(L+L^{-(s-1)}\right)=\ln L+\ln\left(1+L^{-s}\right)\leqslant \ln L+L^{-s},$$

therefore

$$\begin{aligned}\rho&\geqslant\frac{\ln L-L^{-1}-L^{-2}}{(s-1)(\ln L+L^{-s})}\\&\geqslant\frac{1}{s-1}\left(1-\frac{1}{L\ln L}-\frac{1}{L^2\ln L}-\frac{1}{L^s\ln L}+\frac{1}{L^{s+1}(\ln L)^2}\right)\\&\geqslant\frac{1}{s-1}\left(1-\frac{2}{L\ln L}+\frac{1}{L^{s+3}}\right)\end{aligned}$$

by Lemma 2.6. Hence we can derive from Theorem 2.2 the following

**Theorem 2.10.** *for $n\geqslant s$, we have*

$$\left|\frac{G_{n+j}}{G_n}-\tau^j\right|\leqslant c(\tau)G_n^{-1-\frac{1}{s-1}+\frac{2}{(s-1)L\ln L}-\frac{1}{(s-1)L^{s+3}}}, \quad 1\leqslant j\leqslant s-1. \tag{2.29}$$

For the cases $s=2$ and $s=3$, if we use Lemma 2.7 to replace Lemma 2.6, then we have

**Theorem 2.11.** *The right hand side of (2.29) may be replaced by $c(\tau)G_n^{-2}$ and $c(\tau)G_n^{-\frac{3}{2}}$ for the cases $s=2$ and $s=3$ respectively.*

3. Let $H_n(=H_{s,n})(n=0,1,\cdots)$ be the sequence of integers defined by the recurrance relation

$$H_0 = H_1 = \cdots = H_{s-2} = 0, \quad H_{s-1} = 1,$$

$$H_{n+s} = s^2 r^{s-1} H_{n+s-1} + (-1)^{s-1} A_{s-2} r^{s-2} H_{n+s-2} + \cdots + A_1 r H_{n+1} + H_n, \quad n \geqslant 0.$$

Let

$$\rho = -\frac{\ln|\omega^{(s)}|}{\ln \omega}.$$

Then by Lemma 2.8, we have

$$\rho = \frac{\ln\left(4r\left(\sin\frac{\pi}{s}\right)^2\right) + O(r^{-s/2})}{\ln(s^2 r^{s-1}) + O(r^{-s/2})}$$

$$= \frac{\ln r + 2\ln\left(2\sin\frac{\pi}{s}\right) + O(r^{-s/2})}{(s-1)\ln r + 2\ln s + O(r^{-s/2})}$$

$$= \frac{1}{s-1} + \frac{c_1}{\ln r} + \frac{c_2}{(\ln r)^2} + O\left(\frac{1}{(\ln r)^3}\right). \tag{2.30}$$

where

$$c_1 = \frac{2\ln\left(2\sin\frac{\pi}{s}\right)}{s-1} - \frac{2\ln s}{(s-1)^2},$$

$$c_2 = -\frac{4\ln s\ln\left(2\sin\frac{\pi}{s}\right)}{(s-1)^2} + \frac{4(\ln s)^2}{(s-1)^3}. \tag{2.31}$$

Clearly $H_n$ increases with $n$. We can derive from Theorem 2.2 the following

**Theorem 2.12.** *For $n \geqslant s$, we have*

$$\left|\frac{H_{n+j}}{H_n} - \omega^j\right| \leqslant c(\omega) H_n^{-1-\rho}, \quad 1 \leqslant j \leqslant s-1.$$

*Where $\rho$ is defined by* (2.30) *and* (2.31).

*Remarks.* 1. Concerning the generalization of Fibonacci sequence, except those given here and in § 1.3, Raney [1] also gave a generalization and his result may be obtained from the results of § 2.1 and § 2.2 (Cf. G. N. Raney [1]).

2. Although the errors in rational approximations of $\tau$ and $\omega$ are

better, the sequences of $G_n$ and $H_n$ increase too fast as $n$ increases and so they are not as convenient in practical uses as compared with the sequence $F_n$.

## Notes

The definition of PV number was first introduced by C. Pisot [1] and T. Vijayaraghavan [1] (Cf. J. W. S. Cassels [1]).

Lemma 2.8: Cf. Hua Loo Keng [1].

Theorem 2.4 is due to Xie Ting Fan and Pei Ding Yi [1] which improves a theorem of O. Perron [1] and also a theorem of L. Bernstein [1].

The other results: Cf. Hua Loo Keng and Wang Yuan [6,7,8].

# Chapter 3

# Uniform Distribution

## §3.1. Uniform distribution

We use $G_s$ to denote the $s$-dimensional unit cube

$$0 \leqslant x_i \leqslant 1, \quad 1 \leqslant i \leqslant s.$$

Let $n$ be a positive integer and

$$P_n(k) = (x_1^{(n)}(k), \cdots, x_s^{(n)}(k)), \quad 1 \leqslant k \leqslant n$$

be a set of points in $G_s$, where

$$0 \leqslant x_i^{(n)}(k) < 1, \quad 1 \leqslant i \leqslant s.$$

For any $\boldsymbol{\gamma} = (\gamma_1, \cdots, \gamma_s) \in G_s$, let $N_n(\boldsymbol{\gamma}) = N_n(\gamma_1, \cdots, \gamma_s)$ denote the number of points of $P_n(k)(1 \leqslant k \leqslant n)$ satisfying the inequalities

$$0 \leqslant x_i^{(n)}(k) < \gamma_i, \quad 1 \leqslant i \leqslant s.$$

Then

$$\sup_{\gamma \in G_s} \left| \frac{N_n(\boldsymbol{\gamma})}{n} - |\boldsymbol{\gamma}| \right| = D(n), \quad |\boldsymbol{\gamma}| = \gamma_1 \cdots \gamma_s$$

is called the discrepancy of the set of points $P_n(k)$ $(1 \leqslant k \leqslant n)$. Let $n_1 < n_2 < \cdots$ be a sequence of positive integers. Let

$$P_{n_l}(k) = (x_1^{(n_l)}(k), \cdots, x_s^{(n_l)}(k)), \quad 1 \leqslant k \leqslant n_l$$

be a set of points in $G_s$ with discrepancy $D(n_l)$. If $D(n_l) = o(1)$, then the sequence of sets $P_{n_l}(k)(n_1 < n_2 < \cdots)$ is said to be uniformly distributed with discrepancy $D(n)$. In case $n_l = l, x_1^{(l)}(k) = x_1(k), \cdots, x_s^{(l)}(k) = x_s(k)(k = 1, 2, \cdots)$, the sequence $P(k) = (x_1(k), \cdots, x_s(k))(k = 1, 2, \cdots)$ will be called uniformly distributed in $G_s$.

## § 3.2. Vinogradov's lemma

**Lemma 3.1.** *Let $r$ be a positive integer. Let $\alpha, \beta, \Delta$ be real numbers* *satisfying*

$$0 < \Delta < \frac{1}{2}, \quad \Delta \leqslant \beta - \alpha \leqslant 1 - \Delta.$$

*Then there exists periodic function* $\Psi(x)$ *with period* 1 *such that*

1) $\Psi(x) = 1,$ *if* $\alpha + \frac{1}{2}\Delta \leqslant x \leqslant \beta - \frac{1}{2}\Delta,$

2) $0 \leqslant \Psi(x) \leqslant 1,$ *if* $\alpha - \frac{1}{2}\Delta \leqslant x \leqslant \alpha + \frac{1}{2}\Delta$ *and*

$\beta - \frac{1}{2}\Delta \leqslant x \leqslant \beta + \frac{1}{2}\Delta,$

3) $\Psi(x) = 0,$ *if* $\beta + \frac{1}{2}\Delta \leqslant x \leqslant 1 + \alpha - \frac{1}{2}\Delta,$

4) $\Psi(x)$ *has a Fourier expansion*

$$\Psi(x) = \beta - \alpha + \sum{}' C(m) e^{2\pi i m x},$$

*where* $\sum'$ *denotes a sum with* $m = 0$ *deleted and*

$$|C(m)| \leqslant \min\left(\beta - \alpha, \frac{1}{\pi|m|}, \left(\frac{1}{\pi|m|}\right)^{r+1}\left(\frac{r}{\Delta}\right)^{r}\right).$$

*Proof.* Define a periodic function with period 1

$$\Psi_0(x) = \begin{cases} 1, & \text{if} \quad \alpha < x < \beta, \\ \frac{1}{2}, & \text{if} \quad x = \alpha \quad \text{or} \quad x = \beta, \\ 0 & \text{if} \quad \beta < x < 1 + \alpha. \end{cases}$$

Then $\Psi_0(x)$ has the Fourier expansion

$$\Psi_0(x) = C_0^{(0)} + \sum{}' C_m^{(0)} e^{2\pi i m x},$$

where

$$C_0^{(0)} = \int_0^1 \Psi_0(x)\,dx = \beta - \alpha,$$

$$C_m^{(0)} = \int_0^1 \Psi_0(x) e^{-2\pi i m x}\,dx = \int_\alpha^\beta e^{-2\pi i m x}\,dx$$

$$= \frac{e^{-2\pi i m \alpha} - e^{-2\pi i m \beta}}{2\pi i m} \quad (m \neq 0),$$

and so

$$|C_m^{(0)}| \leqslant \min\left(\beta - \alpha, \frac{1}{\pi|m|}\right).$$

Let $\Delta = 2r\delta$ and

$$\Psi_\rho(x) = \frac{1}{2\delta}\int_{-\delta}^{\delta} \Psi_{\rho-1}(x+z)dz,$$

where $\rho = 1, 2, \cdots, r$. Then we may prove by induction that

1)′ $\Psi_\rho(x) = 1$, if $\alpha + \rho\delta < x < \beta - \rho\delta$,

2)′ $0 \leqslant \Psi_\rho(x) \leqslant 1$, if $\alpha - \rho\delta \leqslant x \leqslant \alpha + \rho\delta$ and $\beta - \rho\delta \leqslant x \leqslant \beta + \rho\delta$,

3)′ $\Psi_\rho(x) = 0$, if $\beta + \rho\delta < x < 1 + \alpha - \rho\delta$,

4)′ $\Psi_\rho(x)$ has a Fourier expansion

$$\Psi_\rho(x) = C_0^{(\rho)} + \sum{}' C_m^{(\rho)} e^{2\pi i m x},$$

where

$$C_0^{(\rho)} = \beta - \alpha,$$

$$|C_m^{(\rho)}| \leqslant \min\left(\beta - \alpha, \frac{1}{\pi|m|}, \left(\frac{1}{\pi|m|}\right)^{\rho+1}\left(\frac{r}{\Delta}\right)^\rho\right).$$

In fact, suppose that $\Psi_{\rho-1}(x)$ satisfies 1)′—4)′. Then $\Psi_\rho(x)$ satisfies 1)′—3)′ obviously. Now we prove that $\Psi_\rho(x)$ satisfies 4)′ as follows:

$$C_0^{(\rho)} = \int_0^1 \Psi_\rho(x)\,dx = \frac{1}{2\delta}\int_{-\delta}^{\delta} dz \int_0^1 \Psi_{\rho-1}(x+z)dx$$

$$= C_0^{(\rho-1)} = \beta - \alpha,$$

$$C_m^{(\rho)} = \int_0^1 \Psi_\rho(x) e^{-2\pi i m x} dx = \frac{1}{2\delta}\int_0^1 e^{-2\pi i m x} dx \int_{-\delta}^{\delta} \Psi_{\rho-1}(x+z)dz$$

$$= \frac{C_m^{(\rho-1)}}{2\delta}\int_{-\delta}^{\delta} e^{2\pi i m z} dz = C_m^{(\rho-1)}\left(\frac{e^{2\pi i m\delta} - e^{-2\pi i m\delta}}{4\pi i m\delta}\right)$$

$$= C_m^{(0)}\left(\frac{e^{2\pi i m\delta} - e^{-2\pi i m\delta}}{4\pi i m\delta}\right)^\rho.$$

Hence

$$|C_m^{(\rho)}| \leqslant \min\left(\beta - \alpha, \frac{1}{\pi|m|}, \left(\frac{1}{\pi|m|}\right)^{\rho+1}\left(\frac{r}{\Delta}\right)^\rho\right).$$

Take $\Psi(x) = \Psi_r(x)$. Then we have the lemma.

## § 3.3. The exponential sum and the discrepancy

We use the notations $\bar{x} = \max(1, |x|)$, $\|\boldsymbol{\gamma}\| = \bar{\gamma}_1 \cdots \bar{\gamma}_s$ and $(\boldsymbol{\alpha}, \boldsymbol{\beta}) = \sum_{i=1}^{s} \alpha_i \beta_i$ the scalar product of $\boldsymbol{\alpha}$ and $\boldsymbol{\beta}$.

**Theorem 3.1.** *Let $r, h$ be positive integers such that $h > r/\eta$, where $\eta$ satisfies $0 < \eta < 1/6$. Then for any $\boldsymbol{\gamma} \in G_s$, we have*

$$\left| \frac{1}{n} N_n(\boldsymbol{\gamma}) - |\boldsymbol{\gamma}| \right| < D(n)$$

*where*

$$D(n) = \sum_{\|\mathbf{m}\| \leqslant h}' \frac{1}{\|\pi \mathbf{m}\|} \left| \frac{1}{n} \sum_{k=1}^{n} e^{2\pi i (\mathbf{m}, P_n(k))} \right| + (5s+6)\eta$$
$$+ \frac{s 2^s r^{r-1}}{\pi^{s+r} \eta^r h^r} (\ln 64h)^{s-1},$$

*in which $\sum'$ denotes a sum with $\mathbf{m} = \mathbf{0} = (0, \cdots, 0)$ deleted.*

**Lemma 3.2.** *Let $n, l$ be integers $> 1$. Then*

$$\sum_{m=1}^{n} \frac{1}{m} < 1 + \ln n$$

*and*

$$\sum_{m=n+1}^{\infty} \frac{1}{m^l} < \frac{1}{l-1} n^{-l+1}.$$

*Proof.* Since

$$\frac{1}{m} < \int_{m-1}^{m} \frac{dt}{t},$$

hence

$$\sum_{m=1}^{n} \frac{1}{m} < 1 + \int_1^2 \frac{dt}{t} + \cdots + \int_{n-1}^{n} \frac{dt}{t} = 1 + \int_1^n \frac{dt}{t} = 1 + \ln n.$$

Similarly

$$\sum_{m=n+1}^{\infty} \frac{1}{m^l} < \int_n^{\infty} \frac{dt}{t^l} = \frac{n^{-l+1}}{l-1}.$$

The lemma is proved.

The proof of Theorem 1. First Suppose that $\boldsymbol{\gamma} \in G_s$, where $\gamma_i'$s satisfy

$$3\eta \leqslant \gamma_i \leqslant 1 - 3\eta, \quad 1 \leqslant i \leqslant s.$$

Let

$$G_x(y) = \begin{cases} 1, & \text{if } 0 \leqslant y < x, \\ 0, & \text{if } x \leqslant y < 1. \end{cases}$$

Then

$$\frac{1}{n} N_n(\boldsymbol{\gamma}) = \frac{1}{n} \sum_{k=1}^{n} \sum_{\substack{x_\nu^{(n)}(k) < \gamma_\nu \\ 1 \leqslant \nu \leqslant s}} 1 = \frac{1}{n} \sum_{k=1}^{n} \prod_{\nu=1}^{s} G_{\gamma_\nu}(x_\nu^{(n)}(k)). \tag{3.1}$$

For $3\eta \leqslant x \leqslant 1 - 3\eta$, we construct two auxiliary functions $G_x^{(1)}(y)$ and $G_x^{(2)}(y)$, where $G_x^{(1)}(y)$ satisfies

1) $G_x^{(1)}(y) = 1$, if $-\eta \leqslant y \leqslant x$,

2) $0 \leqslant G_x^{(1)}(y) \leqslant 1$, if $-2\eta \leqslant y \leqslant -\eta$ and $x \leqslant y \leqslant x + \eta$,

3) $G_x^{(1)}(y) = 0$, if $x + \eta \leqslant y \leqslant 1 - 2\eta$,

4) $G_x^{(1)}(y)$ has the Fourier expansion

$$G_x^{(1)}(y) = x + 2\eta + \sum{}' C_1(m) e^{2\pi i m y},$$

in which

$$|C_1(m)| \leqslant \min\left(x + 2\eta, \frac{1}{\pi|m|}, \frac{r^r}{\pi^{r+1}\eta^r|m|^{r+1}}\right),$$

and where $G_x^{(2)}(y)$ satisfies

1)′ $G_x^{(2)}(y) = 1$, if $2\eta \leqslant y \leqslant x - \eta$,

2)′ $0 \leqslant G_x^{(2)}(y) \leqslant 1$, if $\eta \leqslant y \leqslant 2\eta$ and $x - \eta \leqslant y \leqslant x$,

3)′ $G_x^{(2)}(y) = 0$, if $x \leqslant y \leqslant 1 + \eta$,

4)′ $G_x^{(2)}(y)$ has the Fourier expansion

$$G_x^{(2)}(y) = x - 2\eta + \sum{}' C_2(m) e^{2\pi i m y},$$

in which

$$|C_2(m)| \leqslant \min\left(x - 2\eta, \frac{1}{\pi|m|}, \frac{r^r}{\pi^{r+1}\eta^r|m|^{r+1}}\right).$$

Then

$$G_x^{(2)}(y) \leqslant G_x(y) \leqslant G_x^{(1)}(y). \tag{3.2}$$

By Lemma 3.2

$$\begin{aligned} G_x^{(1)}(y) &= \sum_{|m| \leqslant h} C_1(m) e^{2\pi i m y} + \theta \sum_{|m| > h} \frac{r^r}{\pi^{r+1}\eta^r|m|^{r+1}} \\ &= \sum_{|m| \leqslant h} C_1(m) e^{2\pi i m y} + \frac{2\theta r^{r-1}}{\pi^{r+1}\eta^r h^r}, \end{aligned} \tag{3.3}$$

where $C_1^{(0)} = x + 2\eta$ and we use $\theta$ to denote a number satisfying $0 \leqslant |\theta| \leqslant 1$ but not always with the same value. From (3.1), (3.2), (3.3) and Lemma 3.2, we have

$$\frac{1}{n} N_n(\boldsymbol{\gamma}) \leqslant \frac{1}{n} \sum_{k=1}^{n} \prod_{\gamma=1}^{s} G_{\gamma_v}^{(1)}(x_v^{(n)}(k))$$

$$= \frac{1}{n} \sum_{k=1}^{n} \prod_{\gamma=1}^{s} \left( \sum_{|m| \leqslant h} C_1(m) e^{2\pi i m x_v^{(n)}(k)} + \frac{2\vartheta r^{r-1}}{\pi^{r+1} \eta^r h^r} \right) \tag{3.4}$$

and

$$1 + \sum_{|m| \leqslant h} |C_1(m)| \leqslant 2 + \frac{2}{\pi} \sum_{m=1}^{h} \frac{1}{m}$$

$$\leqslant 2 + \frac{2}{\pi} + \frac{2}{\pi} \ln h < \frac{2}{\pi} \ln 64h. \tag{3.5}$$

Since

$$(\gamma_1 + 2\eta) \cdots (\gamma_s + 2\eta) \leqslant \gamma_1(\gamma_2 + 2\eta) \cdots (\gamma_s + 2\eta)$$
$$+ 2\eta \leqslant \cdots \leqslant |\boldsymbol{\gamma}| + 2s\eta, \tag{3.6}$$

so from (3.4), (3.5), (3.6), we have

$$\frac{1}{n} N_n(\boldsymbol{\gamma}) - |\boldsymbol{\gamma}| \leqslant \left| \sum_{|m_i| \leqslant h}' C_1(m) \cdots C_1(m_s) \frac{1}{n} \sum_{k=1}^{n} e^{2\pi i (\mathbf{m}, P_n(k))} \right|$$
$$+ 2s\eta + \frac{2sr^{r-1}}{\pi^{r+1} \eta^r h^r} \left( \frac{2}{\pi} \right)^{s-1} (\ln 64h)^{s-1}.$$

Using $G_x^{(2)}(y)$ to instead of $G_x^{(1)}(y)$, we obtain a similar lower estimate for $\frac{1}{n} N_n(\boldsymbol{\gamma}) - |\boldsymbol{\gamma}|$. Hence

$$\left| \frac{1}{n} N_n(\boldsymbol{\gamma}) - |\boldsymbol{\gamma}| \right| \leqslant \sum_{|m_i| \leqslant h}' \frac{1}{\|\pi \mathbf{m}\|} \left| \frac{1}{n} \sum_{k=1}^{n} e^{2\pi i (\mathbf{m}, P_n(k))} \right| + 2s\eta$$
$$+ \frac{s 2^s r^{r-1}}{\pi^{r+s} \eta^r h^r} (\ln 64h)^{s-1} = \Phi \text{ (say)}.$$

Next, suppose that there are $t$ components of $\boldsymbol{\gamma}$ satisfying $\gamma_i < 3\eta$ and the rest satisfying $3\eta \leqslant \gamma_i \leqslant 1 - 3\eta$. Define $\boldsymbol{\gamma}' = (\gamma_1', \cdots, \gamma_s')$, where $\gamma_i' = 3\eta$ if $\gamma_i < 3\eta$, and $\gamma_i' = \gamma_i$ otherwise. Then $N_n(\boldsymbol{\gamma}') \geqslant N_n(\boldsymbol{\gamma})$ and

$$\left| \frac{N_n(\boldsymbol{\gamma})}{n} - |\boldsymbol{\gamma}| \right| \leqslant \frac{N_n(\boldsymbol{\gamma}')}{n} + |\boldsymbol{\gamma}| \leqslant \left| \frac{N_n(\boldsymbol{\gamma}')}{n} - |\boldsymbol{\gamma}'| \right|$$
$$+ |\boldsymbol{\gamma}'| + |\boldsymbol{\gamma}| \leqslant \Phi + 6\eta. \tag{3.7}$$

Finally, suppose that there are $t$ components of $\boldsymbol{\gamma}$ equal to 1 and the rest are not greater than $1-3\eta$. Then the problem is reduced to the $s-t$ dimensional case. Hence we have (3.7) too. For any $\boldsymbol{\gamma}\in G_s$, define $\boldsymbol{\gamma}'=(\gamma_1',\cdots,\gamma_s')$ and $\boldsymbol{\gamma}''=(\gamma_1'',\cdots,\gamma_s'')$ as follows:

$$\gamma_i'=\begin{cases}1-3\eta, & \text{if } \gamma_i>1-3\eta,\\ \gamma_i, & \text{if } \gamma_i\leqslant 1-3\eta\end{cases}$$

and

$$\gamma_i''=\begin{cases}1, & \text{if } \gamma_i>1-3\eta,\\ \gamma_i, & \text{if } \gamma_i\leqslant 1-3\eta.\end{cases}$$

Hence

$$\left|\frac{N_n(\boldsymbol{\gamma})}{n}-|\boldsymbol{\gamma}|\right|\leqslant\max\left(\left|\frac{N_n(\boldsymbol{\gamma}')}{n}-|\boldsymbol{\gamma}|\right|,\left|\frac{N_n(\boldsymbol{\gamma}'')}{n}-|\boldsymbol{\gamma}|\right|\right)$$
$$\leqslant\Phi+6\eta+3s\eta.$$

The theorem follows.

## § 3.4. The number of solutions to the congruence

We shall always use $M$ to denote a number $\geqslant 1$, $n$ an integer $\geqslant 2$ and $\mathbf{a}=(a_1,\cdots,a_s)$ an integral vector.

**Lemma 3.3.** *Suppose that the congruence*

$$(\mathbf{a},\mathbf{m})=\sum_{i=1}^{s}a_im_i\equiv 0(\operatorname{mod} n) \tag{3.8}$$

*has no solution in the domain*

$$\|\mathbf{m}\|\leqslant M,\quad \mathbf{m}\neq\mathbf{0}.$$

*Then the number of solutions $T_{l,M}$ of (3.8) in the domain*

$$\|\mathbf{m}\|<lM \tag{3.9}$$

*satisfies*

$$T_{l,M}\leqslant c(\varepsilon)^s l^{1+\varepsilon}M^{\varepsilon},$$

*where $l$ is a positive integer.*

We use $P_{s,M}$ to denote the $s$-dimensional parallelepiped with edges parallel to coordinate axes and with volume $\leqslant M$. To prove Lemma 3.3, we shall need

**Lemma 3.4.** *The domain (3.9) can be covered by at most* $c(\varepsilon)^s l^{1+\varepsilon} M^\varepsilon$ *parallelepipeds of the type* $P_{s,M}$.

*Proof.* Take

$$c(\varepsilon) = 2^{2+\varepsilon} \sum_{j=0}^{\infty} (\bar{j}^{-(1+\varepsilon)} + 2^{-\varepsilon j}).$$

For $s = 1$, the domain (3.9) is an open interval $(-lM, lM)$, hence it can be covered by

$$\frac{2lM}{M} = 2l$$

intervals of the type $[c, c + M]$, where $c$ is a constant. The lemma holds.

Suppose that $k$ is a positive integer and the lemma holds for $s = 1, 2, \cdots, k$. Now we proceed to prove that the lemma holds for $s = k + 1$ too.

Divide the domain

$$\bar{m}_1 \cdots \bar{m}_{k+1} < lM$$

into $2[\log_2 M] + 3$ parts

$$m_{k+1} = j, \quad |j| \leqslant l, \tag{3.10}$$

and

$$2^i l < |m_{k+1}| \leqslant 2^{i+1} l, \quad 0 \leqslant i \leqslant [\log_2 M]. \tag{3.11}$$

Suppose that $m_{k+1} = j$. Then

$$\bar{m}_1 \cdots \bar{m}_k < \frac{lM}{\bar{j}} < \left(\left[\frac{l}{\bar{j}}\right] + 1\right) M,$$

so it can be covered by at most

$$Q = c(\varepsilon)^k \left(\left[\frac{l}{\bar{j}}\right] + 1\right)^{1+\varepsilon} M^\varepsilon$$

parallelepipeds of the type $P_{k,M}$ from the induction hypothesis. Using $P_{k,M}$ as basis and 1 as height, we obtain a $P_{k+1,M}$. Hence the subdomain $m_{k+1} = j$ can be covered by at most $Q$ parallelepipeds of the type $P_{k+1,M}$ and so the domain (3.10) can be covered by at most

$$2 \sum_{j=0}^{l} c(\varepsilon)^k \left(\left[\frac{l}{\bar{j}}\right] + 1\right)^{1+\varepsilon} M^\varepsilon \tag{3.12}$$

parallelepipeds of the type $P_{k+1,M}$.

Consider the subdomain of (3.11)

$$2^i l < m_{k+1} \leqslant 2^{i+1} l. \tag{3.13}$$

Then we have

$$\overline{m}_1 \cdots \overline{m}_k < \frac{M}{2^i}$$

for $m_{k+1} = 2^i l + 1$ and it can be covered by at most

$$c(\varepsilon)^k \left(\frac{M}{2^i}\right)^\varepsilon$$

parallelepipeds of the type $P_{k,\frac{M}{2^i}}$ from the induction hypothesis. Since we may obtain a $P_{k+1,M}$ by the use of $P_{k,\frac{M}{2^i}}$ as basis and $2^i$ as height and $2^i l + 2^i l + 1 > 2^{i+1} l$, the domain (3.13) can be covered by at most

$$c(\varepsilon)^k l \left(\frac{M}{2^i}\right)^\varepsilon$$

parallelepipeds of the type $P_{k+1,M}$ and so the domain (3.11) is covered by at most

$$2c(\varepsilon)^k l \sum_{i=0}^{[\log_2 M]} \left(\frac{M}{2^i}\right)^\varepsilon \tag{3.14}$$

parallelepipeds of the type $P_{k+1,M}$.

It follows by (3.13) and (3.14) that the domain (3.9) can be covered by at most

$$c(\varepsilon)^k l^{1+\varepsilon} M^\varepsilon \sum_{j=0}^{\infty} \left(\frac{2^{2+\varepsilon}}{\bar{j}^{1+\varepsilon}} + \frac{2}{2^{\varepsilon j}}\right) \leqslant c(\varepsilon)^{k+1} l^{1+\varepsilon} M^\varepsilon$$

parallelepipeds of the type $P_{k+1,M}$. Hence the lemma follows by induction.

The proof of Lemma 3.3. By Lemma 3.4, it is sufficient to prove that the congruence (3.8) has at most 1 solution in any parallelepiped of the type $P_{s,M}$. Suppose that (3.8) has two solutions $\mathbf{m}'$ and $\mathbf{m}''$ in a $P_{s,M}$, where $\mathbf{m}' \neq \mathbf{m}''$. Let $\mathbf{m} = \mathbf{m}' - \mathbf{m}''$. Then $\|\mathbf{m}\| \leqslant M, \mathbf{m} \neq \mathbf{0}$ and

$$(\mathbf{a}, \mathbf{m}) = (\mathbf{a}, \mathbf{m}') - (\mathbf{a}, \mathbf{m}'') \equiv 0 (\operatorname{mod} n)$$

which leads to a contradiction. The lemma follows.

**Lemma 3.5.** *Under the assumption of Lemma* 3.3,

$$T_{l,M} \leqslant c(s) l (\ln 3lM)^{s-1}.$$

*Proof.* We may prove easily by induction that the conclusion $c(\varepsilon)^s l^{1+\varepsilon} M^\varepsilon$ in Lemma 3.4 can be replaced by $c(s)\, l(\ln 3lM)^{s-1}$ and so the lemma follows.

## § 3.5. The solutions of the congruence and the discrepancy

**Theorem 3.2.** *Under the assumption of Lemma* 3.3, *the set*

$$\left(\left\{\frac{a_1 k}{n}\right\}, \cdots, \left\{\frac{a_s k}{n}\right\}\right), \quad 1 \leqslant k \leqslant n$$

*has discrepancy*

$$D(n) \leqslant c(s) M^{-1} (\ln 3M)^s.$$

**Lemma 3.6.**

$$\frac{1}{n} \sum_{k=1}^{n} e^{2\pi i m k/n} = \begin{cases} 1, & \text{if } n \mid m, \\ 0, & \text{if } n \nmid m. \end{cases}$$

(Cf. Hua Loo Keng [2], Chap. 7).

The proof of Theorem 3.2. Let $h \geqslant 7$. Then by Lemmas 3.5 and 3.6,

$$\sum_{|m_i| \leqslant h}' \frac{1}{\|\pi \mathbf{m}\|} \left| \frac{1}{n} \sum_{k=1}^{n} e^{2\pi i (\mathbf{a}, \mathbf{m}) k/n} \right| \leqslant \sum_{\substack{|m_i| \leqslant h \\ (\mathbf{a}, \mathbf{m}) \equiv 0 (\mathrm{mod}\ n)}}' \frac{1}{\|\mathbf{m}\|}$$

$$\leqslant \sum_{l=1}^{h^s} \frac{(T_{l+1,M} - T_{l,M})}{lM} \leqslant M^{-1} \sum_{l=1}^{h^s} T_{l+1,M} \left(\frac{1}{l} - \frac{1}{l+1}\right) + \frac{T_{h^s+1,M}}{(h^s+1)M}$$

$$\leqslant c(s) M^{-1} \sum_{l=1}^{h^s} \frac{(\ln 3lM)^{s-1}}{l} + c(s) M^{-1} (\ln hM)^{s-1}$$

$$\leqslant c(s) M^{-1} (\ln hM)^s.$$

Take $r = 1$, $\eta = \dfrac{1}{7M}$ and $h = 7([M]+1)^2$ in Theorem 3.1. Then we have the theorem.

## § 3.6. The partial summation formula

**Lemma 3.7.** *Let* $g(\mathbf{m})$ *be a non-negative function of* $\mathbf{m}$. *Then*

$$\sum_{|m_i|\leqslant h}{}' \frac{g(\mathbf{m})}{\|\mathbf{m}\|} \leqslant \sum_{l=0}^{s} \sum_{\mathbf{i}} \frac{1}{h^l} \sum_{m_{i_{l+1}}=1}^{h} \cdots \sum_{m_{i_s}=1}^{h} \frac{1}{m_{i_{l+1}}^2 \cdots m_{i_s}^2}$$

$$\times \sum_{|k_{i_1}|\leqslant h} \cdots \sum_{|k_{i_l}|\leqslant h} \sum_{|k_{i_{l+1}}|\leqslant m_{i_{l+1}}} \cdots \sum_{|k_{i_s}|\leqslant m_{i_s}}{}' g(\mathbf{k}),$$

*where* $\sum_{\mathbf{i}}$ *denotes a sum in which* $\mathbf{i} = (i_1, \cdots, i_s)$ *runs over all the permutations of* $(1, \cdots, s)$.

*Proof.* Since

$$\sum_{|m|\leqslant h} \frac{g(m)}{\bar{m}} = g(0) + \sum_{m=1}^{h} \frac{1}{m}(g(m) + g(-m))$$

$$= g(0) + \sum_{m=1}^{h} \left(\frac{1}{m} - \frac{1}{m+1}\right) \sum_{1\leqslant k\leqslant m} (g(k) + g(-k))$$

$$+ \frac{1}{h+1} \sum_{|k|\leqslant h}{}' g(k)$$

$$\leqslant \sum_{m=1}^{h} \frac{1}{m^2} \sum_{|k|\leqslant m} g(k) + \frac{1}{h} \sum_{|k|\leqslant h} g(k),$$

hence

$$\sum_{|m_i|\leqslant h}{}' \frac{g(\mathbf{m})}{\|\mathbf{m}\|} \leqslant \sum_{\substack{|m_i|\leqslant h \\ 1\leqslant i\leqslant s-1}}{}' \frac{1}{\bar{m}_1 \cdots \bar{m}_{s-1}} \left( \sum_{m_s=1}^{h} \frac{1}{m_s^2} \sum_{|k_s|\leqslant m_s} g(m_1, \cdots, m_{s-1}, k_s) \right.$$

$$\left. + \frac{1}{h} \sum_{|k_s|\leqslant h} g(m_1, \cdots, m_{s-1}, k_s) \right)$$

$$+ \sum_{\substack{|m_i|\leqslant h \\ 1\leqslant i\leqslant s-1}} \frac{1}{\bar{m}_1 \cdots \bar{m}_{s-1}} \left( \sum_{m_s=1}^{h} \frac{1}{m_s^2} \sum_{|k_s|\leqslant m_s}{}' g(m_1, \cdots, m_{s-1}, k_s) \right.$$

$$\left. + \frac{1}{h} \sum_{|k_s|\leqslant h}{}' g(m_1, \cdots, m_{s-1}, k_s) \right) \leqslant \cdots$$

$$\leqslant \sum_{l=0}^{s} \sum_{\mathbf{i}} \frac{1}{h^l} \sum_{m_{i_{l+1}}=1}^{h} \cdots \sum_{m_{i_s}=1}^{h} \frac{1}{m_{i_{l+1}}^2 \cdots m_{i_s}^2}$$

$$\cdot \sum_{|k_{i_1}|\leqslant h} \cdots \sum_{|k_{i_l}|\leqslant h} \sum_{|k_{i_{l+1}}|\leqslant m_{i_{l+1}}} \cdots \sum_{|k_{i_s}|\leqslant m_{i_s}}{}' g(\mathbf{k}).$$

The lemma is proved.

## §3.7. The comparison of discrepancies

**Lemma 3.8.** *Suppose that the sets* $P_n(k) = (x_1^{(n)}(k), \cdots, x_s^{(n)}(k))$ $(1 \leqslant k \leqslant n)$ *and* $Q_n(k) = (y_1^{(n)}(k), \cdots, y_s^{(n)}(k))$ $(1 \leqslant k \leqslant n)$ *have discrepancies* $D(n)$ *and* $E(n)$ *respectively and that*

$$|x_i^{(n)}(k) - y_i^{(n)}(k)| \leqslant \delta, \quad 1 \leqslant k \leqslant n, 1 \leqslant i \leqslant s. \tag{3.15}$$

*Then*

$$|D(n) - E(n)| \leqslant s\delta.$$

*Proof.* For any $\boldsymbol{\gamma} \in G_s$, set $\boldsymbol{\gamma}' = (\gamma_1', \cdots, \gamma_s')$ and $\boldsymbol{\gamma}'' = (\gamma_1'', \cdots, \gamma_s'')$ where

$$\gamma_i' = \begin{cases} \gamma_i - \delta, & \text{if } \gamma_i - \delta \geqslant 0, \\ 0, & \text{if } \gamma_i - \delta < 0. \end{cases}$$

and

$$\gamma_i'' = \begin{cases} \gamma_i + \delta, & \text{if } \gamma_i + \delta \leqslant 1, \\ 1, & \text{if } \gamma_i + \delta > 1. \end{cases}$$

Let $N_n(\boldsymbol{\gamma})$ and $M_n(\boldsymbol{\gamma})$ denote the numbers of $P_n(k)(1 \leqslant k \leqslant n)$ and $Q_n(k)(1 \leqslant k \leqslant n)$ falling into the region

$$0 \leqslant x_i < \gamma_i, \quad 1 \leqslant i \leqslant s$$

respectively. Then by (3.15),

$$M_n(\boldsymbol{\gamma}') \leqslant N_n(\boldsymbol{\gamma}) \leqslant M_n(\boldsymbol{\gamma}''). \tag{3.16}$$

Let

$$\sigma_1 = \left|\frac{M_n(\boldsymbol{\gamma}')}{n} - |\boldsymbol{\gamma}|\right| \quad \text{and} \quad \sigma_2 = \left|\frac{M_n(\boldsymbol{\gamma}'')}{n} - |\boldsymbol{\gamma}|\right|.$$

Then by (3.16),

$$\left|\frac{N_n(\boldsymbol{\gamma})}{n} - |\boldsymbol{\gamma}|\right| \leqslant \max(\sigma_1, \sigma_2). \tag{3.17}$$

Since

$$0 \leqslant |\boldsymbol{\gamma}''| - |\boldsymbol{\gamma}| \leqslant \delta + \gamma_1\left(\prod_{i=2}^{s} \gamma_i'' - \prod_{i=2}^{s} \gamma_i\right) \leqslant \cdots \leqslant s\delta,$$

therefore

$$\sigma_2 \leqslant \left|\frac{M_n(\boldsymbol{\gamma}'')}{n} - |\boldsymbol{\gamma}''|\right| + |\boldsymbol{\gamma}''| - |\boldsymbol{\gamma}| \leqslant E(n) + s\delta.$$

$\sigma_1$ satisfies the same inequality. Hence by (3.17),

$$D(n) \leqslant E(n) + s\delta.$$

Similarly,

$$E(n) \leqslant D(n) + s\delta.$$

The lemma is proved.

## § 3.8. Rational approximation and the solutions of the congruence

We use $\mathbf{h} = (h_0, h_1, \cdots, h_s)$, $\mathbf{m} = (m_1, \cdots, m_s)$ and $\mathbf{m}^{(0)} = (m_0, m_1, \cdots, m_s)$ to denote the vectors with integral components, where $h_0 = 1$.

**Lemma 3.9.** *Suppose that*

$$\langle(\mathbf{r}, \mathbf{m})\rangle \geqslant b\|\mathbf{m}\|^{-a} \tag{3.18}$$

*holds for any* $\mathbf{m} \neq \mathbf{0}$, *where* $a, b$ *are constants satisfying* $s \geqslant a \geqslant 1, b > 0$, *and that*

$$\left|\frac{h_i}{n} - \gamma_i\right| \leqslant dn^{-1-g}, \quad 1 \leqslant i \leqslant s, \tag{3.19}$$

*where* $d, g$ *are constants satisfying* $d > 0, 0 \leqslant g \leqslant 1/s$. *Then there exists a positive constant* $c(b, d, s)(<1)$ *such that the congruence*

$$(\mathbf{h}, \mathbf{m}^{(0)}) = \sum_{i=0}^{s} h_i m_i \equiv 0 \pmod{n} \tag{3.20}$$

*has no solution in the domain*

$$\|\mathbf{m}^{(0)}\| \leqslant c(b, d, s) n^{\frac{1+g}{1+a}}, \quad \mathbf{m}^{(0)} \neq \mathbf{0}. \tag{3.21}$$

*Proof.* Suppose that $\mathbf{m}^{(0)}(\neq \mathbf{0})$ is a solution of (3.20). If $\mathbf{m} = \mathbf{0}$, then $m_0 \neq 0$. By (3.20), we have $m_0 \equiv 0 \pmod{n}$. Hence $\|\mathbf{m}^{(0)}\| \geqslant n$ and so $\mathbf{m}^{(0)}$ not belongs to the domain (3.21). Consequently, we may suppose that $\mathbf{m} \neq \mathbf{0}$. If

$$\|\mathbf{m}\| \geqslant \left(\frac{b}{2ds}\right)^{\frac{1}{1+a}} n^{\frac{1+g}{1+a}},$$

then

$$\|\mathbf{m}^{(0)}\| \geqslant \left(\frac{b}{2ds}\right)^{\frac{1}{1+a}} n^{\frac{1+g}{1+a}}$$

and so we have the theorem. Now, suppose that

$$\|\mathbf{m}\| < \left(\frac{b}{2ds}\right)^{\frac{1}{1+a}} n^{\frac{1+g}{1+a}}.$$

Since

$$\langle \alpha - \beta \rangle \geqslant \langle \alpha \rangle - \langle \beta \rangle,$$

we have

$$\frac{|m_0|}{n} \geqslant \left\langle \frac{m_0}{n} \right\rangle = \left\langle \frac{1}{n} \sum_{i=1}^{s} h_i m_i \right\rangle \geqslant \langle (\boldsymbol{\gamma}, \mathbf{m}) \rangle - \left\langle \sum_{i=1}^{s} \left( \frac{h_i}{n} - \gamma_i \right) m_i \right\rangle$$

$$\geqslant \frac{b}{\|\mathbf{m}\|^a} - \frac{ds}{n^{1+g}} \|\mathbf{m}\|$$

by (3.18) and (3.19). Hence

$$\|\mathbf{m}^{(0)}\| \geqslant \frac{nb}{\|\mathbf{m}\|^{a-1}} - \frac{ds}{n^g} \|\mathbf{m}\|^2 \geqslant b^{\frac{2}{1+a}} (2ds)^{\frac{-1+a}{1+a}} n^{1+\frac{(1-a)(1+g)}{1+a}}$$

$$- \left( \frac{b}{2} \right)^{\frac{2}{1+a}} (ds)^{\frac{-1+a}{1+a}} n^{\frac{2(1+g)}{1+a} - g} \geqslant \frac{1}{2} (2ds)^{\frac{-1+a}{1+a}} b^{\frac{2}{1+a}} n^{\frac{1+g}{1+a}}.$$

The lemma is proved.

## § 3.9. The rational approximation and the discrepancy

In this section, we shall prove the following three theorems.

**Theorem 3.3.** *Suppose that* (3.18) *holds for* $\mathbf{m} \neq \mathbf{0}$, *where* $a, b$ *are constants satisfying* $1 \leqslant a \leqslant 1 + \frac{1}{2s}$ *and* $b > 0$. *Then the set*

$$P_n(k) = (\{\gamma_1 k\}, \cdots, \{\gamma_s k\}), \quad 1 \leqslant k \leqslant n$$

*has discrepancy*

$$D(n) \leqslant c(b, s) n^{-1+2s(a-1)} (\ln n)^{1+s\delta_{1,a}}.$$

**Theorem 3.4.** *Suppose that* (3.19) *holds, where* $d, g$ *are constants satisfying* $d > 0$ *and* $0 \leqslant g \leqslant \frac{1}{s}$. *Then under the assumption of Theorem* 3.3, *the set*

$$\left( \left\{ \frac{k}{n} \right\}, \left\{ \frac{h_1 k}{n} \right\}, \cdots, \left\{ \frac{h_s k}{n} \right\} \right), \quad 1 \leqslant k \leqslant n$$

*has discrepancy*

$$D(n) \leqslant c(b, d, s) n^{-\frac{-1+g}{1+a}} (\ln n)^{s+1}.$$

**Theorem 3.5.** *Suppose that $q$ is an integer satisfying $1 \leqslant q \leqslant n^{\frac{1+g}{2-2s(a-1)}}$. Then under the assumption of Theorem 3.4, the set*

$$\left(\left\{\frac{h_1 k}{n}\right\}, \cdots, \left\{\frac{h_s k}{n}\right\}\right), \quad 1 \leqslant k \leqslant q \tag{3.22}$$

*has discrepancy*

$$D(q) \leqslant c(b, d, s) q^{-1+2s(a-1)} (\ln 3q)^{1+s\delta_{1,a}}.$$

To prove these theorems, we shall need

**Lemma 3.10.** *Let $\delta$ be a real number. Then*

$$\left| \sum_{k=1}^{n} e^{2\pi i \delta k} \right| \leqslant \min\left(n, \frac{1}{2\langle\delta\rangle}\right)$$

(Cf. Hua Loo Keng [2], Chap. 7).

**Lemma 3.11.**

$$\sum_{m=1}^{n} \frac{1}{m^\alpha} \leqslant \begin{cases} \dfrac{n^{1-\alpha}}{1-\alpha}, & \text{if } 0 \leqslant \alpha < 1, \\ \dfrac{(n+1)^{1-\alpha}}{1-\alpha}, & \text{if } \alpha < 0. \end{cases}$$

*Proof.* For $0 \leqslant \alpha < 1$,

$$\sum_{m=1}^{n} \frac{1}{m^\alpha} \leqslant 1 + \int_1^n \frac{dt}{t^\alpha} = 1 + \frac{n^{1-\alpha}}{1-\alpha} - \frac{1}{1-\alpha} \leqslant \frac{n^{1-\alpha}}{1-\alpha}.$$

The other inequality may be proved similarly.

**Lemma 3.12.** *Let $Q = [2^{sa}\|\mathbf{m}\|^a b^{-1}] + 1$, where $\mathbf{m} \neq \mathbf{0}$. Then under the assumption of Theorem 3.3, there is at most 1 point $(\mathbf{k}, \boldsymbol{\gamma}) = \sum_{i=1}^{s} k_i \gamma_i$ in any interval of the type $(P, P+Q^{-1}]$, where $k$ is a vector with integral components which satisfy $|k_i| \leqslant |m_i| (1 \leqslant i \leqslant s)$.*

*Proof.* If there are two points $(\mathbf{k}', \boldsymbol{\gamma})$ and $(\mathbf{k}'', \boldsymbol{\gamma})$ in the interval $(P, P + Q^{-1}]$, where $\mathbf{k}' \neq \mathbf{k}''$, $|k_i'| \leqslant |m_i|$ and $|k_i''| \leqslant |m_i| (1 \leqslant i \leqslant s)$, then

$$\langle(\mathbf{k}' - \mathbf{k}'', \boldsymbol{\gamma})\rangle \leqslant Q^{-1}$$

On the other hand from (3.18),

$$\langle(\mathbf{k}' - \mathbf{k}'', \boldsymbol{\gamma})\rangle > b\|\mathbf{k}' - \mathbf{k}''\|^{-a} \geqslant 2^{-sa} b \|\mathbf{m}\|^{-a} > Q^{-1}$$

which gives a contradiction. The lemma is proved.

**Lemma 3.13.** *Under the assumption of Lemma* 3.12,

$$\sum_{|k_i| \leqslant |m_i|}' \frac{1}{\langle(\mathbf{k}, \boldsymbol{\gamma})\rangle} \leqslant 4Q \ln 3Q.$$

*Proof.* Divide the interval $(0, 1]$ into $Q$ subintervals

$$I_l = \left(\frac{l}{Q}, \frac{l+1}{Q}\right], \quad l = 0, 1, \cdots, Q-1.$$

None of the points $(\mathbf{k}, \boldsymbol{\gamma})$ is contained in $I_0$, where $\mathbf{k} \neq \mathbf{0}$ and $|k_i| \leqslant |m_i| (1 \leqslant i \leqslant s)$. Otherwise by (3.18),

$$Q^{-1} \geqslant \langle(\mathbf{k}, \boldsymbol{\gamma})\rangle \geqslant b\|\mathbf{k}\|^{-a} \geqslant b\|\mathbf{m}\|^{-a} > Q^{-1}$$

which gives a contradiction. It follows by Lemma 3.12 that there is at most 1 point $(\mathbf{k}, \boldsymbol{\gamma})$ in any interval $I_l$, where $l \geqslant 1$. Hence

$$\sum_{|k_i| \leqslant |m_i|}' \frac{1}{\langle(\mathbf{k}, \boldsymbol{\gamma})\rangle} < 4 \sum_{k=1}^{Q-1} \frac{Q}{k} < 4Q(1 + \ln Q) < 4Q \ln 3Q.$$

The lemma follows.

**Lemma 3.14.** *Let $h$ be an integer* $\geqslant 2$. *Then under the assumption of Theorem* 3.3,

$$\sum_{|m_i| \leqslant h}' \frac{1}{\|\mathbf{m}\| \langle(\mathbf{m}, \boldsymbol{\gamma})\rangle} \leqslant c(b, s) h^{s(a-1)} (\ln h)^{1+s\delta_{1,a}}.$$

*Proof.* By Lemmas 3.2, 3.7, 3.11 and 3.13,

$$\sum_{|m_i| \leqslant h}' \frac{1}{\|\mathbf{m}\| \langle(\mathbf{m}, \boldsymbol{\gamma})\rangle} \leqslant \sum_{l=0}^{s} \sum_{\mathbf{i}} \frac{1}{h^l} \sum_{m_{i_{l+1}}=1}^{h} \cdots \sum_{m_{i_s}=1}^{h} \frac{1}{m_{i_{l+1}}^2 \cdots m_{i_s}^2}$$

$$\times \sum_{|k_{i_1}|\leqslant h} \cdots \sum_{|k_{i_l}|\leqslant h} \sum_{|k_{i_{l+1}}|\leqslant m_{i_{l+1}}} \cdots \sum_{|k_{i_s}|\leqslant m_{i_s}}{}' \frac{1}{\langle(\mathbf{k}, \boldsymbol{\gamma})\rangle}$$

$$\leqslant \sum_{l=0}^{s} \sum_{\mathbf{i}} \frac{1}{h^l} \sum_{m_{i_{l+1}}=1}^{h} \cdots \sum_{m_{i_s}=1}^{h} \frac{1}{m_{i_{l+1}}^2 \cdots m_{i_s}^2} c(b, s) h^{la} (m_{i_{l+1}} \cdots m_{i_s})^a \ln h$$

$$\leqslant c(b, s) h^{s(a-1)} (\ln h)^{1+s\delta_{1,a}}.$$

The lemma is proved.

The proof of Theorem 3.3. By Lemmas 3.10 and 3.14,

$$\sum_{|m_i|\leqslant h}{}' \frac{1}{\|\mathbf{m}\|} \left| \frac{1}{n} \sum_{k=1}^{n} e^{2\pi i(\mathbf{m}, \gamma)k} \right|$$

$$\leqslant \frac{1}{2n} \sum_{|m_i|\leqslant h}{}' \frac{1}{\|\mathbf{m}\| \langle(\mathbf{m}, \boldsymbol{\gamma})\rangle}$$

$$\leqslant c(b, s) n^{-1} h^{s(a-1)} (\ln h)^{1+s\delta_{1,a}}.$$

Take $r = 1, \eta = \frac{1}{7n}$ and $h = 8n^2$ in Theorem 3.1. Then we have the theorem.

The proof of Theorem 3.5. From Theorem 3.3, the set

$$(\{\gamma_1 k\}, \cdots, \{\gamma_s k\}), \quad 1 \leqslant k \leqslant q$$

has discrepancy $\leqslant c(b, s) q^{-1+2s(a-1)} (\ln 3q)^{1+s\delta_{1,a}}$. Hence by (3.19) and Lemma 3.8. The discrepancy $D(q)$ of the set (3.22) satisfies

$$D(q) \leqslant c(b, s) q^{-1+2s(a-1)} (\ln 3q)^{1+s\delta_{1,a}} + sdqn^{-1-g}$$

$$\leqslant c(b, d, s) q^{-1+2s(a-1)} (\ln 3q)^{1+s\delta_{1,a}}.$$

The theorem is proved.

Theorem 3.4 may be derived from Theorem 3.2 and Lemma 3.9 immediately.

## § 3.10. The lower estimate of discrepancy

**Theorem 3.6.** *Suppose that* $s \geqslant 2$. *Then for any set of points* $P_n(k)$ $(1 \leqslant k \leqslant n)$,

$$D(n) > 2^{-2s-4} (s-1)^{-\frac{s-1}{2}} n^{-1} (\log_2 n)^{\frac{s-1}{2}}.$$

Evidently, Theorem 3.6 is a consequence of the following:

**Theorem 3.7.** *Suppose that* $s \geqslant 2$. *Then for any set of points* $P_n(k)$ $(1 \leqslant k \leqslant n)$,

$$\int_{G_s} (N_n(\mathbf{x}) - n|\mathbf{x}|)^2 d\mathbf{x} > 2^{-4s-8}(s-1)^{-(s-1)} (\log_2 n)^{s-1}.$$

We may suppose that $s = 2$, since the proof of the case $s > 2$ is completely similar. Denote

$$P_n(k) = (X_k, Y_k), \quad 1 \leqslant k \leqslant n.$$

Any point $x$ in the interval $0 \leqslant x \leqslant 1$ can be represented uniquely by the series

$$x = \frac{x_1}{2} + \frac{x_2}{2^2} + \cdots, \quad x_i = 0 \quad \text{or} \quad 1.$$

Let

$$\tau_r(x) = (-1)^{x_r}, \quad r = 1, 2, \cdots.$$

Let $m$ be an integer $> 2$ which will be determined later. For any $(x, y) \in G_2$, we define $F_r(x, y)$ $(r = 1, \cdots, m-1)$ as follows. If there is a $k (1 \leqslant k \leqslant n)$ such that

$$x_1(X_k) = x_1(x), \cdots, x_{r-1}(X_k) = x_{r-1}(x),$$
$$y_1(Y_k) = y_1(y), \cdots, y_{m-r-1}(Y_k) = y_{m-r-1}(y), \tag{3.23}$$

then $F_r(x, y) = 0$, otherwise

$$F_r(x, y) = \tau_r(x)\tau_{m-r}(y).$$

**Lemma 3.15.** *Suppose that* $u$ *is a non-negative integer. Then*

$$\int_{u2^{-r+1}}^{(u+1)2^{-r+1}} F_r(x, y) dx = 0 \tag{3.24}$$

*and*

$$\int_{u2^{-m+r+1}}^{(u+1)2^{-m+r+1}} F_r(x, y) dy = 0. \tag{3.25}$$

*Proof.* For any given $y$ and $x \in [u2^{-r+1}, (u+1)2^{-r+1})$, the $x_1, \cdots, x_{r-1}$ in the binary representation of $x$ are the same, but

$$x_r = \begin{cases} 0, & \text{if} \quad x \in [u2^{-r+1}, u2^{-r+1} + 2^{-r}), \\ 1, & \text{if} \quad x \in [u2^{-r+1} + 2^{-r}, (u+1)2^{-r+1}). \end{cases}$$

Hence we have (3.24). The proof of (3.25) is similar.

**Lemma 3.16.** *Let*

$$F(x, y) = \sum_{r=1}^{m-1} F_r(x, y). \tag{3.26}$$

*Then*

$$\int_0^1 \int_0^1 xyF(x, y)dxdy \geqslant (m-1)2^{-2m}(2^{m-2} - n).$$

*Proof.* It is sufficient to prove that

$$\int_0^1 \int_0^1 xyF_r(x, y)dxdy \geqslant 2^{-2m}(2^{m-2} - n), \quad r = 1, 2, \cdots, m-1.$$

Divide the intervals of the integrals of $x$ and $y$ into $2^{r-1}$ and $2^{m-r-1}$ equal parts respectively. Then the above integral is equal to

$$\Sigma^* \int_{-2^{-r}}^{2^{-r}} \int_{-2^{-m+r}}^{2^{-m+r}} (\xi + x)(\eta + y)\,(\text{sign}\, x)\,(\text{sign}\, y)dxdy, \tag{3.27}$$

where

$$\xi = \frac{x_1}{2} + \cdots + \frac{x_{r-1}}{2^{r-1}} + \frac{1}{2^r},$$

$$\eta = \frac{y_1}{2} + \cdots + \frac{y_{m-r-1}}{2^{m-r-1}} + \frac{1}{2^{m-r}}$$

and $\Sigma^*$ denotes a sum of $x_1, \cdots, x_{r-1}, y_1, \cdots, y_{m-r-1}$ such that (3.23) is not satisfied for any $k (1 \leqslant k \leqslant n)$.

Since

$$\int_{-a}^{a} (\xi + x)\,\text{sign}\, x dx = \int_{-a}^{a} x \,\text{sign}\, x \; dx = a^2,$$

(3.27) is equal to

$$\Sigma^* 2^{-2m}.$$

Since the total number of the sets of $x_1, \cdots, x_{r-1}, y_1, \cdots, y_{m-r-1}$ is $2^{m-2}$ in which the number of sets such that (3.23) are satisfied is at most $n$, hence the total number of terms of $\Sigma^*$ is no less than $2^{m-2} - n$. The lemma follows.

**Lemma 3.17.**

$$\int_0^1 \int_0^1 F(x, y)^2 dxdy \leqslant m - 1.$$

*Proof.* By (3.26),

$$\int_0^1\int_0^1 F(x,y)^2dxdy = \sum_{r=1}^{m-1}\int_0^1\int_0^1 F_r(x,y)^2dxdy$$
$$+ 2\sum_{1\leqslant r<s\leqslant m-1}\int_0^1\int_0^1 F_r(x,y)F_s(x,y)dxdy.$$

Since $|F_r(x,y)|\leqslant 1$, the first term of the right hand side does not exceed $m-1$. Now we proceed to prove that the second term is equal to zero. For any given $y$, divide the interval of the integral of $x$ into $2^{s-1}$ equal parts. Similar to (3.24), the integrals over these subintervals are all equal to zero. Hence we have the lemma.

**Lemma 3.18.**

$$\int_{X_k}^1\int_{Y_k}^1 F(x,y)dxdy = 0,\quad 1\leqslant k\leqslant n.$$

*Proof.* It is sufficient to prove that

$$\int_{X_k}^1\int_{Y_k}^1 F_r(x,y)dxdy = 0,\quad r=1,\cdots,m-1.$$

Divide the integral into four parts

$$\int_{X_k}^X\int_{Y_k}^Y + \int_X^1\int_{Y_k}^Y + \int_{X_k}^X\int_Y^1 + \int_X^1\int_Y^1, \tag{3.28}$$

where $X$ is the least integer $\geqslant X_k$ and also a multiple of $2^{-r+1}$ and where $Y$ is the least integer $\geqslant Y_k$ and also a multiple of $2^{-m+r+1}$. Since (3.23) holds in the rectangle

$$X_k\leqslant x<X,\quad Y_k\leqslant y<Y,$$

the first integral of (3.28) is equal to zero. It follows by Lemma 3.15 that the other integrals are all equal to zero too. The lemma is proved.

The proof of Theorem 3.7. From Lemma 3.18,

$$\int_0^1\int_0^1 N_n(x,y)F(x,y)dxdy = \int_0^1\int_0^1 F(x,y)\left(\sum_{\substack{k=1\\ X_k<x,\ Y_k<y}}^{n} 1\right)dxdy$$
$$= \sum_{k=1}^n\int_{X_k}^1\int_{Y_k}^1 F(x,y)dxdy = 0.$$

Hence by Lemma 3.16,

$$\int_0^1\int_0^1 (nxy - N_n(x, y))F(x, y)dxdy$$
$$= n\int_0^1\int_0^1 xyF(x, y)dxdy \geqslant n(m-1)2^{-2m}(2^{m-2}-n).$$

Let $m$ be the integer such that

$$8n < 2^m \leqslant 16n.$$

Then

$$\int_0^1\int_0^1 (nxy - N_n(x, y))^2 dxdy$$
$$\geqslant \left(\int_0^1\int_0^1 (n\ x\ y - N_n(x, y))\ F(x, y)\, dxdy\right)^2 \left(\int_0^1\int_0^1 F(x, y)^2 dxdy\right)^{-1}$$
$$\geqslant (m-1)(n2^{-2m}(2^{m-2}-n))^2$$

by Schwarz inequality and Lemma 3.17. Since

$$2^{m-2} - n > n, \quad n^2 2^{-2m} \geqslant 2^{-8}, \quad m > \log_2 n + 3,$$

hence

$$\int_0^1\int_0^1 (N_n(x, y) - n\ x\ y)^2 dxdy > 2^{-16}\log_2 n.$$

The theorem is proved.

If the sequence of sets $P_{n_l}(k)(n_1 < n_2 < \cdots)$ have discrepancy

$$D(n) = O(n^{-1+\varepsilon}),$$

where the constant implied by the symbol "$O$" depends only on $\varepsilon$, the sequence $P_{n_l}(k)(n_1 < n_2 < \cdots)$ (or simply the set $P_n(k)$) is said to be best uniformly distributed.

## Notes

The definition of uniform distribution was first given by H. Weyl [1].

Lemma 3.1: Cf. I. M. Vinogradov [1].

Theorem 3.1 in a slightly different form was proved by P. Erdös, and P. Turán [1] for $s=1$ and J. F. Koksma [2] for $s>1$ (Cf. also E. Hlawka [2] and Hua Loo Keng and Wang Yuan [7]).

Lemmas 3.3, 3.4 and 3.5: Cf N. S. Bahvalov [1] and also Hua Loo Keng and Wang Yuan [7].

Theorem 3.3 was proved independently by Hua Loo Keng and Wang Yuan [7] and H. Niederreiter [1] in a slightly different form.

Concerning the lower estimation of the discrepancy, T. Van Aardenne-Ehrenfest [1] first proved that $D(n) > \frac{c \ln \ln n}{n \ln \ln \ln n}$. Theorem 3.6 is due to K. F. Roth [1] whose result was improved to $D(n) > \frac{c \ln n}{n}$ for $s = 2$ by W. M. Schmidt [3].

*Chapter 4*

# Estimation of Discrepancy

## § 4.1. The set of equi-distribution

Suppose that $s \geqslant 2, m_1, \cdots, m_s$ are $s$ positive integers, $n = m_1 \cdots m_s$ and $m = \min(m_1, \cdots, m_s)$. The set

$$\left(\frac{l_1}{m_1}, \cdots, \frac{l_s}{m_s}\right), \quad 0 \leqslant l_i < m_i, 1 \leqslant i \leqslant s \tag{4.1}$$

is called the set of equi-distribution.

**Theorem 4.1.** *The set* (4.1) *has discrepancy*

$$D(n) \leqslant 2^s m^{-1}.$$

*Proof.* Let $\boldsymbol{\gamma} \in G_s$. Since the number of $l_i$ satisfying

$$\frac{l_i}{m} < \gamma_i, \quad l_i = 0, 1, \cdots$$

is equal to $[m_i \gamma_i]$ or $[m_i \gamma_i] + 1$, therefore

$$N_n(\boldsymbol{\gamma}) = ([m_1 \gamma_1] + \theta_1) \cdots ([m_s \gamma_s] + \theta_s),$$

where ${}_i\theta = 0$ or $1$ $(1 \leqslant i \leqslant s)$. Hence

$$0 \leqslant \frac{N_n(\boldsymbol{\gamma})}{n} - |\boldsymbol{\gamma}| \leqslant \prod_{i=1}^{s} \frac{(m_i \gamma_i + 1)}{m_i} - \prod_{i=1}^{s} \gamma_i$$

$$= \prod_{i=1}^{s} \left(\gamma_i + \frac{1}{m_i}\right) - \prod_{i=1}^{s} \gamma_i \leqslant 2^s m^{-1}.$$

The theorem is proved.

**Theorem 4.2.** *The discrepancy of the set* (4.1) *satisfies*

$$D(n) \geqslant \frac{1}{2m} \geqslant 2^{-1}n^{-1/s}.$$

*Proof.* Without loss of generality, we may suppose that $m = m_1$. Set $\boldsymbol{\gamma} = \left(\frac{1}{2m}, 1, \cdots, 1\right)$. Then

$$N_n(\boldsymbol{\gamma}) = m_2 \cdots m_s = \frac{n}{m}.$$

Hence

$$\left|\frac{N_n(\boldsymbol{\gamma})}{n} - |\boldsymbol{\gamma}|\right| = \frac{1}{m} - \frac{1}{2m} = \frac{1}{2m} \geqslant 2^{-1}n^{-1/s}.$$

The theorem is proved.

Clearly, the best case for the equi-distribution is to choose $m_1 = \cdots = m_s = m$, i.e.,

$$\left(\frac{l_1}{m}, \cdots, \frac{l_s}{m}\right), \quad 0 \leqslant l_1, \cdots, l_s < m.$$

And it follows from Theorem 4.2 that the discrepancy $D(n)$ of the set of equi-distribution increases rapidly when $s$ increases.

## § 4.2. The Halton theorem

**Lemma 4.1.** *The number of solutions of the congruence*

$$x \equiv a \pmod{m}, \quad 1 \leqslant x \leqslant n$$

*is equal to* $\left[\frac{n}{m}\right]$ *or* $\left[\frac{n}{m}\right] + 1$.

*Proof.* Since the congruence has exactly 1 solution in any $m$ consecutive integers, the lemma follows.

**Lemma 4.2.** *Suppose that* $m_1, \cdots, m_s$ *are positive integers which are relatively prime to each other. Then the system of congruences*

$$x \equiv a_i \pmod{m_i}, \quad 1 \leqslant i \leqslant s$$

*has a unique solution* mod $m_1 \cdots m_s$ (Cf. Hua Loo Keng [2], Chap. 2).

Let $r$ be an integer $> 1$. Then any positive integer $k$ can be represented uniquely as

$$k = k_0 + k_1 r + \cdots + k_M r^M, \quad 0 \leqslant k_i \leqslant r-1.$$

Which is denoted in digits by

$$k = k_M \cdots k_1 k_0$$

and any number $h$ in the interval $[0, 1]$ can be represented uniquely by

$$h = \frac{h_0}{r} + \frac{h_1}{r^2} + \cdots + \frac{h_M}{r^{M+1}} + \cdots, \quad 0 \leqslant h_i \leqslant r-1$$

which is denoted by

$$h = 0.h_0 h_1 \cdots h_M \cdots.$$

For any given positive integer $k = k_0 + k_1 r + \cdots + k_M r^M$, where it corresponds to a number

$$\varphi_r(k) = \frac{k_0}{r} + \frac{k_1}{r^2} + \cdots + \frac{k_M}{r^{M+1}}.$$

If $k_M \neq 0$, then

$$r^M \leqslant k < r^{M+1}$$

and so

$$M = \left[\frac{\ln k}{\ln r},\right]$$

where $M+1$ is called the number of digits of $k$.

**Lemma 4.3.** *Suppose that* $n > r$. *Then the set*

$$\varphi_r(k), \quad k = 1, 2, \cdots, n$$

*has discrepancy*

$$D(n) < \left(\frac{r \ln rn}{\ln r}\right) n^{-1}.$$

*Proof.* Let $\alpha$ satisfy $0 < \alpha \leqslant 1$. Then $\alpha$ may be written as

$$\alpha = 0.a_0 a_1 \cdots a_M \cdots.$$

Without loss of generality, we may suppose that $\alpha$ does not terminate. Otherwise if $\alpha = 0.a_0 a_1 \cdots a_M$, where $a_M \neq 0$, then $\alpha$ may be written as $\alpha = 0.a_0 a_1 \cdots a_{M-1} a'_M a'_{M+1} \cdots$, where $a'_M = a_M - 1$, $a'_i = r - 1 (i = M+1,$

$M+2, \cdots$). If $\varphi_r(k) < \alpha$, then the integer $k$ satisfies one of the following conditions:

$$1)\quad k_0 < a_0,$$

$$2)\quad k_0 = a_0, k_1 < a_1,$$

$$3)\quad k_0 = a_0, k_1 = a_1, k_2 < a_2,$$

$$\cdots$$

Since the number of digits of $k$ does not exceed $\left[\frac{\ln n}{\ln r}\right] + 1$ for $1 \leqslant k \leqslant n$, we may write $M = \left[\frac{\ln n}{\ln r}\right]$. The final two steps are

$$M+1)\quad k_0 = a_0, k_1 = a_1, \cdots, k_{M-1} = a_{M-1}, k_M < a_M,$$

$$M+2)\quad k_0 = a_0, k_1 = a_1, \cdots, k_{M-1} = a_{M-1}, k_M = a_M.$$

Write these formulas in the forms of congruences

$$1)\quad k \equiv k_0 \pmod r, 0 \leqslant k_0 < a_0,$$

$$2)\quad k \equiv a_0 + k_1 r \pmod{r^2}, 0 \leqslant k_1 < a_1,$$

$$3)\quad k \equiv a_0 + a_1 r + k_2 r^2 \pmod{r^3}, 0 \leqslant k_2 < a_2,$$

$$\cdots$$

$$M+1)\quad k \equiv a_0 + a_1 r + \cdots + a_{M-1} r^{M-1} + k_M r^M \pmod{r^{M+1}}, \quad 0 \leqslant k_M < a_M,$$

$$M+2)\quad k \equiv a_0 + a_1 r + \cdots + a_{M-1} r^{M-1} + a_M r^M \pmod{r^{M+2}}.$$

By Lemma 4.1, The numbers of solutions of the above congruences for $1 \leqslant k \leqslant n$ are equal to

$$a_0\left(\left[\frac{n}{r}\right] + \theta\right)$$

$$a_1\left(\left[\frac{n}{r^2}\right] + \theta\right),$$

$$\cdots\cdots$$

$$a_M\left(\left[\frac{n}{r^{M+1}}\right] + \theta\right),$$

$$\left(\left[\frac{n}{r^{M+2}}\right] + \theta\right) = \theta$$

respectively, where $\theta$ satisfies $0 \leqslant \theta \leqslant 1$. (But not always having the same value.)

Let $N_n(\alpha)$ denote the number of $k$ satisfying $\varphi_r(k) < \alpha (1 \leqslant k \leqslant n)$. Then

$$N_n(\alpha) = a_0\left(\left[\frac{n}{r}\right] + \theta\right) + a_1\left(\left[\frac{n}{r^2}\right] + \theta\right) + \cdots + a_M\left(\left[\frac{n}{r^{M+1}}\right] + \theta\right) + \theta$$

and

$$\begin{aligned}
&\left|N_n(\alpha) - a_0\frac{n}{r} - a_1\frac{n}{r^2} - \cdots - a_M\frac{n}{r^{M+1}}\right| \\
&\leqslant a_0\left|\frac{n}{r} - \left[\frac{n}{r}\right] - \theta\right| + a_1\left|\frac{n}{r^2} - \left[\frac{n}{r^2}\right] - \theta\right| + \cdots \\
&\quad + a_M\left|\frac{n}{r^{M+1}} - \left[\frac{n}{r^{M+1}}\right] - \theta\right| + \theta \\
&\leqslant a_0 + a_1 + \cdots + a_M + 1.
\end{aligned}$$

Hence

$$|N_n(\alpha) - \alpha n| \leqslant a_0 + a_1 + \cdots + a_M + 1 + \left(\frac{a_{M+1}}{r^{M+2}} + \frac{a_{M+2}}{r^{M+3}} + \cdots\right)n. \tag{4.2}$$

Since

$$\begin{aligned}
\frac{a_{M+1}}{r^{M+2}} + \frac{a_{M+2}}{r^{M+3}} + \cdots &\leqslant (r-1)\left(\frac{1}{r^{M+2}} + \frac{1}{r^{M+3}} + \cdots\right) \\
&= \frac{r-1}{r^{M+2}}\left(1 - \frac{1}{r}\right)^{-1} = \frac{1}{r^{M+1}} < \frac{1}{n},
\end{aligned}$$

we have

$$\begin{aligned}
|N_n(\alpha) - \alpha n| &\leqslant (M+1)(r-1) + 2 \\
&\leqslant \left(\frac{\ln n}{\ln r} + 1\right)(r-1) + 2 \leqslant \frac{r \ln rn}{\ln r}
\end{aligned}$$

for $n > r$. The lemma is proved.

**Theorem 4.3.** *Suppose that $r_i (1 \leqslant i \leqslant s)$ are $s$ integers $> 1$ which are relatively prime to each other and that $n > \max(r_1, \cdots, r_s)$. Then the set*

$$(\varphi_{r_1}(k), \cdots, \varphi_{r_s}(k)), \quad 1 \leqslant k \leqslant n \tag{4.3}$$

*has discrepancy*

$$D(n) \leqslant \left(\prod_{i=1}^{s} \frac{r_i \ln r_i n}{\ln r_i}\right) n^{-1}.$$

*Proof.* We may suppose that $s=2$, since the proof is similar for the case $s>2$. Suppose that $r$ and $t$ are integers $>1$ and $(r,t)=1$ and that $0\leqslant\beta<1$ and its expansion in the scale of $t$ is

$$\beta=0.b_0b_1\cdots.$$

Let $N_n(\alpha,\beta)$ denote the number of integers $k$ satisfying

$$\varphi_r(k)<\alpha,\quad \varphi_t(k)<\beta,\quad 1\leqslant k\leqslant n.$$

Let $L=\left[\dfrac{\ln n}{\ln t}\right]$. Then the integers in the interval $1\leqslant k\leqslant n$ can be represented uniquely by

$$k=l_0+l_1t+\cdots+l_Lt^L,\quad 0\leqslant l_i\leqslant t-1.$$

Similar to the congruences $1), 2), \cdots, M+2)$, we have a system of congruences

$$1)'\quad k\equiv l_0\ (\mathrm{mod}\ t),\ 0\leqslant l_0<b_0,$$

$$2)'\quad k\equiv b_0+l_1t\ (\mathrm{mod}\ t^2),\ 0\leqslant l_1<b_1,$$

$$\cdots$$

$$L+1)'\quad k\equiv b_0+b_1t+\cdots+b_{L-1}t^{L-1}+l_Lt^L\ (\mathrm{mod}\ t^{L+1}),\ 0\leqslant l_L<b_L,$$

$$L+2)'\quad k\equiv b_0+b_1t+\cdots+b_{L-1}t^{L-1}+b_Lt^L\ (\mathrm{mod}\ t^{L+2}).$$

Since it follows by Lemmas 4.1 and 4.2 that the number of integers in $1\leqslant k\leqslant n$ which satisfy $m)$ and $l)'$ is

$$a_{m-1}b_{l-1}\left(\left[\frac{n}{r^mt^l}\right]+\theta\right),$$

so $N_n(\alpha,\beta)$ is equal to

$$\sum_{m=1}^{M+1}\sum_{l=1}^{L+1}a_{m-1}b_{l-1}\left(\left[\frac{n}{r^mt^l}\right]+\theta\right)+\sum_{m=1}^{M+1}a_{m-1}\theta+\sum_{l=1}^{L+1}b_{l-1}\theta+\theta$$

and so

$$\left|N_n(\alpha,\beta)-\sum_{m=1}^{M+1}\sum_{l=1}^{L+1}a_{m-1}b_{l-1}\frac{n}{r^mt^l}\right|$$

$$\leqslant\sum_{m=1}^{M+1}\sum_{l=1}^{L+1}a_{m-1}b_{l-1}+\sum_{m=1}^{M+1}a_{m-1}+\sum_{l=1}^{L+1}b_{l-1}+1$$

$$=\left(\sum_{m=1}^{M+1}a_{m-1}+1\right)\left(\sum_{l=1}^{L+1}b_{l-1}+1\right).$$

Consequently, we have

$$|N_n(\alpha,\beta)-\alpha\beta n| \leqslant \Big(\sum_{m=1}^{M+1} a_{m-1}+1\Big)\Big(\sum_{l=1}^{L+1} b_{l-1}+1\Big)$$

$$+n\sum_{m=M+2}^{\infty}\frac{a_{m-1}}{r^m}+n\sum_{l=L+2}^{\infty}\frac{b_{l-1}}{t^l}+n^{-1}$$

$$\leqslant ((r-1)(M+1)+1)((t-1)(L+1)+1)+3$$

$$\leqslant ((r-1)(M+1)+2)((t-1)(L+1)+2)$$

$$\leqslant \left(\frac{r\ln rn}{\ln r}\right)\left(\frac{t\ln tn}{\ln t}\right). \tag{4.4}$$

The theorem is proved.

**Lemma 4.4.** *Suppose that $n>r^2$. Then under the assumption of Lemma* 4.3,

$$\frac{1}{q}\sum_{l=1}^{q}\left|\frac{1}{n}N_n\left(\frac{l}{q}\right)-\frac{l}{q}\right| \leqslant \left(\frac{\ln rn}{2\ln r}\right)n^{-1}.$$

*Proof.* Let

$$1-\alpha=0.a_0'a_1'\cdots a_M'\cdots.$$

Then

$$a_\nu+a_\nu'=r-1,\quad \nu=0,1,\cdots$$

and by (4.2),

$$|N_n(\alpha)-\alpha n|+|N_n(1-\alpha)-(1-\alpha)n|$$

$$\leqslant (r-1)(M+1)+2+(r-1)n\sum_{l=M+2}^{\infty}\frac{1}{r^l}$$

$$\leqslant (r-1)(M+1)+2+\frac{n}{r^{M+1}}$$

$$\leqslant (r-1)(M+1)+3\leqslant \frac{r\ln rn}{\ln r}.$$

Hence

$$\frac{1}{q}\sum_{l=1}^{q}\left|\frac{1}{n}N_n\left(\frac{l}{q}\right)-\frac{l}{q}\right| \leqslant \left(\frac{r\ln rn}{2\ln r}\right)n^{-1}.$$

The lemma is proved.

**Theorem 4.4.** *Suppose that* $n > \max(r_1^4, \cdots, r_s^4)$. *Then under the assumption of Theorem* 4.3,

$$\frac{1}{q^s}\sum_{l=1}^{q}\cdots\sum_{l_s=1}^{q} q^{\delta_{l_1,q}+\cdots+\delta_{l_s,q}}\left|\frac{1}{n}N_n\left(\frac{l_1}{q},\cdots,\frac{l_s}{q}\right)-\frac{l_1\cdots l_s}{q^s}\right|$$
$$\leqslant\left(\prod_{i=1}^{s}\frac{r_i\ln r_i n}{2\ln r_i}\right)n^{-1}.$$

*Proof.* We may suppose that $s=2$ as in the proof of theorem 4.3. Let

$$1-\beta = 0.b_0'b_1'\cdots b_L'\cdots.$$

Then

$$b_\nu + b_\nu' = t-1, \quad \nu = 0, 1, \cdots.$$

Hence from (4.4), we have

$$\begin{aligned}
&|N_n(\alpha,\beta)-\alpha\beta n| + |N_n(1-\alpha,\beta)-(1-\alpha)\beta n|\\
&\quad + |N_n(\alpha,1-\beta)-\alpha(1-\beta)n|\\
&\quad + |N_n(1-\alpha,1-\beta)-(1-\alpha)(1-\beta)n|\\
&\leqslant((r-1)(M+1)+2)((t-1)(L+1)+2)\\
&\quad + 2n\sum_{m=M+2}^{\infty}\frac{r-1}{r^m} + 2n\sum_{l=L+2}^{\infty}\frac{t-1}{t^l}+\frac{4}{n}\\
&\leqslant((r-1)(M+1)+2)((t-1)(L+1)+2)+5\\
&\leqslant((r-1)(M+1)+3)((t-1)(L+1)+3)
\end{aligned}$$

and

$$\begin{aligned}
&\frac{1}{q^2}\sum_{l=1}^{q-1}\sum_{m=1}^{q-1}\left|\frac{1}{n}N_n\left(\frac{l}{q},\frac{m}{q}\right)-\frac{lm}{q^2}\right| + \frac{1}{q}\sum_{l=1}^{q-1}\left|\frac{1}{n}N_n\left(\frac{l}{q},1\right)-\frac{l}{q}\right|\\
&\quad+\frac{1}{q}\sum_{m=1}^{q-1}\left|\frac{1}{n}N_n\left(1,\frac{m}{q}\right)-\frac{m}{q}\right|\\
&\leqslant\frac{1}{4n}((r-1)(M+1)+3)((t-1)(L+1)+3)\\
&\quad+\frac{1}{2n}((r-1)(M+1)+3)+\frac{1}{2n}((t-1)(L+1)+3)\\
&\leqslant\frac{1}{4n}((r-1)(M+1)+5)((t-1)(L+1)+5)\\
&<\left(\frac{r\ln rn}{2\ln r}\right)\left(\frac{t\ln tn}{2\ln t}\right)n^{-1}.
\end{aligned}$$

The theorem is proved.

**Theorem 4.5.** *Suppose that* $n > \max(r_1, \cdots, r_{s-1})$. *Then the set*

$$\left(\frac{k}{n}, \varphi_{r_1}(k), \cdots, \varphi_{r_{s-1}}(k)\right), \quad 1 \leqslant k \leqslant n \tag{4.5}$$

*has discrepancy*

$$D(n) \leqslant \left(\prod_{i=1}^{s-1} \frac{r_i \ln r_i n}{\ln r_i}\right) n^{-1}. \tag{4.6}$$

*Proof.* Let $\boldsymbol{\gamma} \in G_s$. If $n\gamma_1 \leqslant \max(r_1, \cdots, r_{s-1}) = r$ (say), then we derive from $\frac{k}{n} < \gamma_1$ that $N_n(\boldsymbol{\gamma}) \leqslant r$. Hence the left hand side of (4.6) does not exceed $\frac{r}{n}$, but the right hand side of (4.6) is greater than $\frac{r}{n}$, so we have the theorem. Now, suppose that $n\gamma_1 > r$. Obviously $N_n(\boldsymbol{\gamma})$ equals the number of integers $k$ satisfying

$$\varphi_{r_1}(k) < \gamma_2, \cdots, \varphi_{r_{s-1}}(k) < \gamma_s, \quad 1 \leqslant k < n\gamma_1.$$

We may also suppose that $n\gamma_1$ is an integer. Otherwise we can suppose that $m < n\gamma_1 < m+1$, where $m$ is an integer. Then $N_n(\boldsymbol{\gamma})$ is unchanged, if $\gamma_1$ is replaced by $\frac{m+1}{n}$. Hence by Theorem 4.3,

$$\left|\frac{N_n(\boldsymbol{\gamma})}{n\gamma_1} - \gamma_2 \cdots \gamma_s\right| \leqslant \gamma_1^{-1} \left(\prod_{i=1}^{s-1} \frac{r_i \ln r_i \gamma_1 n}{\ln r_i}\right) n^{-1}$$
$$\leqslant \gamma_1^{-1} \left(\prod_{i=1}^{s-1} \frac{r_i \ln r_i n}{\ln r_i}\right) n^{-1}.$$

The theorem follows.

Similar to Theorem 4.4, we have

**Theorem 4.6.** *Suppose that* $n > \max(r_1^4, \cdots, r_{s-1}^4)$. *Then under the assumption of Theorem* 4.5,

$$\frac{1}{q^s} \sum_{l_1=1}^{q} \cdots \sum_{l_s=1}^{q} q^{\delta_{l_1,q} + \cdots + \delta_{l_s,q}} \left| N_n\left(\frac{l_1}{q}, \cdots, \frac{l_s}{q}\right) - \frac{l_1 \cdots l_s}{q^s}\right|$$
$$\leqslant \left(\prod_{i=1}^{s-1} \frac{r_i \ln r_i n}{2 \ln r_i}\right) n^{-1}.$$

From Theorems 4.3 and 4.5, we know that the sets (4.3) and (4.5) have discrepancies $D(n)=O\left(\frac{(\ln n)^s}{n}\right)$ and $D(n)=O\left(\frac{(\ln n)^{s-1}}{n}\right)$ respectively. Hence they are best uniformly distributed.

*Remark.* For practical use, we may take $r_i=p_i(1\leqslant i\leqslant s)$, where $p_i$ denotes the $i$-th prime number.

## § 4.3. The $p$ set

As usual, we use $p$ to denote a prime number (Cf. §1.3). The sets

$$\left(\left\{\frac{k}{p}\right\},\left\{\frac{ak}{p}\right\},\cdots,\left\{\frac{a^{s-1}k}{p}\right\}\right),\quad 1\leqslant a,k\leqslant p, \tag{4.7}$$

$$\left(\left\{\frac{k}{p^2}\right\},\left\{\frac{k^2}{p^2}\right\},\cdots,\left\{\frac{k^s}{p^2}\right\}\right),\quad 1\leqslant k\leqslant p^2 \tag{4.8}$$

and

$$\left(\left\{\frac{k}{p}\right\},\left\{\frac{k^2}{p}\right\},\cdots,\left\{\frac{k^s}{p}\right\}\right),\quad 1\leqslant k\leqslant p \tag{4.9}$$

are called the $p$ sets. The discrepancies of these sets can be evaluated with the aid of the estimates for exponential sums (Cf. Hua Loo Keng [4.5]).

**Theorem 4.7.** *The set* (4.7) *has discrepancy*

$$D(p^2)<c(s)p^{-1}(\ln p)^s. \tag{4.10}$$

**Theorem 4.8.** *The set* (4.8) *has discrepancy*

$$D(p^2)<c(s)p^{-1}(\ln p)^s.$$

**Theorem 4.9.** *The set* (4.9) *has discrepancy*

$$D(p)<c(s)p^{-\frac{1}{2}}(\ln p)^s.$$

**Lemma 4.5.** *Let* $a_0,a_1,\cdots,a_s$ *be a set of integers such that their great common divisor is not divisible by* $p$. *Then the number of solutions of the congruence*

$$a_sx^s+\cdots+a_1x+a_0\equiv 0\ (\mathrm{mod}\ p),\ 1\leqslant x\leqslant p$$

*is at most* $s$ (Cf. Hua Loo Keng [2], Chap. 2).

**Lemma 4.6.** *Let*

$$S = \sum_{x=1}^{p^2} e^{2\pi i f(x)/p^2},$$

*where* $f(x) = a_s x^s + \cdots + a_1 x$ *and at least one of the coefficients is not divisible by* $p$. *Then*

$$|S| \leqslant (s-1)p.$$

*Proof.* Since

$$f(x+py) \equiv f(x) + f'(x)py \pmod{p^2}$$

and

$$S = \sum_{x=1}^{p} \sum_{y=1}^{p} e^{2\pi i f(x+py)/p^2} = \sum_{x=1}^{p} e^{2\pi i f(x)/p^2} \sum_{y=1}^{p} e^{2\pi i f'(x)y/p},$$

so from Lemmas 3.6 and 4.5,

$$|S| \leqslant \sum_{x=1}^{p} \left| \sum_{y=1}^{p} e^{2\pi i f'(x)y/p} \right| = p \sum_{\substack{x=1 \\ f'(x)\equiv 0 \pmod p}}^{p} 1 \leqslant (s-1)p.$$

The lemma is proved.

**Lemma 4.7.** *Under the assumptions of Lemma* 4.6,

$$\left| \sum_{x=1}^{p} e^{2\pi i f(x)/p} \right| \leqslant (s-1)\sqrt{p}$$

(Cf. A. Weil [1]).

The proof of Theorem 4.7. Let $\mathbf{a} = (1, a, \cdots, a^{s-1})$ and $\mathbf{m} = (m_1, \cdots, m_s)$. (Notice that we also use $(m_1, \cdots, m_s)$ to denote the great common divisor of $m_1, \cdots, m_s$. Please don't be confused with the notation for a vector). By Lemmas 3.6 and 4.5,

$$\Sigma(\mathbf{m}) = \frac{1}{p^2} \sum_{a=1}^{p} \sum_{k=1}^{p} e^{2\pi i (\mathbf{a},\mathbf{m})k/p} = p^{-1} \sum_{\substack{a=1 \\ (\mathbf{a},\mathbf{m})\equiv 0 \pmod p}}^{p} 1$$

$$\leqslant \begin{cases} 1, & \text{if } p \mid (m_1, \cdots, m_s), \\ \dfrac{s-1}{p}, & \text{if } p \nmid (m_1, \cdots, m_s). \end{cases}$$

Hence by Lemma 3.2,

$$\sum_{|m_i|\leqslant p^2}' \frac{1}{\|\pi\mathbf{m}\|}|\Sigma(\mathbf{m})| \leqslant \sum_{\substack{|m_i|\leqslant p^2 \\ p|(m_1,\cdots,m_s)}}' \frac{1}{\|\pi\mathbf{m}\|} + \frac{s-1}{p} \sum_{\substack{|m_i|\leqslant p^2 \\ p\nmid(m_1,\cdots,m_s)}}' \frac{1}{\|\pi\mathbf{m}\|}$$

$$\leqslant \frac{s}{p}\left(1+\frac{2}{\pi}\sum_{m=1}^{p^2}\frac{1}{m}\right)^s < c(s)p^{-1}(\ln p)^s.$$

Take $r=1, \eta=p^{-1}$ and $h=p^2$ in Theorem 3.1 for $p>6$. Then we have the Theorem. The theorem is obvious for $p\leqslant 6$.

The proof of Theorem 4.8. Let $\mathbf{k}=(k, k^2, \cdots, k^s)$. Then from Lemma 4.6,

$$\sum_{|m_i|\leqslant p^2}' \frac{1}{\|\pi\mathbf{m}\|}\left|\frac{1}{p^2}\sum_{k=1}^{p^2} e^{2\pi i(\mathbf{k},\mathbf{m})/p^2}\right| \leqslant \sum_{\substack{|m_i|\leqslant p^2 \\ p|(m_1,\cdots,m_s)}}' \frac{1}{\|\pi\mathbf{m}\|}$$

$$+ \frac{s-1}{p} \sum_{\substack{|m_i|\leqslant p^2 \\ p\nmid(m_1,\cdots,m_s)}} \frac{1}{\|\pi\mathbf{m}\|} < c(s)p^{-1}(\ln p)^s.$$

Put $r=1, \eta=p^{-1}$ and $h=p^2$ in Theorem 3.1 for $p>6$. The theorem follows. For the case $p\leqslant 6$, the theorem is obvious.

The proof of Theorem 4.9. From Lemma 4.7,

$$\sum_{|m_i|<p}' \frac{1}{\|\pi\mathbf{m}\|}\left|\frac{1}{p}\sum_{k=1}^{p} e^{2\pi i(\mathbf{k},\mathbf{m})/p}\right| \leqslant \frac{s-1}{\sqrt{p}} \sum_{|m_i|<p}' \frac{1}{\|\pi\mathbf{m}\|}$$

$$\leqslant \frac{s-1}{\sqrt{p}}\left(1+\frac{2}{\pi}\sum_{m=1}^{p-1}\frac{1}{m}\right)^s < c(s)p^{-\frac{1}{2}}(\ln p)^s.$$

Put $r=1, \eta=p^{-\frac{1}{2}}$ and $h=p-1$ in Theorem 3.1 for $p\geqslant 37$. Then we have the theorem. For the case $p<37$, the theorem is obvious.

Now we study the lower estimate for the discrepancy of the set (4.7). Let $p\geqslant 3$. Put $x_1=1, x_2=\dfrac{2}{p}$ and $x_3=\cdots=x_s=1$. Since the congruences

$$ak\equiv 0 \text{ or } 1 \pmod{p}, \quad 1\leqslant a, k\leqslant p$$

have $3p-2$ pairs of solutions $a, k$,

$$N_{p^2}\left(1, \frac{2}{p}, 1, \cdots, 1\right) = 3p-2$$

and

$$\left|\frac{1}{p^2}N_{p^2}\left(1, \frac{2}{p}, 1, \cdots, 1\right) - \frac{2}{p}\right| = \frac{3p-2}{p^2} - \frac{2}{p} = \frac{p-2}{p^2} > \frac{1}{2p}.$$

And so the discrepancy of the set (4.7) satisfies $D(p^2) > \frac{1}{2p}$, i.e., the factor $p^{-1}$ in the right hand side of (4.10) does not admit further improvement.

*Remark.* For any given $\varepsilon > 0$, take $\eta = p^{-1}$, $h = [p^{1+r^{-1}}]$ and $r = r(\varepsilon)$ sufficiently large in Theorem 3.1. Then the discrepancies of the sets (4.7) and (4.8) may be given by

$$D(p^2) < \left(\left(\frac{2}{\pi}\right)^s s + \varepsilon\right) p^{-1} (\ln p)^s$$

for $p > c(s, \varepsilon)$. And if we take $\eta = p^{-\frac{1}{2}}, h = [p^{\frac{1}{2}+\frac{1}{r}}]$ and $r = r(\varepsilon)$ sufficiently large in Theorem 3.1, then the discrepancy of the set (4.9) may be given by

$$D(p) < \left(\left(\frac{1}{\pi}\right)^s (s-1) + \varepsilon\right) p^{-\frac{1}{2}} (\ln p)^s$$

for $p > c(s, \varepsilon)$.

## § 4.4. The *gp* set

Let $\boldsymbol{\gamma} \in G_s$. If the set of the form

$$P(k) = (\{\gamma_1 k\}, \cdots, \{\gamma_s k\}), \quad 1 \leqslant k \leqslant n \tag{4.11}$$

has discrepancy

$$D(n) \leqslant c(\boldsymbol{\gamma}, \varepsilon) n^{-1+\varepsilon}, \tag{4.12}$$

then the set (4.11) is called a *gp* set and $\boldsymbol{\gamma}$ a good point. Hence the sequence of sets so obtained is best uniformly distributed.

We know from Theorem 3.3 that if

$$\langle(\boldsymbol{\gamma}, \mathbf{m})\rangle \geqslant c(\boldsymbol{\gamma}, \varepsilon) \|\mathbf{m}\|^{-1-\varepsilon} \tag{4.13}$$

holds for any integral vector $\mathbf{m} \neq \mathbf{0}$, then $\boldsymbol{\gamma}$ is a good point and (4.11) is a *gp* set. Hence the problem of the existence and construction of a *gp* set is reduced to the problem of the existence and construction of a vector $\boldsymbol{\gamma}$ satisfying (4.13).

**Theorem 4.10.** *The measure of the points* $\boldsymbol{\gamma} \in G_s$ *such that* (4.13) *holds for any integral vector* $\mathbf{m} \neq \mathbf{0}$ *is equal to* 1.

We derive immediately

**Theorem 4.11.** *The measure of the points* $\boldsymbol{\gamma} \in G_s$ *such that the set* (4.11) *has discrepancy* (4.12) *equals* 1.

To proof Theorem 4.10, we shall need

**Lemma 4.8.** *Suppose that* $m$ *is a non-zero integer and* $c$ *is a real number. Then the measure of the points satisfying*

$$\langle c+mx\rangle \leqslant \varepsilon, \quad x \in [0,1]$$

*is* $\leqslant 2\varepsilon$.

*Proof.* Clearly, we may suppose that $\varepsilon < \frac{1}{2}$ and $m > 0$.

Suppose that $c = 0$. Then the set of points such that

$$\langle mx\rangle \leqslant \varepsilon$$

holds, is given by the intervals

$$0 \leqslant x \leqslant \frac{\varepsilon}{m}, \quad \frac{1-\varepsilon}{m} \leqslant x \leqslant \frac{1+\varepsilon}{m}, \quad \frac{2-\varepsilon}{m} \leqslant x \leqslant \frac{2+\varepsilon}{m}, \cdots,$$

$$\frac{m-1-\varepsilon}{m} \leqslant x \leqslant \frac{m-1+\varepsilon}{m}, \quad \frac{m-\varepsilon}{m} \leqslant x \leqslant 1.$$

So its measure is $\leqslant 2\varepsilon$.

Suppose that $c \neq 0$. Then

$$c+mx = m\left(x+\frac{c}{m}\right) = my,$$

so it is reduced to the case $c = 0$. The lemma follows.

Theorem 4.10 is a consequence of the following

**Lemma 4.9.** *If* $\psi(\bar{n}) > 0$ *and*

$$\sum_{n=-\infty}^{\infty} \frac{1}{\bar{n}\psi(\bar{n})} < \infty,$$

*then the inequality*

$$\langle(\boldsymbol{\gamma}, \mathbf{m})\rangle > \frac{c(\boldsymbol{\gamma}, \psi)}{\prod_{i=1}^{s} \bar{m}_i \psi(\bar{m}_i)}, \quad \mathbf{m} \neq \mathbf{0}$$

*is satisfied for almost all* $\boldsymbol{\gamma} \in G_s$.

*Proof.* We may easily prove by Lemma 4.8 and mathematical induction that the measure of points $\boldsymbol{\gamma}$ satisfying

$$\langle(\boldsymbol{\gamma}, \mathbf{m})\rangle \leqslant \varepsilon$$

is $\leqslant 2\varepsilon$, if $\mathbf{m} \neq \mathbf{0}$. Hence the measure of the set $\sigma_\eta$ of points $\boldsymbol{\gamma}$ such that the inequality

$$\langle(\boldsymbol{\gamma}, \mathbf{m})\rangle \leqslant \frac{\eta}{\prod_{i=1}^{s} \bar{m}_i \psi(\bar{m}_i)}, \quad \eta > 0$$

holds for any $\mathbf{m} \neq \mathbf{0}$ does not exceed

$$2\eta \sum{}' \frac{1}{\prod_{i=1}^{s} \bar{m}_i \psi(\bar{m}_i)} = 2\eta \left( \left( \sum_{m=-\infty}^{\infty} \frac{1}{\bar{m}\psi(\bar{m})} \right)^s - \frac{1}{\psi(1)^s} \right)$$

$$= c(\psi)\eta \text{ (say)}.$$

Now we shall prove that for any $\tau > 0$, the measure of the set $\sigma$ of $\boldsymbol{\gamma}$ such that the inequality

$$\langle(\boldsymbol{\gamma}, \mathbf{m})\rangle > \frac{\tau}{\prod_{i=1}^{s} \bar{m}_i \psi(\bar{m}_i)}$$

can not be satisfied by all $\mathbf{m} \neq \mathbf{0}$ is equal to zero. Otherwise, suppose that $\sigma$ has the measure $\delta > 0$. Then for any $\eta > 0, \sigma$ is the subset of $\sigma_\eta$. Take $\eta = \frac{\delta}{2c(\psi)}$. Then the measure of $\sigma$ does not exceed the measure of $\sigma_\eta$, i.e., $\frac{\delta}{2}$. This gives a contradiction. The lemma is proved.

## § 4.5. The construction of good points

Theorem 4.11 means that the measure of the set of good points in $G_s$ equals 1, but Theorem 4.11 is not constructive. Two constructive results were obtained by W. M. Schmidt and A. Baker respectively.

**Theorem 4.12.** *Let* $\boldsymbol{\alpha}=(\alpha_1, \cdots, \alpha_s)$, *where* $\alpha_i (1 \leqslant i \leqslant s)$ *is a set of real algebraic numbers such that* $1, \alpha_1, \cdots, \alpha_s$ *are linearly independent over* $Q$. *Then*

$$\langle(\boldsymbol{\alpha}, \mathbf{m})\rangle > c(\boldsymbol{\alpha}, \varepsilon)\|\mathbf{m}\|^{-1-\varepsilon}$$

*holds for any integral vector* $\mathbf{m} \neq \mathbf{0}$ (Cf. W. M. Schmidt [2, 4]).

**Theorem 4.13.** *Let* $\boldsymbol{\beta}=(\beta_1, \cdots, \beta_s)$, *where* $\beta_i = e^{r_i} (1 \leqslant i \leqslant s)$ *with* $r_i (1 \leqslant i \leqslant s)$ *denoting a set of different non-zero rational numbers. Then*

$$\langle(\boldsymbol{\beta}, \mathbf{m})\rangle > c(\boldsymbol{\beta}, \varepsilon)\|\mathbf{m}\|^{-1-\varepsilon}$$

*holds for any integral vector* $\mathbf{m} \neq \mathbf{0}$ (Cf. A. Baker [1]).

From Theorem 4.12 and 4.13, we derive immediately

**Theorem 4.14.** *The set*

$$P(k)=(\{\alpha_1 k\}, \cdots, \{\alpha_s k\}), \quad 1 \leqslant k \leqslant n$$

*has discrepancy*

$$D(n) \leqslant c(\boldsymbol{\alpha}, \varepsilon) n^{-1+\varepsilon}.$$

**Theorem 4.15.** *The set*

$$P(k)=(\{\beta_1 k\}, \cdots, \{\beta_s k\}), \quad 1 \leqslant k \leqslant n$$

*has discrepancy*

$$D(n) \leqslant c(\boldsymbol{\beta}, \varepsilon) n^{-1+\varepsilon}.$$

*Remark.* The constant in Theorem 4.13 is effective (Cf. K. Mahler [1]).

## § 4.6. The $\mathscr{R}_s$ set

Let $h_1, \cdots, h_s, n(>0)$ be a set of integers. If the set of rational points

$$\left(\left\{\frac{h_1 k}{n}\right\}, \cdots, \left\{\frac{h_s k}{n}\right\}\right), \quad 1 \leqslant k \leqslant n \tag{4.14}$$

has discrepancy

$$D(n) \leqslant c(s, \varepsilon) n^{-1+\varepsilon},$$

the set (4.14) is called a *glp* set and **h** an optimal coefficient mod $n$ or a good lattice point mod $n$.

Notice that the definition of optimal coefficient or good lattice point given here is different from the original definition due to Korobov and Hlawka (Cf. N. M. Kopobov [2] and E. Hlawka [3]).

Let $\mathscr{F}_s$ be a real algebraic number field of degree $s$ and $\omega_1, \cdots, \omega_s$ be an integral basis of $\mathscr{F}_s$, where $\omega_2, \cdots, \omega_s$ are irrational numbers. Let $(n_l, h_{l1}, \cdots, h_{ls})(l = 1, 2, \cdots)$ be a sequence of sets of integers satisfying

$$\left|\frac{h_{lj}}{n_l} - \omega_j\right| \leqslant c(\mathscr{F}_s) n_l^{-1-\frac{1}{s-1}}, \quad 2 \leqslant j \leqslant s. \tag{4.15}$$

For simplicity, we omit the index $l$.

**Theorem 4.16.** *The set*

$$\left(\left\{\frac{k}{n}\right\}, \left\{\frac{h_2 k}{n}\right\}, \cdots, \left\{\frac{h_s k}{n}\right\}\right), \quad 1 \leqslant k \leqslant n \tag{4.16}$$

*has discrepancy*

$$D(n) \leqslant c(\mathscr{F}_s, \varepsilon) n^{-\frac{1}{2}-\frac{1}{2(s-1)}+\varepsilon}.$$

**Theorem 4.17.** *Suppose that* $1 \leqslant q \leqslant n^{\frac{1}{2}+\frac{1}{2(s-1)}}$. *Then the set*

$$\left(\left\{\frac{h_2 k}{n}\right\}, \cdots, \left\{\frac{h_s k}{n}\right\}\right), \quad 1 \leqslant k \leqslant q \tag{4.17}$$

*has discrepancy*

$$D(q) \leqslant c(\mathscr{F}_s, \varepsilon) q^{-1+\varepsilon}.$$

Since $1, \omega_2, \cdots, \omega_s$ is a basis of $\mathscr{F}_s$, they are linearly independent over the rational field Q. Hence Theorems 4.16 and 4.17 follow by (4.15) and Theorems 3.4, 3.5 and 4.12.

The sets (4.16) and (4.17) are called the $\mathscr{F}_s$ sets, Especially, for the real cyclotomic field $\mathscr{R}_s = Q\left(\cos \frac{2\pi}{m}\right)$, where $s = \varphi(m)/2$, we have

$$\left|\frac{c_j}{n} - 2\cos\frac{2\pi(j-1)}{m}\right| < c(\mathscr{R}_s, \varepsilon) n^{-1-\frac{1}{s-1}}, \quad 2 \leqslant j \leqslant s$$

(Cf. §1.3) and so the following

**Theorem 4.18.** *The set*

$$\left(\left\{\frac{c_1 k}{n}\right\}, \cdots, \left\{\frac{c_s k}{n}\right\}\right), \quad 1 \leqslant k \leqslant n$$

*has discrepancy*

$$D(n) \leqslant c(\mathscr{R}_s, \varepsilon) n^{-\frac{1}{2} - \frac{1}{2(s-1)} + \varepsilon}.$$

**Theorem 4.19.** *Suppose that* $1 \leqslant q \leqslant n^{\frac{1}{2} + \frac{1}{2(s-1)}}$. *Then the set*

$$\left(\left\{\frac{c_2 k}{n}\right\}, \cdots, \left\{\frac{c_s k}{n}\right\}\right), \quad 1 \leqslant k \leqslant q$$

*has discrepancy*

$$D(q) \leqslant c(\mathscr{R}_s, \varepsilon) q^{-1+\varepsilon}.$$

## § 4.7. The $\eta$ set

Let $\alpha$ be a PV number of degree $s$, i.e., $\alpha > 1$ and its conjugates satisfy

$$|\alpha^{(2)}| \leqslant \cdots \leqslant |\alpha^{(s)}| < 1.$$

Let

$$\rho = -\frac{\ln|\alpha^{(s)}|}{\ln \alpha}.$$

Let $\alpha$ satisfy the irreducible polynomial

$$x^s - a_{s-1}x^{s-1} - \cdots - a_1 x - a_0 = 0$$

and $Q_n (n = 0, 1, \cdots)$ be a set of integers defined by the recurrence relation

$$Q_0 = Q_1 = \cdots = Q_{s-2} = 0, \quad Q_{s-1} = 1,$$

$$Q_n = a_{s-1}Q_{n-1} + \cdots + a_1 Q_{n-s+1} + a_0 Q_{n-s}, \quad n \geqslant s.$$

Further let

$$Q_n(j) = Q_{n+j-1} - a_{s-1}Q_{n+j-2} - \cdots - a_{s-j+2}Q_{n+1} - a_{s-j+1}Q_n,$$
$$2 \leqslant j \leqslant s.$$

Then

$$\left| \frac{Q_n(j)}{Q_n} - \omega_j \right| < c(\alpha)|Q_n|^{-1-\rho}, \quad 2 \leqslant j \leqslant s, n > c_1(\alpha), \tag{4.18}$$

where

$$\omega_j = \alpha^{j-1} - a_{s-1}\alpha^{j-2} - \cdots - a_{s-j+2}\alpha - a_{s-j+1} \quad (2 \leqslant j \leqslant s) \text{ (Cf. §2.2).}$$

**Theorem 4.20.** *The set*

$$\left( \left\{ \frac{k}{Q_n} \right\}, \left\{ \frac{Q_n(2)}{Q_n} k \right\}, \cdots, \left\{ \frac{Q_n(s)}{Q_n} k \right\} \right), \quad 1 \leqslant k \leqslant |Q_n| \tag{4.19}$$

*has discrepancy*

$$D(Q_n) \leqslant c(\alpha, \varepsilon)|Q_n|^{-\frac{1}{2}-\frac{\rho}{2}+\varepsilon}.$$

**Theorem 4.21.** *Suppose that* $1 \leqslant q \leqslant |Q_n|^{\frac{1}{2}+\frac{\rho}{2}}$. *Then the set*

$$\left( \left\{ \frac{Q_n(2)}{Q_n} k \right\}, \cdots, \left\{ \frac{Q_n(s)}{Q_n} k \right\} \right), \quad 1 \leqslant k \leqslant q \tag{4.20}$$

*has discrepancy*

$$D(q) \leqslant c(\alpha, \varepsilon) q^{-1+\varepsilon}.$$

Since $1, \omega_2, \cdots, \omega_s$ form a basis of $Q(\alpha)$, hence Theorems 4.20 and 4.21 follow by (4.18), and Theorems 3.4, 3.5 and 4.12.

The sets (4.19) and (4.20) are called the $\alpha$ set. Especially, we take the $\eta$ set, where $\eta$ is the greatest real root of the equation

$$x^s - x^{s-1} - \cdots - x - 1 = 0.$$

Let $F_n(=F_{s,n})$ be a set of integers defined by the recurrence relation

$$F_0 = F_1 = \cdots = F_{s-2} = 0, \quad F_{s-1} = 1,$$

$$F_n = F_{n-1} + \cdots + F_{n-s+1} + F_{n-s}, \quad n \geqslant s.$$

Further let

$$F_n(j) = F_{n+j-1} - F_{n+j-2} - \cdots - F_n, \quad 2 \leqslant j \leqslant s.$$

Then

$$\left| \frac{F_n(j)}{F_n} - \omega_j \right| \leqslant c(\eta) F_n^{-1-\frac{1}{2^s \ln 2}-\frac{1}{2^{2s+1}}}, \quad 2 \leqslant j \leqslant s,$$

where $\omega_j = \eta^{j-1} - \eta^{j-2} - \cdots - \eta - 1$ (Cf. §2.8). Hence we have

**Theorem 4.22.** *The set*

$$\left( \left\{ \frac{k}{F_n} \right\}, \left\{ \frac{F_n(2)}{F_n} k \right\}, \cdots, \left\{ \frac{F_n(s)}{F_n} k \right\} \right), \quad 1 \leqslant k \leqslant F_n \tag{4.21}$$

*has discrepancy*

$$D(F_n) \leqslant c(\eta) F_n^{-\frac{1}{2}-\frac{1}{2^{s+1}\ln 2}-\frac{1}{2^{2s+3}}}. \tag{4.22}$$

**Theorem 4.23.** *Suppose that* $1 \leqslant q \leqslant F_n^{\frac{1}{2}+\frac{1}{2^{s+1}\ln 2}+\frac{1}{2^{2s+2}}}$. *Then the set*

$$\left( \left\{ \frac{F_n(2)}{F_n} k \right\}, \cdots, \left\{ \frac{F_n(s)}{F_n} k \right\} \right), \quad 1 \leqslant k \leqslant q$$

*has discrepancy*

$$D(q) \leqslant c(\eta, \varepsilon) q^{-1+\varepsilon}.$$

For real quadratic irrational $\alpha$, we have

$$\langle \alpha m \rangle > \frac{c(\alpha)}{|m|}. \tag{4.23}$$

(Cf. Hua Loo Keng [2], Chap. 10). If we use (4.23) and Theorem 2.9 to instead of Theorem 4.12 and Theorem 2.8 respectively, then we have

**Theorem 4.24.** *For* $s=2$, *the right hand side of* (4.22) *may be replaced by* $c(\eta) F_n^{-1} (\ln F_n)^2$.

Hence the set (4.21) is a *glp* set for $s=2$. For $s=3$, we may use Theorem 2.9 instead of theorem 2.8. Then we have

**Theorem 4.25.** *For $s=3$, the right hand side of (4.22) and the range of $q$ in Theorem 4.23 may be replaced by $c(\eta,\varepsilon)F_n^{-3/4+\varepsilon}$ and $q\leqslant F_n^{3/4}$ respectively.*

## § 4.8. The case $s=2$

Let

$$a_3, a_4, \cdots$$

be a set of positive integers such that $a_n \leqslant M (n=3,4,\cdots)$. Let $q_1, q_2$ be two positive integers such that $q_1 \leqslant q_2$ and $(q_1, q_2)=1$. Further let

$$q_n = a_n q_{n-1} + q_{n-2}, \quad n \geqslant 3. \tag{4.24}$$

Then $q_n (n=1,2,\cdots)$ form an increasing sequence of integers and

$$n-1 \leqslant q_n \leqslant (M+1) q_{n-1}, \quad n \geqslant 3. \tag{4.25}$$

**Theorem 4.26.** *There exists a constant $c(q_1, q_2, M)$ such that the solution of the equation*

$$\begin{gathered} x_1 + q_{n-1}x_2 = q_n y, \quad 0 < |x_1| \leqslant q_n/2, \\ 0 < |x_2| \leqslant q_n/2, \quad y \neq 0 \end{gathered} \tag{4.26}$$

*satisfies*

$$|x_1 x_2| \geqslant c q_n, \quad |x_1 y| \geqslant c q_{n-1}.$$

*Proof.* The theorem holds for $n=2,3$ obviously. Suppose that $n>3$ and that the theorem holds for any integer $<n$. Now we proceed to prove that the theorem holds for $n$ also.

Clearly, $x_2$ and $y$ have the same sign. Otherwise

$$\frac{q_n}{2} \geqslant |x_1| = |q_n y - q_{n-1} x_2| \geqslant q_n + q_{n-1}.$$

This leads to a contradiction.

If $|x_1| \leqslant \frac{1}{2} q_{n-1}$, $|y| \leqslant \frac{1}{2} q_{n-1}$, then by (4.24) and (4.26),

$$x_1 + q_{n-1}x_2 = y(a_nq_{n-1} + q_{n-2}),$$

$$x_1 - q_{n-2}y = q_{n-1}(a_ny - x_2). \tag{4.27}$$

If $a_ny = x_2$, then by (4.27), $x_1 = q_{n-2}y$, hence from (4.25),

$$x_1x_2 = a_nq_{n-2}y^2 \geqslant q_{n-2} \geqslant \frac{1}{M+1} q_{n-1} \geqslant \frac{1}{(M+1)^2} q_n,$$

$$x_1y = q_{n-2}y^2 \geqslant q_{n-2} \geqslant \frac{1}{M+1} q_{n-1}.$$

Take $c \leqslant \frac{1}{(M+1)^2}$. The theorem follows.

Now, suppose that $a_ny \neq x_2$. Then it follows from the induction hypothesis that there exists a constant $c = c(q_1, q_2, M)$ such that the solution of the equation (4.26) satisfies

$$|x_1y| \geqslant cq_{n-1}, \quad |x_1(a_ny - x_2)| \geqslant cq_{n-2}. \tag{4.28}$$

Since $y$ and $x_2 - a_ny$ have the same sign, we have

$$|x_1x_2| = |x_1(x_2 - a_ny) + x_1a_ny| \geqslant cq_{n-2} + ca_nq_{n-1} = cq_n$$

by (4.28).

If $|x_1| \geqslant \frac{1}{2} q_{n-1}$, then by (4.25),

$$|x_1x_2| \geqslant \frac{1}{2} q_{n-1} \geqslant \frac{1}{2(M+1)} q_n,$$

$$|x_1y| \geqslant \frac{1}{2} q_{n-1}.$$

If $|y| \geqslant \frac{1}{2} q_{n-1}$, then

$$|x_1y| \geqslant \frac{1}{2} q_{n-1},$$

$$\frac{1}{2} q_{n-1}q_n \leqslant |y|q_n \leqslant |x_1| + q_{n-1}|x_2| \leqslant \frac{1}{2} q_n + q_{n-1}|x_2|,$$

$$|x_2| \geqslant \frac{1}{2} q_n\left(1 - \frac{1}{q_{n-1}}\right) \geqslant \frac{1}{4} q_n.$$

So

$$|x_1x_2| \geqslant \frac{1}{4} q_n.$$

Hence the theorem holds for $n$ too. The theorem follows by induction.

**Theorem 4.27.** *There exists a constant $c = c(q_1, q_2, M)$ such that the congruence*

$$x_1 + q_{n-1}x_2 \equiv 0 \pmod{q_n} \tag{4.29}$$

*has no solution in the domain*

$$\bar{x}_1\bar{x}_2 < cq_n, \quad (x_1, x_2) \neq (0, 0).$$

*Proof.* Since $(q_1, q_2) = 1$, we have $(q_{n-1}, q_n) = 1 (n = 3, 4, \cdots)$ by (4.24). If $(x_1, x_2) \neq (0, 0)$ is a solution of (4.29), then $q_n | x_2$ if $x_1 = 0$ and $q_n | x_1$ if $x_2 = 0$. Hence $\bar{x}_1\bar{x}_2 \geqslant q_n$. If $x_1 \neq 0$ and $x_2 \neq 0$, then it follows by Theorem 4.26 that the solution of (4.29) satisfies $\bar{x}_1\bar{x}_2 \geqslant cq_n$. The theorem is proved.

From Theorem 3.2, we derive

**Theorem 4.28.** *The set*

$$\left(\left\{\frac{k}{q_n}\right\}, \left\{\frac{q_{n-1}k}{q_n}\right\}\right), \quad 1 \leqslant k \leqslant q_n$$

*has discrepancy*

$$D(q_n) \leqslant c(q_1, q_2, M) q_n^{-1} (\ln 3q_n)^2.$$

Take $q_1 = q_2 = 1$ and $a_n = 1 (n = 3, 4, \cdots)$. Then the sequence $q_n (n = 1, 2, \cdots)$ is the usual Fibonacci sequence $F (= F_{2,n}) (n = 1, 2, \cdots)$ of dimension 2. Hence Theorem 4.24 may be derived by Theorem 4.28 also.

## § 4.9. The *glp* set

**Theorem 4.29.** *There exists an integral vector $\mathbf{a} (= \mathbf{a}(p))$ such that the set*

$$\left(\left\{\frac{a_1k}{p}\right\}, \cdots, \left\{\frac{a_sk}{p}\right\}\right), \quad 1 \leqslant k \leqslant p \tag{4.30}$$

*has discrepancy*

$$D(p) < c(s)p^{-1}(\ln p)^s.$$

**Theorem 4.30.** *Suppose that $n$ is an integer satisfying $1 \leqslant n \leqslant p$. Then there exists an integral vector $\mathbf{a}(=\mathbf{a}(p))$ such that the set*

$$\left(\left\{\frac{a_1 k}{p}\right\}, \cdots, \left\{\frac{a_s k}{p}\right\}\right), \quad 0 \leqslant k < n \tag{4.31}$$

*has discrepancy*

$$D(n) < c(s)n^{-1}(\ln p)^{s+1}.$$

To prove these theorems, we shall need

**Lemma 4.10.** *Let $\mathbf{a}$ be an integral vector and $q$ an integer $> 1$. I $(a_i, q) = 1 (1 \leqslant i \leqslant s)$, then for any positive integer $r$,*

$$\sum_{\substack{(\mathbf{a},\mathbf{m})\equiv 0 \,(\mathrm{mod}\, q)\\ |m_i|\leqslant qr}}' \frac{1}{\|\pi\mathbf{m}\|} - \sum_{\substack{(\mathbf{a},\mathbf{m}^{(0)})\equiv 0 \,(\mathrm{mod}\, q)\\ -q/2<m_i^{(0)}\leqslant q/2}}' \frac{1}{\|\pi\mathbf{m}^{(0)}\|} < \frac{s2^s(\ln 20qr)^s}{\pi^s q}.$$

*Proof.* If $\mathbf{m}^{(0)}$ is a solution of

$$(\mathbf{a}, \mathbf{m}^{(0)}) = a_1 m_1^{(0)} + \cdots + a_s m_s^{(0)} \equiv 0 \pmod q$$

$$-q/2 < m_i^{(0)} \leqslant q/2, \quad 1 \leqslant i \leqslant s.$$

Then

$$m_1 = m_1^{(0)} + ql_1, \cdots, m_s = m_s^{(0)} + ql_s \tag{4.32}$$

is also a solution of the congruence

$$(\mathbf{a}, \mathbf{m}) \equiv 0 \pmod q. \tag{4.33}$$

On the other hand, any solution of (4.33) may be represented by (4.32). If any $s-1$ integers of $m_1^{(0)}, \cdots, m_s^{(0)}$ are given, then the remaining one can be determined uniquely, since $(a_i, q) = 1 (1 \leqslant i \leqslant s)$. By Lemma 3.2, we have

$$\sum_{-q/2<m\leqslant q/2} \sum_{l=-r}^{r} \frac{1}{\pi(m+lq)} \leqslant 1 + \frac{2}{\pi} \sum_{m=1}^{[q(r+\frac{1}{2})]} \frac{1}{m} < \frac{2}{\pi} \ln 20qr$$

and

$$\sum_{l=-r}^{r}{}' \frac{1}{\pi(m+lq)} \leqslant \frac{2}{\pi} \sum_{l=1}^{r} \frac{1}{q\left(l-\frac{1}{2}\right)} < \frac{2}{\pi q} \ln 20r$$

for $-q/2 < m \leqslant q/2$. Hence

$$\left| \sum_{\substack{(\mathbf{a},\mathbf{m})\equiv 0 \pmod q \\ |m_i|\leqslant qr}}' \frac{1}{\|\pi\mathbf{m}\|} - \sum_{\substack{(\mathbf{a},\mathbf{m}^{(0)})\equiv 0 \pmod q \\ -q/2<m_i^{(0)}\leqslant q/2}}' \frac{1}{\|\pi\mathbf{m}^{(0)}\|} \right|$$

$$\leqslant \sum_{\nu=1}^{s} \sum_{\substack{(\mathbf{a},\mathbf{m}^{(0)})\equiv 0 \pmod q \\ -q/2<m_\nu^{(0)}\leqslant q/2}} \left( \sum_{l_\nu=-r}^{r}{}' \frac{1}{\pi(m_\nu^{(0)}+l_\nu q)} \right) \prod_{\substack{\mu=1 \\ \mu\neq\nu}}^{s} \left( \sum_{l_\mu=-r}^{r} \frac{1}{\pi(m_\mu^{(0)}+l_\mu q)} \right)$$

$$< s\left(\frac{2}{\pi}\right)^s q^{-1} (\ln 20qr)^s.$$

The lemma is proved.

**Lemma 4.11.** *There exists an integral vector* $\mathbf{a}(=\mathbf{a}(p))$ *such that*

$$\sum_{\substack{(\mathbf{a},\mathbf{m})\equiv 0 \pmod p \\ |m_i|<p/2}}' \frac{1}{\|\pi\mathbf{m}\|} < c(s)p^{-1}(\ln p)^s.$$

*Proof*. Let $\mathbf{a}=(1, a, \cdots, a^{s-1})$, where $a$ is an integer. Let

$$\Lambda(a) = \sum_{\substack{(\mathbf{a},\mathbf{m})\equiv 0 \pmod p \\ |m_i|<p/2}}' \frac{1}{\|\pi\mathbf{m}\|}.$$

Then by Lemma 4.5,

$$\min_{1\leqslant a\leqslant p} \Lambda(a) \leqslant \frac{1}{p}\sum_{a=1}^{p} \Lambda(a) = p^{-1} \sum_{|m_i|<p/2}' \frac{1}{\|\pi\mathbf{m}\|} \sum_{\substack{1\leqslant a\leqslant p \\ (\mathbf{a},\mathbf{m})\equiv 0 \pmod p}} 1$$

$$\leqslant (s-1)p^{-1}\left( \sum_{|m|<p/2} \frac{1}{\pi\bar{m}} \right)^s < c(s)p^{-1}(\ln p)^s.$$

Hence there exists an integer $a$ such that

$$\Lambda(a) < c(s)p^{-1}(\ln p)^s.$$

The lemma is proved.

The proof of Theorem 4.29. Take $\mathbf{a}$ satisfying Lemma 4.11. Then by Lemmas 4.10 and 4.11,

$$\sum_{|m_i|\leqslant p^2}' \frac{1}{\|\pi\mathbf{m}\|} \left| \frac{1}{p} \sum_{k=1}^{p} e^{2\pi i(\mathbf{a},\mathbf{m})k/p} \right| = \sum_{\substack{(\mathbf{a},\mathbf{m})\equiv 0 \pmod p \\ |m_i|\leqslant p^2}}' \frac{1}{\|\pi\mathbf{m}\|}$$

$$< \sum_{\substack{(\mathbf{a},\mathbf{m}^{(0)})\equiv 0 \pmod p \\ |m_i^{(0)}|<p/2}}' \frac{1}{\|\pi\mathbf{m}^{(0)}\|} + \frac{s2^{2s}(\ln 8p)^s}{\pi^s p}$$

$$< c(s)p^{-1}(\ln p)^s.$$

Take $r=1, \eta=p^{-1}$ and $h=p^2$ in Theorem 3.1 for $p>6$. Then we have the result. The theorem is obvious for $p\leqslant 6$.

The proof of Theorem 4.30. Let $\mathbf{a}=(1, a, \cdots, a^s)$ be a vector such that Theorem 4.29 holds. Then for $\boldsymbol{\gamma}\in G_s$, the number $N_n(\boldsymbol{\gamma})$ of integers $k$ satisfying

$$\left\{\frac{ak}{p}\right\}<\gamma_1, \cdots, \left\{\frac{a^s k}{p}\right\}<\gamma_s, \quad 0\leqslant k<n$$

is equal to the number of integers $k$ satisfying

$$\left\{\frac{k}{p}\right\}<\frac{n}{p}, \left\{\frac{ak}{p}\right\}<\gamma_1, \cdots, \left\{\frac{a^s k}{p}\right\}<\gamma_s, \quad 1\leqslant k\leqslant p.$$

Hence from Theorem 4.29,

$$\left|p^{-1}N_n(\boldsymbol{\gamma})-\frac{n}{p}|\boldsymbol{\gamma}|\right|<c(s)p^{-1}(\ln p)^{s+1}$$

the theorem follows.

The vector $\mathbf{a}$ in Theorem 4.29 is given by the integer $a$ such that $\Lambda(a)$ is minimal. Now we shall give an explicit expression for $\Lambda(a)$.

**Lemma 4.12.** *Let $q$ be a positive integer and $x$ satisfy $0<x<1$. Then*

$$1-\frac{2}{\pi}\ln(2\sin\pi x)=\sum_{|m|<q}\frac{e^{2\pi i m x}}{\pi \overline{m}}+\frac{\psi}{\pi q\langle x\rangle}.$$

*Hereafter $\psi$ denotes a number satisfying $|\psi|\leqslant 1$.*

*Proof.* For $0\leqslant r<1$.

$$\sum_{m=1}^{\infty}\frac{r^m e^{2\pi i m x}}{m}=-\ln(1-re^{2\pi i x}).$$

Since the series $\sum_{m=1}^{\infty}\frac{e^{2\pi i m x}}{m}$ is convergent, so

$$\sum_{m=1}^{\infty} \frac{e^{2\pi imx}}{m} = -\ln\left(1 - e^{2\pi ix}\right)$$

and

$$\sum_{m=-\infty}^{\infty} \frac{e^{2\pi imx}}{\pi \overline{m}} = 1 + \pi^{-1}\sum_{m=1}^{\infty} \frac{e^{2\pi imx}}{m} + \pi^{-1}\sum_{m=1}^{\infty} \frac{e^{-2\pi imx}}{m}$$

$$= 1 - \pi^{-1}(\ln\left(1 - e^{2\pi ix}\right) + \ln\left(1 - e^{-2\pi ix}\right))$$

$$= 1 - \pi^{-1}\ln\left(2 - 2\cos 2\pi x\right) = 1 - \frac{2}{\pi}\ln\left(2\sin \pi x\right).$$

Hence

$$1 - \frac{2}{\pi}\ln\left(2\sin \pi x\right) = \sum_{|m|<q} \frac{e^{2\pi imx}}{\pi \overline{m}} + R,$$

where

$$R = \sum_{m=q}^{\infty} \frac{e^{2\pi imx}}{\pi m} + \sum_{m=q}^{\infty} \frac{e^{-2\pi imx}}{\pi m}.$$

From the identity

$$\frac{e^{2\pi imx}}{m} = \frac{1}{e^{2\pi ix} - 1}\left(\frac{e^{2\pi i(m+1)x}}{m+1} - \frac{e^{2\pi imx}}{m} + \frac{e^{2\pi i(m+1)x}}{m(m+1)}\right),$$

we have

$$\left|\sum_{m=q}^{\infty} \frac{e^{2\pi imx}}{\pi m}\right| = \frac{1}{\pi\left|e^{2\pi ix} - 1\right|}\left|-\frac{e^{2\pi iqx}}{q} + \sum_{m=q}^{\infty} \frac{e^{2\pi i(m+1)x}}{m(m+1)}\right|$$

$$\leqslant \frac{1}{2\pi\sin \pi x}\left(\frac{1}{q} + \sum_{m=q}^{\infty} \frac{1}{m(m+1)}\right) = \frac{1}{q\pi\sin \pi x}$$

$$\leqslant \frac{1}{2\pi q\langle x\rangle}.$$

Hence

$$R = -\frac{\phi}{\pi q\langle x\rangle}.$$

The lemma is proved.

Since

$$\left|1 - \frac{2}{\pi}\ln\left(2\sin \pi \left\{\frac{a^{\nu-1}k}{p}\right\}\right)\right| \leqslant 1 + \frac{2}{\pi}\left|\ln\left(2\sin \frac{\pi}{p}\right)\right|$$

$$\leqslant 1 + \frac{2}{\pi}\ln p < \frac{2}{\pi}\ln 8p$$

for $1 \leqslant a, k \leqslant p$, then by Lemma 4.12,

$$\prod_{\nu=1}^{s}\left(1-\frac{2}{\pi}\ln\left(2\sin\pi\left\{\frac{a^{\nu-1}k}{p}\right\}\right)\right)=\sum_{|m_i|<p/2}\frac{e^{2\pi i(\mathbf{a},\mathbf{m})k/p}}{\|\pi\mathbf{m}\|}$$

$$+\psi\left(\frac{2}{\pi}\right)^{s-1}(\ln 8p)^{s-1}\frac{2}{\pi(p+1)}\sum_{\nu=1}^{s}\frac{1}{\left\langle\dfrac{a^{\nu-1}k}{p}\right\rangle}.$$

Since

$$\sum_{k=1}^{p-1}\frac{1}{\left\langle\dfrac{a^{\nu-1}k}{p}\right\rangle}=2p\sum_{k=1}^{\frac{p-1}{2}}\frac{1}{k}<2p\ln 8p$$

then, by Lemma 3.2,

$$\Lambda(a)=p^{-1}\sum_{k=1}^{p}\sum_{|m_i|<p/2}\frac{e^{2\pi i(\mathbf{a},\mathbf{m})k/p}}{\|\pi\mathbf{m}\|}-1$$

$$=p^{-1}\sum_{k=1}^{p-1}\sum_{|m_i|<p/2}\frac{e^{2\pi i(\mathbf{a},\mathbf{m})k/p}}{\|\pi\mathbf{m}\|}-1+p^{-1}\sum_{|m_i|<p/2}\frac{1}{\|\pi\mathbf{m}\|}$$

$$=p^{-1}\sum_{k=1}^{p-1}\prod_{\nu=1}^{s}\left(1-\frac{2}{\pi}\ln\left(2\sin\pi\left\{\frac{a^{\nu-1}k}{p}\right\}\right)\right)$$

$$+\frac{\psi 2^{s+1}}{\pi^s}p^{-1}(\ln 8p)^s-1+\psi\frac{2^s}{\pi^s}p^{-1}(\ln 8p)^s$$

$$=p^{-1}\sum_{k=1}^{p-1}\prod_{\nu=1}^{s}\left(1-\frac{2}{\pi}\ln\left(2\sin\pi\left\{\frac{a^{\nu-1}k}{p}\right\}\right)\right)$$

$$-1+3\psi\left(\frac{2}{\pi}\right)^s p^{-1}(\ln 8p)^s.$$

For $a=1,2,\cdots,p-1$, we may find the integer $a$ such that $\Lambda(a)$ assumes a minimum or satisfies

$$\sum_{k=1}^{p-1}\prod_{\nu=1}^{s}\left(1-\frac{2}{\pi}\ln\left(2\sin\pi\left\{\frac{a^{\nu-1}k}{p}\right\}\right)\right)\leqslant p+O((\ln p)^s).$$

Hence we have the vector $\mathbf{a}=(1, a, \cdots, a^{s-1})$ satisfying Theorem 4.29.

The set (4.30) is a *glp* set which has not only a precise discrepancy but is also very convenient for use. However, it has the disadvantage that it requires heavy computations to obtain the integer $a$. Roughly speaking, it requires $O(p^2)$ elementary operations (Cf. N. M. Korobov [5, 7]).

*Remark.* For $\varepsilon > 0$, take $r = r(\varepsilon)$ sufficiently large, $\eta = p^{-1}$ and $h = [p^{1+r^{-1}}]$. Then the discrepancies of the sets (4.30) and (4.31) may be replaced by

$$D(p) < \left(\frac{(2s-1)2^s}{\pi^s} + \varepsilon\right) p^{-1} (\ln p)^s$$

and

$$D(n) < \left(\frac{(2s+1)2^{s+1}}{\pi^{s+1}} + \varepsilon\right) n^{-1} (\ln p)^{s+1}$$

respectively for $p > c(s, \varepsilon)$.

## Notes

J. G. Van der Corput [1] first gave the best uniformly distributed set $\left(\frac{k}{n}, \varphi_2(k)\right)(1 \leqslant k \leqslant n)$. J. M. Hammersley [1] and J. H. Halton [1] suggested the generalizations $\left(\frac{k}{n}, \varphi_{p_1}(k), \cdots \varphi_{p_{s-1}}(k)\right)(1 \leqslant k \leqslant n)$ and $(\varphi_{p_1}(k), \cdots, \varphi_{p_s}(k))(1 \leqslant k \leqslant n)$ respectively. Theorem 4.3 was proved by Halton.

The $p$ sets $\left(\left\{\frac{k}{p^2}\right\}, \cdots, \left\{\frac{k^s}{p^2}\right\}\right)(1 \leqslant k \leqslant p^2)$ and $\left(\left\{\frac{k}{p}\right\}, \cdots, \left\{\frac{k^s}{p}\right\}\right)(1 \leqslant k \leqslant p)$ were proposed by N. M. Korobov [1,7] and the $p$ set $\left(\left\{\frac{k}{p}\right\}, \left\{\frac{ak}{p}\right\}, \cdots, \left\{\frac{a^{s-1}k}{p}\right\}\right)(1 \leqslant a, k \leqslant p)$ was given by Hua Loo Keny and Wang Yuan [3.7].

Theorems 4.8 and 4.9: Cf. E. Hlawka [4] and N. M. Kolobov [8].

Theorem 4.10: Cf. A. Y. Khintchine [1] and N. S. Bahvalov [1].

Theorem 4.11: Cf. also W. M. Schmidt [1] and Wang Yuan [5].

The $\eta$ set of dimension 2 was proposed by N. S. Bahvalov [1] and Hua Loo Keng and Wang Yuan [1,2] independently. (Cf. also S. Haber and C. F. Osgood [1], S. K. Zaremba [1]). Concerning Theorem 4.24, S. K. Zaremba [1] proved a more precise result $D(F_n) = O(F_n^{-1} \ln F_n)$.

The $\mathscr{R}_s$ set and $\eta$ set of dimension $>2$ were given by Hua Loo Keng and Wang Yuan [1, 4, 5, 6, 7].

Theorem 4.29: Cf. N. M. Korobov [7] and E. Hlawka [4].

The *glp* set $\left(\left\{\frac{a_1 k}{p}\right\}, \cdots, \left\{\frac{a_s k}{p}\right\}\right)(1 \leqslant k \leqslant p)$ was first introduced by N. M. Korobov [2] and E. Hlawka [3]. Later, Korobov [4] pointed out that the *glp* set may take the form $\left(\left\{\frac{k}{p}\right\}, \left\{\frac{ak}{p}\right\}, \cdots, \left\{\frac{a^{s-1}k}{p}\right\}\right)(1 \leqslant k \leqslant p)$. H. Niederreiter [3,4] proved that the prime number $p$ in Theorem 4.29 may be replaced by the composite integer $m$ and that for any prime $p$, there exists a primitive root $g$ mod $p$ such that the set $\left(\left\{\frac{k}{p}\right\}, \left\{\frac{gk}{p}\right\}, \cdots, \left\{\frac{g^{s-1}k}{p}\right\}\right)$ $(1 \leqslant k \leqslant p)$ has discrepancy $D(p) = O(p^{-1} (\ln p)^s \ln \ln p)$.

*Chapter 5*

# Uniform Distribution and Numerical Integration

## § 5.1. The function of bounded variation

Let

$$(\sigma)\qquad \begin{aligned} 0 &= x_0 < x_1 < \cdots < x_l = 1, \\ 0 &= y_0 < y_1 < \cdots < y_m = 1 \end{aligned}$$

be any division of $G_2$. Let $f(x,y)$ be a function defined on $G_2$ and

$$\begin{aligned} \Delta_{10} f(x_i, y) &= f(x_{i+1}, y) - f(x_i, y), \\ \Delta_{01} f(x, y_j) &= f(x, y_{j+1}) - f(x, y_j), \\ \Delta_{11} f(x_i, y_j) &= f(x_i, y_j) - f(x_{i+1}, y_j) \\ &\quad - f(x_i, y_{j+1}) + f(x_{i+1}, y_{j+1}). \end{aligned}$$

If the variation

$$V_\sigma = \sum_{i=0}^{l-1} \sum_{j=0}^{m-1} |\Delta_{11} f(x_i, y_j)| + \sum_{i=0}^{l-1} |\Delta_{10} f(x_i, 1)| + \sum_{j=0}^{m-1} |\Delta_{01} f(1, y_j)|$$

has an upper bound which is independent of the choice of $(\sigma)$, then $f$ is called a function of bounded variation in the sense of Harday and Krause. The least upper bound of $V_\sigma$ is called the total variation of $f$ and is denoted by $V(f)$. The class of these functions is denoted by $B_2$. Similarly, we may define $B_s (s > 2)$.

If

1. $f(x', y) - f(x, y)$ has the same sign or equals zero,

2. $f(x, y') - f(x, y)$ has the same sign or equals zero and

3. $f(x, y) - f(x', y) - f(x, y') + f(x', y')$ has the same sign or equals zero for any given $x, y, x', y'$ satisfying $0 \leqslant x < x' \leqslant 1$ and $0 \leqslant y < y' \leqslant 1$, then $f$ is called a generalized monotonic function and the class of these functions is denoted by $M_2$. Similarly, we may define $M_s (s > 2)$.

If $f \in M_2$, then

$$V_\sigma = \left|\sum_{i=0}^{l-1}\sum_{j=0}^{m-1}\Delta_{11}f(x_i,y_j)\right| + \left|\sum_{i=0}^{l-1}\Delta_{10}f(x_i,1)\right| + \left|\sum_{j=0}^{m-1}\Delta_{01}f(1,y_j)\right|$$
$$= |f(0,0) - f(0,1) - f(1,0) + f(1,1)| + |f(1,1) - f(0,1)| + |f(1,1) - f(1,0)|.$$

Hence $f \in B_2$ and so $M_2 \subset B_2$. Similarly, we may prove that $M_s \subset B_s (s > 2)$.

**Theorem 5.1.** *Every function of $B_s$ can be represented as a difference of two functions of $M_s$.*

*Proof.* We prove the theorem only for the case $s = 2$, since the proof is similar for $s > 2$.

For any division $(\sigma)$ of $G_2$, consider a part of $V_\sigma$

$$\sum_{i=1}^{l-1}\sum_{j=0}^{m-1}|\Delta_{11}f(x_i,y_j)|.$$

Change the notations $x_0, y_0$ to $x, y$ respectively. We use $P(x,y)$ and $N(x,y)$ to denote the sums of those terms in above formula satisfying $\Delta_{11}f(x_i,y_j) \geqslant 0$ and $\Delta_{11}f(x_i,y_j) < 0$ respectively. Then

$$P(x,y) - N(x,y) = \sum_{i=0}^{l-1}\sum_{j=0}^{m-1}\Delta_{11}f(x_i,y_j)$$
$$= f(1,1) - f(x,1) - f(1,y) + f(x,y),$$
$$f(x,y) = P(x,y) - N(x,y) + f(x,1) + f(1,y) - f(1,1). \quad (5.1)$$

Since the functions $P(x,y)$ and $N(x,y)$ satisfy

$$\begin{array}{lll} \Delta_{10}P \leqslant 0, & \Delta_{01}P \leqslant 0, & \Delta_{11}P \geqslant 0, \\ \Delta_{10}N \leqslant 0 & \Delta_{01}N \leqslant 0, & \Delta_{11}N \geqslant 0, \end{array} \quad (5.2)$$

the functions $P$ and $N$ are all generalized monotonic functions.

Since $f(x,1)$ and $f(1,y)$ are functions of bounded variation of a single variable, they are differences of two monotonic functions, i.e.,

$$\left.\begin{aligned} f(x,1) &= P_1(x,1) - N_1(x,1), \\ f(1,y) &= P_2(1,y) - N_2(1,y), \end{aligned}\right\} \tag{5.3}$$

where $P_1, N_1, P_2, N_2$ are monotonic decreasing functions which may be defined as before.

Substituting (5.3) into (5.1), we have

$$f = F - G,$$

where

$$F = -(N + N_1 + N_2) - f(1,1), \quad G = -(P + P_1 + P_2).$$

From (5.2) and (5.3), we have

$$\Delta_{10}F \geqslant 0, \quad \Delta_{01}F \geqslant 0, \quad \Delta_{11}F \leqslant 0,$$
$$\Delta_{10}G \geqslant 0, \quad \Delta_{01}G \geqslant 0, \quad \Delta_{11}G \leqslant 0,$$

i.e., $F$ and $G$ are generalized monotonic functions. The theorem is proved.

If there exists a positive constant $L$ such that

1. $|f(x',1) - f(x,1)| \leqslant L(x' - x)$,
2. $|f(1,y') - f(1,y)| \leqslant L(y' - y)$,

and

3. $|f(x,y) - f(x',y) - f(x,y') + f(x',y')| \leqslant L(x' - x)(y' - y)$

hold for any given $x, y, x', y'$ satisfying $0 \leqslant x < x' \leqslant 1$ and $0 \leqslant y < y' \leqslant 1$, then $f$ is said to be a function satisfying the generalized Lipschitz condition and the class of these functions is denoted by $L_2$. Similarly, we may define $L_s (s > 2)$.

If $f \in L_2$, then

$$\begin{aligned} V_\sigma &= \left|\sum_{i=0}^{l-1}\sum_{j=0}^{m-1} \Delta_{11} f(x_i, y_j)\right| + \left|\sum_{i=0}^{l-1} \Delta_{10} f(x_i, 1)\right| + \left|\sum_{j=0}^{m-1} \Delta_{01} f(1, y_j)\right| \\ &\leqslant L \sum_{i=0}^{l-1}\sum_{j=0}^{m-1} (x_{i+1} - x_i)(y_{j+1} - y_j) + L \sum_{i=0}^{l-1} (x_{i+1} - x_i) \\ &\quad + L \sum_{j=0}^{m-1} (y_{j+1} - y_j) = 3L \end{aligned}$$

and so $f \in B_2$ and $L_2 \subset B_2$. Similarly, we may prove $L_s \subset B_s (s > 2)$.

Similar to Theorem 5.1, we have

**Theorem 5.2.** *Any function of $L_s$ may be represented as a difference of two generalized monotonic functions which satisfy the generalized Lipschitz condition.*

If $f(\mathbf{x})$ satisfies

$$|f| \leqslant L, \quad \left|\frac{\partial f}{\partial x_i}\right| \leqslant L(1 \leqslant i \leqslant s), \quad \left|\frac{\partial^2 f}{\partial x_i \, \partial x_j}\right| \leqslant L(1 \leqslant i < j \leqslant s),$$

$$\cdots, \left|\frac{\partial^s f}{\partial x_1 \cdots \partial x_s}\right| \leqslant L, \tag{5.4}$$

then $f \in L_s$ evidently.

## § 5.2. Uniform distribution and numerical integration

**Theorem 5.3.** *Let $P_n(k)$ $(1 \leqslant k \leqslant n)$ be a set with discrepancy $D(n)$. If $f \in B_s$, then*

$$\left|\int_{G_s} f(\mathbf{x}) d\mathbf{x} - \frac{1}{n} \sum_{k=1}^{n} f(P_n(k))\right| \leqslant V(f) D(n). \tag{5.5}$$

*Proof.* We will prove the theorem only for the case $s = 2$, since the proof is similar for $s > 2$.

Since $f \in B_2$, it follows by Theorem 5.1 that $f$ can be represented as

$$f = F - G,$$
$$F = -(N + N_1 + N_2) - f(1, 1),$$
$$G = -(P + P_1 + P_2),$$

where

$$\Delta_{10}F \geqslant 0, \Delta_{01}F \geqslant 0, \Delta_{11}F \leqslant 0, \Delta_{10}G \geqslant 0, \Delta_{01}G \geqslant 0, \Delta_{11}G \leqslant 0,$$
$$\Delta_{10}P_2 = \Delta_{10}N_2 = \Delta_{01}P_1 = \Delta_{01}N_1 = 0,$$
$$\Delta_{11}P_1 = \Delta_{11}N_1 = \Delta_{11}P_2 = \Delta_{11}N_2 = 0,$$
$$P(x, 1) = N(x, 1) = P(1, y) = N(1, y) = 0,$$
$$P_1(1, 1) = N_1(1, 1) = P_2(1, 1) = N_2(1, 1) = 0. \tag{5.6}$$

The number of points $P_n(k)(1 \leqslant k \leqslant n)$ which fall in the rectangle

$$\frac{i-1}{q} \leqslant x < \frac{i}{q}, \quad \frac{j-1}{q} \leqslant y < \frac{j}{q} \tag{5.7}$$

is equal to

$$N_n\left(\frac{i}{q},\frac{j}{q}\right)-N_n\left(\frac{i-1}{q},\frac{j}{q}\right)-N_n\left(\frac{i}{q},\frac{j-1}{q}\right)$$
$$+N_n\left(\frac{i-1}{q},\frac{j-1}{q}\right).$$

It follows from (5.6) that $F(x,y)\leqslant F\left(\frac{i}{q},\frac{j}{q}\right)$ for $(x,y)$ belonging to (5.7). Hence

$$\begin{aligned}
S_1 &= \frac{1}{n}\sum_{k=1}^{n} F(x_1^{(n)}(k),\,x_2^{(n)}(k))\\
&\leqslant \frac{1}{n}\sum_{i=1}^{q}\sum_{j=1}^{q}\left(N_n\left(\frac{i}{q},\frac{j}{q}\right)-N_n\left(\frac{i-1}{q},\frac{j}{q}\right)\right.\\
&\qquad \left.-N_n\left(\frac{i}{q},\frac{j-1}{q}\right)+N_n\left(\frac{i-1}{q},\frac{j-1}{q}\right)\right)F\left(\frac{i}{q},\frac{j}{q}\right)\\
&= \frac{1}{n}\sum_{i=1}^{q-1}\sum_{j=1}^{q-1} N_n\left(\frac{i}{q},\frac{j}{q}\right)\left(F\left(\frac{i}{q},\frac{j}{q}\right)-F\left(\frac{i+1}{q},\frac{j}{q}\right)\right.\\
&\qquad \left.-F\left(\frac{i}{q},\frac{j+1}{q}\right)+F\left(\frac{i+1}{q},\frac{j+1}{q}\right)\right)\\
&\qquad +\frac{1}{n}\sum_{i=1}^{q-1} N_n\left(\frac{i}{q},1\right)\left(F\left(\frac{i}{q},1\right)-F\left(\frac{i+1}{q},1\right)\right)\\
&\qquad +\frac{1}{n}\sum_{j=1}^{q-1} N_n\left(1,\frac{j}{q}\right)\left(F\left(1,\frac{j}{q}\right)-F\left(1,\frac{j+1}{q}\right)\right)\\
&\qquad +F(1,1).
\end{aligned}$$

Since

$$N_n(x,y)=xyn+\phi D(n)n,$$

where we use $\phi$ to denote a number with absolute value $\leqslant 1$, then

$$\begin{aligned}
S_1 \leqslant &\sum_{i=1}^{q-1}\sum_{j=1}^{q-1}\left(\frac{ij}{q^2}+\phi D(n)\right)\left(F\left(\frac{i}{q},\frac{j}{q}\right)-F\left(\frac{i+1}{q},\frac{j}{q}\right)\right.\\
&\left.-F\left(\frac{i}{q},\frac{j+1}{q}\right)+F\left(\frac{i+1}{q},\frac{j+1}{q}\right)\right)\\
&+\sum_{i=1}^{q-1}\left(\frac{i}{q}+\phi D(n)\right)\left(F\left(\frac{i}{q},1\right)-F\left(\frac{i+1}{q},1\right)\right)
\end{aligned}$$

$$+\sum_{j=1}^{q-1}\left(\frac{j}{q}+\psi D(n)\right)\left(F\left(1,\frac{j}{q}\right)-F\left(1,\frac{j+1}{q}\right)\right)+F(1,1)$$

$$\leqslant\frac{1}{q^2}\sum_{i=1}^{q}\sum_{j=1}^{q}F\left(\frac{i}{q},\frac{j}{q}\right)+V(f)D(n).$$

Let $q\to\infty$. Then

$$S_1\leqslant\int_0^1\int_0^1 F(x_1,x_2)dx_1dx_2+V(F)D(n).$$

Similarly, we may prove

$$S_1\geqslant\int_0^1\int_0^1 F(x_1,x_2)dx_1dx_2-V(F)D(n).$$

Hence

$$S_1=\int_0^1\int_0^1 F(x_1,x_2)dx_1dx_2+\psi V(F)D(n). \tag{5.8}$$

Similarly,

$$S_2=\frac{1}{n}\sum_{k=1}^{n}G(x_1^{(n)}(k),x_2^{(n)}(k))$$

$$=\int_0^1\int_0^1 G(x_1,x_2)dx_1dx_2+\psi V(G)D(n). \tag{5.9}$$

For any given division $(\sigma)$, it follows from (5.6) that

$$\sum_{i=0}^{l-1}\sum_{j=0}^{m-1}|\Delta_{11}f(x_i,y_j)|=P(0,0)-P(0,1)-P(1,0)$$

$$+P(1,1)+N(0,0)-N(0,1)-N(1,0)+N(1,1)$$

$$=\sum_{i=0}^{l-1}\sum_{j=0}^{m-1}|\Delta_{11}P(x_i,y_j)|+\sum_{i=0}^{l-1}\sum_{j=0}^{m-1}|\Delta_{11}N(x_i,y_j)|$$

$$=\sum_{i=0}^{l-1}\sum_{j=0}^{m-1}|\Delta_{11}F(x_i,y_j)|+\sum_{i=0}^{l-1}\sum_{j=0}^{m-1}|\Delta_{11}G(x_i,y_j)|.$$

Similarly

$$\sum_{i=0}^{l-1}|\Delta_{10}f(x_i,1)|=\sum_{i=0}^{l-1}|\Delta_{10}F(x_i,1)|+\sum_{i=0}^{l-1}|\Delta_{10}G(x_i,1)|$$

and

$$\sum_{j=0}^{m-1}|\Delta_{01}f(1,y_j)|=\sum_{j=0}^{m-1}|\Delta_{01}F(1,y_j)|+\sum_{j=0}^{m-1}|\Delta_{01}G(1,y_j)|.$$

Hence

$$V(f) = V(F) + V(G)$$

and so from (5.8) and (5.9),

$$\left| \int_0^1 \int_0^1 f(x_1, x_2) dx_1 dx_2 - \frac{1}{n} \sum_{k=1}^{n} f(x_1^{(n)}(k), x_2^{(n)}(k)) \right| \leqslant V(f) D(n).$$

The theorem is proved.

**Theorem 5.4.** *If (5.5) holds for all $f \in B_s$, then $P_n(k)(1 \leqslant k \leqslant n)$ is a set with discrepancy $D(n)$.*

*Proof.* Let $f(\mathbf{x})$ be the characteristic function of the domain

$$(\mathscr{R}) \qquad 0 \leqslant x_1 < \gamma_1, \cdots, 0 \leqslant x_s < \gamma_s, \quad \boldsymbol{\gamma} \in G_s,$$

i.e.,

$$f(\mathbf{x}) = \begin{cases} 1, & \text{if } \mathbf{x} \in \mathscr{R}, \\ 0, & \text{if } \mathbf{x} \bar{\in} \mathscr{R}. \end{cases}$$

Then $V(f) \leqslant 1$,

$$\int_{G_s} f(\mathbf{x}) d\mathbf{x} = \int_{\mathscr{R}} f(\mathbf{x}) d\mathbf{x} = |\boldsymbol{\gamma}|$$

and

$$\frac{1}{n} \sum_{k=1}^{n} f(P_n(k)) = \frac{1}{n} N_n(\boldsymbol{\gamma}).$$

Hence it follows from (5.5) that

$$\left| \frac{1}{n} N_n(\boldsymbol{\gamma}) - |\boldsymbol{\gamma}| \right| \leqslant D(n).$$

The theorem is proved.

If the inequality

$$\frac{1}{q^s} \sum_{l_1=1}^{q} \cdots \sum_{l_s=1}^{q} q^{\delta_{l_1 q} + \cdots + \delta_{l_s q}} \left| \frac{1}{n} N_n \left( \frac{l_1}{q}, \cdots, \frac{l_s}{q} \right) - \frac{l_1 \cdots l_s}{q^s} \right| \leqslant D^*(n)$$

holds for any given integer $q \geqslant 1$, then the set of points $P_n(k)(1 \leqslant k \leqslant n)$ is called a set with average discrepancy $D^*(n)$.

**Theorem 5.5.** *If $f \in L_s$, then*

$$\left|\int_{G_s} f(\mathbf{x})d\mathbf{x} - \frac{1}{n}\sum_{k=1}^{n} f(P_n(k))\right| \leqslant LD^*(n).$$

*Proof.* We still suppose that $s = 2$. Since

$$|f(x', y) - f(x, y)| \leqslant |f(x', y) - f(x, y) - f(x', 1) + f(x, 1)| + |f(x', 1) - f(x, 1)| \leqslant 2L|x' - x|$$

and

$$|f(x, y') - f(x, y)| \leqslant 2L|y' - y|,$$

therefore $f$ is uniformly continuous on $G_2$. Hence for any $\varepsilon > 0$,

$$\begin{aligned} S_1 &= \frac{1}{n}\sum_{k=1}^{n} f(x_1^{(n)}(k), x_2^{(n)}(k)) \\ &= \frac{1}{n}\sum_{i=1}^{q}\sum_{j=1}^{q}\left(N_n\left(\frac{i}{q}, \frac{j}{q}\right) - N_n\left(\frac{i-1}{q}, \frac{j}{q}\right)\right. \\ &\quad \left. - N_n\left(\frac{i}{q}, \frac{j-1}{q}\right) + N_n\left(\frac{i-1}{q}, \frac{j-1}{q}\right)\right)\left(f\left(\frac{i}{q}, \frac{j}{q}\right) + \delta\right), \end{aligned}$$

if $q$ is sufficiently large, where $|\delta| \leqslant \varepsilon/2$. Let $S_2$ be that part of $S_1$ which contains those terms not involving $\delta$. Since

$$\begin{aligned} &\frac{1}{n}\sum_{i=1}^{q}\sum_{j=1}^{q}\left(N_n\left(\frac{i}{q}, \frac{j}{q}\right) - N_n\left(\frac{i-1}{q}, \frac{j}{q}\right)\right. \\ &\quad \left. - N_n\left(\frac{i}{q}, \frac{j-1}{q}\right) + N_n\left(\frac{i-1}{q}, \frac{j-1}{q}\right)\right) = 1, \end{aligned}$$

therefore

$$|S_1 - S_2| < \varepsilon/2. \tag{5.10}$$

By partial summation,

$$\begin{aligned} S_2 &= \frac{1}{n}\sum_{i=1}^{q-1}\sum_{j=1}^{q-1} N_n\left(\frac{i}{q}, \frac{j}{q}\right)\left(f\left(\frac{i}{q}, \frac{j}{q}\right) - f\left(\frac{i+1}{q}, \frac{j}{q}\right)\right. \\ &\quad \left. - f\left(\frac{i}{q}, \frac{j+1}{q}\right) + f\left(\frac{i+1}{q}, \frac{j+1}{q}\right)\right) \\ &\quad + \frac{1}{n}\sum_{i=1}^{q-1} N_n\left(\frac{i}{q}, 1\right)\left(f\left(\frac{i}{q}, 1\right) - f\left(\frac{i+1}{q}, 1\right)\right) \end{aligned}$$

$$+\frac{1}{n}\sum_{j=1}^{q-1} N_n\left(1,\frac{j}{q}\right)\left(f\left(1,\frac{j}{q}\right)-f\left(1,\frac{j+1}{q}\right)\right)$$

$$+ f(1,1).$$

Let

$$S_3=\frac{1}{q^2}\sum_{i=1}^{q}\sum_{j=1}^{q} f\left(\frac{i}{q},\frac{j}{q}\right).$$

Then

$$\begin{aligned}|S_2-S_3| \leqslant &\sum_{i=1}^{q-1}\sum_{j=1}^{q-1}\left|\frac{1}{n}N_n\left(\frac{i}{q},\frac{j}{q}\right)-\frac{ij}{q^2}\right|\left|f\left(\frac{i}{q},\frac{j}{q}\right)\right.\\
&\left.-f\left(\frac{i+1}{q},\frac{j}{q}\right)-f\left(\frac{i}{q},\frac{j+1}{q}\right)+f\left(\frac{i+1}{q},\frac{j+1}{q}\right)\right|\\
&+\sum_{i=1}^{q-1}\left|\frac{1}{n}N_n\left(\frac{i}{q},1\right)-\frac{i}{q}\right|\left|f\left(\frac{i}{q},1\right)-f\left(\frac{i+1}{q},1\right)\right|\\
&+\sum_{j=1}^{q-1}\left|\frac{1}{n}N_n\left(1,\frac{j}{q}\right)-\frac{j}{q}\right|\left|f\left(1,\frac{j}{q}\right)-f\left(1,\frac{j+1}{q}\right)\right|\\
\leqslant &\, LD^*(n)\end{aligned}\tag{5.11}$$

and

$$\left|\int_0^1\int_0^1 f(x_1, x_2)dx_1dx_2 - S_3\right| < \frac{\varepsilon}{2}, \tag{5.12}$$

if $q$ is sufficiently large. Hence from (5.10), (5.11) and (5.12),

$$\begin{aligned}\left|\int_0^1\int_0^1 f(x_1, x_2)dx_1dx_2 - S_1\right| &\leqslant |S_1-S_2| + |S_2-S_3|\\
&+\left|S_3-\int_0^1\int_0^1 f(x_1,x_2)dx_1dx_2\right| \leqslant LD^*(n)+\varepsilon.\end{aligned}$$

Since $\varepsilon$ is arbitrary, the theorem follows.

It follows from these theorems that we may use the arithmetic mean of the values of the function $f(\mathbf{x})$ over a set $P_n(k)(1\leqslant k\leqslant n)$

$$\sum_{k=1}^{n} f(P_n(k))$$

to approximate the definite integral

$$\int_{G_s} f(\mathbf{x})d\mathbf{x}.$$

The difference between them is closely related to the discrepancy of $P_n(k)$ $(1 \leqslant k \leqslant n)$, if $f(\mathbf{x})$ satisfies certain conditions. Hence the problem for finding the best quadrature formula is equivalent to the problem for finding the best uniformly distributed sequence of sets. From the view point of numerical analysis, we demand not only the discrepancy of $P_n(k)(1 \leqslant k \leqslant n)$ should be low but also the $P_n(k)$ should be convenient for computation.

## § 5.3. The lower estimation for the error term of quadrature formula

It follows from Theorems 5.3 and 3.6 that for any given set $P_n(k)(1 \leqslant k \leqslant n)$, the estimate

$$\left|\int_{G_s} f(\mathbf{x})\,d\mathbf{x} - \frac{1}{n}\sum_{k=1}^{n} f(P_n(k))\right|$$
$$\leqslant V(f)2^{-2s-4}(s-1)^{-\frac{s-1}{2}} n^{-1} (\log_2 n)^{\frac{s-1}{2}},$$

cannot hold for all $f \in B_s$.

In this section, we shall prove a more precise result for the lower estimation of the error term of quadrature formula.

Define

$$g(\mathbf{x}) = \delta^{-(2s-1)(q+\lambda)} \prod_{k=1}^{s} ((x_k - a_k)(a_k + \delta - x_k))^{q+\lambda},$$

where

$$a_k \leqslant x_k < a_k + \delta, \quad 1 \leqslant k \leqslant s \tag{5.13}$$

and where $q$ is an integer, $0 \leqslant \lambda \leqslant 1$ and $\delta > 0$. Outside (5.13), we define

$$g(\mathbf{x}) = 0.$$

Evidently, the function has the following properties:

1) $g(\mathbf{x})$ has $q$-th continuous derivatives everywhere,

2) for $1 \leqslant k \leqslant s$, $i_j \geqslant 0$ $(1 \leqslant j \leqslant s)$ and $i_1 + \cdots + i_s = q$, the limit

$$\lim_{y_k \to x_k} \frac{1}{|y_k - x_k|^{\lambda}} \left| \frac{\partial^q f(x_1, \cdots, y_k, \cdots, x_s)}{\partial x_1^{i_1} \cdots \partial y_k^{i_k} \cdots \partial x_s^{i_s}} - \frac{\partial^q g(x_1, \cdots, x_k, \cdots, x_s)}{\partial x_1^{i_1} \cdots \partial x_k^{i_k} \cdots \partial x_s^{i_s}} \right|$$

exists and does not exceed $c(q, \lambda, s)$ and

3) $$\int_{-\infty}^{\infty} \cdots \int_{-\infty}^{\infty} g(\mathbf{x}) d\mathbf{x}$$

$$= \delta^{-(2s-1)(q+\lambda)} \prod_{k=1}^{s} \int_{a_k}^{a_k+\delta} ((x_k - a_k)(a_k + \delta - x_k))^{q+\lambda} dx_k$$

$$= \delta^{s+q+\lambda} \left( \int_0^1 [t(1-t)]^{q+\lambda} dt \right)^s$$

$$= c(q, \lambda, s) \delta^{s+q+\lambda}.$$

Let $n_0 = [(2n)^{1/s}] + 1 (> (2n)^{1/s})$. Divide $G_s$ into $n_0^s (> 2n)$ equal parallelepipeds like (5.13). If the right hand side of (5.13) is 1, the symbol "$<$" should be replaced by "$\leqslant$". For any given $n$ points $P_n(k)$ $(k = 1, \cdots, n)$ of $G_s$, there exists at least $n$ parallelepipeds not containing these points and then they are denoted by

$$Q_1, \cdots, Q_n.$$

We use $R$ to denote the complement of $\bigcup_{i=1}^{n} Q_i$ with respect to $G_s$. Let

$$f(\mathbf{x}) = \begin{cases} n_0^{(2s-1)(q+\lambda)} \prod_{i=1}^{s} ((x_i - a_i^{(k)})(a_i^{(k)} + n_0^{-1} - x_i))^{q+\lambda}, & \text{if } \mathbf{x} \in Q_k, \\ 0, \text{ if } \mathbf{x} \in R. \end{cases}$$

Since $P_n(k) \bar{\in} \bigcup_{i=1}^{n} Q_i$ for any $k (1 \leqslant k \leqslant n)$, hence

$$f(P_n(k)) = 0, \quad 1 \leqslant k \leqslant n. \tag{5.14}$$

Clearly, we have

$$\int_{G_s} f(\mathbf{x}) d\mathbf{x} \geqslant c(q, \lambda, s) n \cdot n_0^{-(s+q+\lambda)} \geqslant c(q, \lambda, s) n^{-\frac{q+\lambda}{s}} \tag{5.15}$$

by 3). Hence we have

**Theorem 5.6.** *For any given $n$ points $P_n(k)$ $(1 \leqslant k \leqslant n)$ of $G_s$, there exists a function $f(\mathbf{x})$ which has the properties* 1) *and* 2) *such that* (5.14) *and* (5.15) *hold.*

Theorem 5.6 means that for any given set of points $P_n(k) (1 \leqslant k \leqslant n)$ and a set of real numbers $\rho_k (1 \leqslant k \leqslant n)$, the difference between the sum

$$\sum_{k=1}^{n} \rho_k f(P_n(k))$$

and the integral

$$\int_{G_s} f(\mathbf{x}) d\mathbf{x}$$

cannot be expected to be less than $c(q, \lambda, s) n^{-\frac{q+\lambda}{s}}$. Especially, if $f$ satisfies (5.4), then the error term of quadrature formula cannot be expected to be better than $O(n^{-1})$.

## § 5.4. The quadrature formulas

Suppose in this section that $f \in B_s$. Denote

$$I(f) = \int_{G_s} f(\mathbf{x}) d\mathbf{x}.$$

Then by the results of Chapter 4 and Theorem 5.3, we have the following quadrature formulas:

$$\left| I(f) - \frac{1}{n} \sum_{l_1=0}^{m-1} \cdots \sum_{l_s=0}^{m-1} f\left(\frac{l_1}{m}, \cdots, \frac{l_s}{m}\right) \right|$$

$$\leqslant V(f) 2^s n^{-\frac{1}{s}}, \quad n = m^s, \qquad \text{(Cf. §4.1)}$$

$$\left| I(f) - \frac{1}{n} \sum_{k=1}^{n} f(\varphi_{p_1}(k), \cdots, \varphi_{p_s}(k)) \right|$$

$$\leqslant V(f) \left( \prod_{i=1}^{s} \frac{p_i \ln p_i n}{\ln p_i} \right) n^{-1}, \qquad \text{(Cf. § 4.2)}$$

$$\left| I(f) - \frac{1}{n} \sum_{k=1}^{n} f\left(\frac{k}{n}, \varphi_{p_1}(k), \cdots, \varphi_{p_{s-1}}(k)\right) \right|$$

$$\leqslant V(f) \left( \prod_{i=1}^{s-1} \frac{p_i \ln p_i n}{\ln p_i} \right) n^{-1}, \qquad \text{(Cf. § 4.2)}$$

$$\left| I(f) - \frac{1}{n} \sum_{a=1}^{p} \sum_{k=1}^{p} f\left(\left\{\frac{k}{p}\right\}, \left\{\frac{ak}{p}\right\}, \cdots, \left\{\frac{a^{s-1}k}{p}\right\}\right) \right|$$

$$\leqslant V(f) c(s) n^{-\frac{1}{2}} (\ln n)^s, \quad n = p^2. \qquad \text{(Cf. § 4.3)}$$

$$\left|I(f)-\frac{1}{n}\sum_{k=1}^{n}f\left(\left\{\frac{k}{n}\right\},\left\{\frac{k^2}{n}\right\},\cdots,\left\{\frac{k^s}{n}\right\}\right)\right|$$
$$\leqslant V(f)c(s)n^{-\frac{1}{2}}(\ln n)^s,\quad n=p^2. \qquad \text{(Cf. § 4.3)}$$

$$\left|I(f)-\frac{1}{p}\sum_{k=1}^{p}f\left(\left\{\frac{k}{p}\right\},\left\{\frac{k^2}{p}\right\},\cdots,\left\{\frac{k^s}{p}\right\}\right)\right|$$
$$\leqslant V(f)p^{-\frac{1}{2}}(\ln p)^s. \qquad \text{(Cf. § 4.3)}$$

$$\left|I(f)-\frac{1}{n}\sum_{k=1}^{n}f(\{\alpha_1 k\},\cdots,\{\alpha_s k\})\right|$$
$$\leqslant V(f)c(\boldsymbol{\alpha},\varepsilon)n^{-1+\varepsilon}. \qquad \text{(Cf. § 4.5)}$$

$$\left|I(f)-\frac{1}{n}\sum_{k=1}^{n}f(\{\beta_1 k\},\cdots,\{\beta_s k\})\right|$$
$$\leqslant V(f)c(\boldsymbol{\beta},\varepsilon)n^{-1+\varepsilon}. \qquad \text{(Cf. § 4.5)}$$

$$\left|I(f)-\frac{1}{n}\sum_{k=1}^{n}f\left(\left\{\frac{c_1 k}{n}\right\},\cdots,\left\{\frac{c_s k}{n}\right\}\right)\right|$$
$$\leqslant V(f)c(\mathscr{R}_s,\varepsilon)n^{-\frac{1}{2}-\frac{1}{2(s-1)}+\varepsilon},$$
$$s=\frac{\varphi(m)}{2},\quad n=n_l,\quad c_i=c_{li}$$
$$(l=1,2,\cdots;\quad i=1,\cdots,s). \qquad \text{(Cf. § 4.6)}$$

$$\left|I(f)-\frac{1}{q}\sum_{k=1}^{q}f\left(\left\{\frac{c_2 k}{n}\right\},\cdots,\left\{\frac{c_{s+1}k}{n}\right\}\right)\right|$$
$$\leqslant V(f)c(\mathscr{R}_{s+1},\varepsilon)q^{-1+\varepsilon},$$
$$s=\frac{\varphi(m)}{2}-1,\quad q=[n^{\frac{1}{2}+\frac{1}{2s}}]. \qquad \text{(Cf. § 4.6)}$$

$$\left|I(f)-\frac{1}{F_n}\sum_{k=1}^{F_n}f\left(\left\{\frac{k}{F_n}\right\},\left\{\frac{F_n(2)}{F_n}k\right\},\cdots,\left\{\frac{F_n(s)}{F_n}k\right\}\right)\right|$$
$$\leqslant V(f)c(\eta)F_n^{-\frac{1}{2}-\frac{1}{2^{s+1}\ln 2}-\frac{1}{2^{2s+3}}},\quad F_n=F_{s,n}. \qquad \text{(Cf. § 4.7)}$$

$$\left|I(f)-\frac{1}{F_n}\sum_{k=1}^{F_n}f\left(\left\{\frac{k}{F_n}\right\},\left\{\frac{F_{n-1}}{F_n}k\right\}\right)\right|$$
$$\leqslant V(f)cF_n^{-1}(\ln F_n)^2,\quad F_n=F_{2,n} \qquad \text{(Cf. § 4.7)}$$

$$\left|I(f) - \frac{1}{q}\sum_{k=1}^{q} f\left(\left\{\frac{F_n(2)}{F_n}k\right\}, \ldots, \left\{\frac{F_n(s+1)}{F_n}k\right\}\right)\right|$$
$$\leqslant V(f)c(\eta, \varepsilon)q^{-1+\varepsilon}, \quad F_n = F_{s+1,n},$$
$$q = [F_n^{\frac{1}{2}+\frac{1}{2^{s+2}\ln 2}+\frac{1}{2^{2s+4}}}]. \qquad \text{(Cf. § 4.7)}$$

$$\left|I(f) - \frac{1}{p}\sum_{k=1}^{p} f\left(\left\{\frac{a_1 k}{p}\right\}, \ldots, \left\{\frac{a_s k}{p}\right\}\right)\right|$$
$$\leqslant V(f)c(s)p^{-1}(\ln p)^s. \qquad \text{(Cf. § 4.9)}$$

$$\left|I(f) - \frac{1}{n}\sum_{k=1}^{n} f\left(\left\{\frac{a_2 k}{p}\right\}, \ldots, \left\{\frac{a_{s+1} k}{p}\right\}\right)\right|$$
$$\leqslant V(f)c(s)n^{-1}(\ln p)^{s+1}, \quad 1 \leqslant n \leqslant p. \qquad \text{(Cf. § 4.9)}$$

## Notes

The definition for a function of bounded variation was given by M. Krause [1] and G. H. Hardy [1] (Cf. also C. R. Adams, and J. A. Clarkson [1,2] and S. K. Zaremba [2]).

Theorem 5.3 was proved by J. F. Koksma [1] for $s = 1$ and generalized to $s > 1$ by E. Hlawka [1] (Cf. also E. M. Sobol [1] for the class of functions $L_s$).

Theorem 5.6: Cf. N. S. Bahvalov [1].

*Chapter 6*

# Periodic Functions

## § 6.1. The classes of functions

The $G_s$ may be regarded as tori. The 1-dimensional torus $G_1$ may be obtained by identifying two end-points of the unit interval $0 \leqslant x_1 \leqslant 1$ and $G_2$ by indentifying 2 opposite sides of the unit square $0 \leqslant x_1 \leqslant 1,\ 0 \leqslant x_2 \leqslant 1$. In general, $G_s$ is obtained by identifying the $2s$ opposite surfaces of the $s$-dimensional unit cube, i.e., the points

$$(x_1, \cdots, x_{\nu-1}, 0, x_{\nu+1}, \cdots, x_s)$$

and

$$(x_1, \cdots, x_{\nu-1}, 1, x_{\nu+1}, \cdots, x_s)$$

are identified, where $1 \leqslant \nu \leqslant s$.

Hereafter, we shall use $G_s$ to denote the $s$-dimensional torus, if not otherwise specified.

If a single-valued function $f(x_1, \cdots, x_s)$ is a periodic function of $s$ variables each with period 1, i.e.,

$$f(x_1, \cdots, x_\nu + 1, \cdots, x_s) = f(x_1, \cdots, x_\nu, \cdots, x_s), \quad 1 \leqslant \nu \leqslant s,$$

then we have a single-valued function over $G_s$. A simple example is $e^{2\pi i(m_1 x_1 + \cdots + m_s x_s)}$, where $m_1, \cdots, m_s$ are integers.

Let $f(x_1, \cdots, x_s)$ be a single-valued function of $G_s$. Let $\alpha = \rho + \beta$, where $\rho$ is a non-negative integer and $0 < \beta \leqslant 1$. Put

$$\delta_{h,k} f = (2i)^{-1}(f(x_1, \cdots, x_k + h, \cdots, x_s) - f(x_1, \cdots, x_k - h, \cdots, x_s)).$$

Suppose that the derivatives

$$\frac{\partial^{\tau_1 + \cdots + \tau_s} f}{\partial x_1^{\tau_1} \cdots \partial x_s^{\tau_s}} = f^{(\tau_1, \cdots, \tau_s)}, \quad 0 \leqslant \tau_1, \cdots, \tau_s \leqslant \rho$$

exist and are the periodic functions of $s$ variables each with period 1. Let

$$\|f^\alpha\| = \sup_{\substack{0<h_k\leqslant 1\\ \mathbf{x}\in G_s}} \left|\left(\left(\prod_{k=1}^{s} h_k^{-\beta}\delta_{h_k,k}\right) f\right)^{(\rho,\cdots,\rho)}\right|.$$

Let $H_s^\alpha(C)$ denote the class of functions $f(\mathbf{x})$ of $G_s$ such that

$$\|f^\alpha\| \leqslant C,$$

where as usual, $C$ denotes the absolute constant and the lower derivatives of $f(\mathbf{x})$ are also bounded by $C$. Let

$$\mu(x) = \begin{cases} \left(\cos\left(\frac{\pi}{2}\log_2|x|\right)\right)^2, & \text{if } \frac{1}{2}\leqslant |x| \leqslant 2,\\ 0, & \text{otherwise,}\end{cases}$$

$$\mu_t(x) = \mu(2^{1-t}x)$$

for any given positive integer $t$ and

$$\mu_0(x) = 1 - \sum_{t=1}^{\infty}\mu_t(x). \tag{6.1}$$

Then

$$\mu(x) + \mu\left(\frac{x}{2}\right) = \left(\cos\left(\frac{\pi}{2}\log_2|x|\right)\right)^2 + \left(\sin\left(\frac{\pi}{2}\log_2|x|\right)\right)^2 = 1$$

for $1\leqslant |x| \leqslant 2$ and so

$$\sum_{t=1}^{\infty}\mu_t(x) = 1$$

or

$$\mu_0(x) = 0$$

for $|x| \geqslant 1$.

Suppose that $f(\mathbf{x})$ has the Fourier expansion

$$f(\mathbf{x}) \sim \sum C(\mathbf{m})e^{2\pi i(\mathbf{m},\mathbf{x})},$$

where $\mathbf{m}$ runs over all the integral vectors. If the series

$$\sum C(\mathbf{m})\lambda(\mathbf{m})e^{2\pi i(\mathbf{m},\mathbf{x})}$$

converges everywhere, then its sum is denoted by $f(\mathbf{x})\odot\lambda(\mathbf{m})$. For any given non-negative integral vector $\mathbf{t} = (t_1,\cdots,t_s)$ (i.e., $t_i\geqslant 0$, $1\leqslant i\leqslant s$),

let

$$t_0 = t_1 + \cdots + t_s$$

and

$$\varphi_{\mathbf{t}}(\mathbf{x}) = f(\mathbf{x}) \odot \prod_{k=1}^{s} \mu_{t_k}(m_k) = \sum C_{\mathbf{t}}(\mathbf{m}) e^{2\pi i(\mathbf{m},\mathbf{x})},$$

where

$$C_{\mathbf{t}}(\mathbf{m}) = C(\mathbf{m}) \mu_{t_1}(m_1) \cdots \mu_{t_s}(m_s).$$

Let $Q_s^\alpha(C)$ denote the class of continuous functions $f(\mathbf{x})$ of $G_s$ satisfying

$$\|\varphi_{\mathbf{t}}\| = \sup_{\mathbf{x} \in G_s} |\varphi_{\mathbf{t}}| \leqslant C 2^{-\alpha t_0}.$$

Let $E_s^\alpha(C)$ denote the set of functions $f(\mathbf{x})$ of $G_s$ such that

$$f(\mathbf{x}) \sim \sum C(\mathbf{m}) e^{2\pi i(\mathbf{m},\mathbf{x})},$$

where

$$|C(\mathbf{m})| \leqslant \frac{C}{\|\mathbf{m}\|^\alpha}.$$

## § 6.2. Several lemmas

**Lemma 6.1.** *Let*

$$\Delta_2 \lambda(n) = \lambda(n+1) - 2\lambda(n) + \lambda(n-1) \tag{6.2}$$

*and*

$$K_\lambda(x) = \sum \lambda(n) e^{2\pi i n x}.$$

*If* $\lambda(n) = 0$ *for* $|n| \geqslant M$, *where* $M$ *is a positive integer, then*

$$\sum \Delta_2 \lambda(n) = 0, \tag{6.3}$$

$$\sum n \Delta_2 \lambda(n) = 0 \tag{6.4}$$

*and*

$$K_\lambda(x) = -\sum \Delta_2 \lambda(n) \frac{e^{2\pi i n x}}{4(\sin \pi x)^2}. \tag{6.5}$$

*Proof.* (6.3) and (6.4) immediately follow from (6.2). Now we proceed to prove (6.5):

$$-\sum \Delta_2 \lambda(n) \frac{e^{2\pi i n x}}{4(\sin \pi x)^2} = -\sum \lambda(n) e^{2\pi i n x} \frac{(e^{2\pi i x} - 2 + e^{-2\pi i x})}{4(\sin \pi x)^2}$$

$$= -\sum \lambda(n) e^{2\pi i n x} \frac{(e^{\pi i x} - e^{-\pi i x})^2}{4(\sin \pi x)^2} = K_\lambda(x).$$

The lemma is proved.

Put

$$\|f(\mathbf{x})\|_1 = \int_{G_s} |f(\mathbf{x})|\, d\mathbf{x}.$$

**Lemma 6.2.** *Under the assumption of Lemma* 6.1, *we have*

$$\|K_\lambda(x)\|_1 \leqslant \frac{\pi M}{2} \sum |\Delta_2 \lambda(n)|.$$

*Proof.* Let

$$\|K_\lambda(x)\|_1 = I_1 + I_2, \tag{6.6}$$

where

$$I_1 = \int_\varepsilon^{1-\varepsilon} |K_\lambda(x)|\, dx, \quad I_2 = \int_{-\varepsilon}^{\varepsilon} |K_\lambda(x)|\, dx,$$

in which $\varepsilon = \dfrac{1}{2\pi M}$. Since $\sin \pi x \geqslant 2x$ for $0 \leqslant x \leqslant \dfrac{1}{2}$, therefore

$$\int_\varepsilon^{1-\varepsilon} \left| \frac{e^{2\pi i n x}}{4(\sin \pi x)^2} \right| dx \leqslant 2 \int_\varepsilon^{1/2} \frac{dx}{4(\sin \pi x)^2} \leqslant \frac{1}{8} \int_\varepsilon^{1/2} \frac{dx}{x^2} < \frac{1}{8\varepsilon} = \frac{\pi M}{4}.$$

Hence

$$I_1 \leqslant \frac{\pi M}{4} \sum |\Delta_2 \lambda(n)| \tag{6.7}$$

by (6.5). Since

$$K_\lambda(x) = -\sum \Delta_2 \lambda(n) \frac{(e^{2\pi i n x} - 1 - 2\pi i n x)}{4(\sin \pi x)^2}$$

by Lemma 6.1 and

$$|e^{i\alpha} - 1 - i\alpha| \leqslant \alpha^2$$

for $-1 \leqslant \alpha \leqslant 1$, therefore

$$\int_{-\varepsilon}^{\varepsilon} \left| \frac{e^{2\pi i n x} - 1 - 2\pi i n x}{4(\sin \pi x)^2} \right| dx \leqslant \pi^2 n^2 \int_{-\varepsilon}^{\varepsilon} \left( \frac{x}{\sin \pi x} \right)^2 dx$$

$$\leqslant \frac{\pi^2 n^2 \varepsilon}{2} = \frac{\pi M}{4}$$

for $|n| \leqslant M$. Hence

$$I_2 \leqslant \frac{\pi M}{4} \sum |\Delta_2 \lambda(n)|. \tag{6.8}$$

The lemma follows from (6.6), (6.7) and (6.8).

**Lemma 6.3.** *Suppose that $f(\mathbf{x})$ has the Fourier expansion*

$$f(\mathbf{x}) \sim \sum C(\mathbf{m}) e^{2\pi i(\mathbf{m},\mathbf{x})}.$$

*Then*

$$\|f(\mathbf{x}) \odot \lambda(\mathbf{m})\| \leqslant \|f(\mathbf{x})\| \prod_{k=1}^{s} \|K_{\lambda_k}(x)\|_1,$$

*where*

$$\lambda(\mathbf{m}) = \lambda_1(m_1)\cdots\lambda_s(m_s).$$

*Proof.* Since

$$\int_0^1 e^{2\pi i n x}\, dx = \begin{cases} 1, & \text{if } n = 0, \\ 0, & \text{if } n \neq 0, \end{cases} \tag{6.9}$$

so

$$\begin{aligned} f(\mathbf{x}) \odot \lambda(\mathbf{m}) &= \sum C(\mathbf{m}) \lambda(\mathbf{m}) e^{2\pi i(\mathbf{m},\mathbf{x})} \\ &= \int_{G_s} \prod_{\nu=1}^{s} K_{\lambda_\nu}(x_\nu - z_\nu) f(\mathbf{z}) d\mathbf{z} \\ &= (-1)^s \int_{G_s} \prod_{\nu=1}^{s} K_{\lambda_\nu}(y_\nu) f(\mathbf{x} - \mathbf{y}) d\mathbf{y} \end{aligned}$$

and

$$\|f(\mathbf{x}) \odot \lambda(\mathbf{m})\| \leqslant \|f(\mathbf{x})\| \prod_{k=1}^{s} \|K_{\lambda_k}(x)\|_1.$$

The lemma is proved.

**Lemma 6.4.**

$$d(\mu) = \int_{-\infty}^{\infty} (|\mu| + 2|\mu'|)dx + V(\mu') < c,$$

*where $c$ is a positive constant and $V(f)$ denotes the total variation of $f(x)$ in the interval $(-\infty, \infty)$.*

*Proof.* Since

$$\mu'(x) = \begin{cases} -\dfrac{\pi}{|x|\ln 2} \cos\left(\dfrac{\pi}{2} \log_2|x|\right) \sin\left(\dfrac{\pi}{2} \log_2|x|\right), \\ \quad \text{if } \dfrac{1}{2} \leqslant x \leqslant 2, \\ 0, \quad \text{otherwise,} \end{cases}$$

$\mu'(x)$ is therefore a product of the functions of bounded variation and so $\mu'(x)$ is a function of bounded variation too. The lemma is proved.

**Lemma 6.5.** *Let*

$$g(\psi) = \max_{\frac{1}{2}\leqslant|x|\leqslant 2} (|\psi|, |\psi'|, |\psi''|).$$

*Then*

$$\|K_{\psi(n2^{1-t})\mu(n2^{1-t})}(x)\|_1 \leqslant 2\pi d(\mu)g(\psi).$$

*Proof.* Since

$$\begin{aligned}
&|\Delta_2(\psi(n2^{1-t})\mu(n2^{1-t}))| \\
&\quad = \left|\int_0^{2^{1-t}} ((\psi\mu)'(n2^{1-t}+z) - (\psi\mu)'((n-1)2^{1-t}+z))dz\right| \\
&\quad \leqslant 2^{1-t} \max_{0\leqslant z\leqslant 2^{1-t}} |(\psi\mu)'(n2^{1-t}+z) - (\psi\mu)'((n-1)2^{1-t}+z)|
\end{aligned}$$

and $\mu(n2^{1-t}) = 0$ for $|n| \geqslant 2^t$, therefore by Lemma 6.2,

$$\begin{aligned}
\|K_{\psi(n2^{1-t})\mu(n2^{1-t})}(x)\|_1 &\leqslant \pi 2^{t-1}\sum|\Delta_2(\psi(n2^{1-t})\mu(n2^{1-t}))| \\
&\leqslant \pi\sum \max_{0\leqslant z\leqslant 2^{1-t}} |(\psi\mu)'(n2^{1-t}+z) - (\psi\mu)'((n-1)2^{1-t}+z)| \\
&\leqslant 2\pi V((\psi\mu)'). \qquad (6.10)
\end{aligned}$$

Since

$$\begin{aligned}
(\psi\mu)'(y) - (\psi\mu)'(x) &= \int_x^y (\psi\mu)''dz \\
&= \int_x^y (\psi\mu'' + 2\psi'\mu' + \psi''\mu)dz,
\end{aligned}$$

so

$$V((\psi\mu)') \leqslant g(\psi)d(\mu) \qquad (6.11)$$

by Lemma 6.4. The lemma follows by substituting (6.11) into (6.10).

## § 6.3. The relations between $H_s^a(C)$, $Q_s^a(C)$ and $E_s^a(C)$

**Theorem 6.1.** $H_s^a(C) \subset Q_s^a(C\cdot c(a)^s)$.

*Proof.* Let

$$\mu_{l,t_k}(n) = \mu_{t_k}(n)((2\pi in)^l \sin 2\pi nh_k)^{-1}$$

and

$$K_{l,t_k}(x) = \sum \mu_{l,t_k}(n) e^{2\pi i n x},$$

where $h_k = \dfrac{1}{5} 2^{-t_k}$. Let $f \in H_s^\alpha(C)$ and

$$f(\mathbf{x}) \sim \sum C(\mathbf{m}) e^{2\pi i (\mathbf{m}, \mathbf{x})}.$$

Since

$$\left(\prod_{k=1}^{s} \delta_{h_k, k}\right) f = \sum C(\mathbf{m}) e^{2\pi i(\mathbf{m}, \mathbf{x})} \prod_{k=1}^{s} (2i)^{-1} (e^{2\pi i m_k h_k} - e^{-2\pi i m_k h_k})$$

$$= f \odot \prod_{k=1}^{s} \sin 2\pi m_k h_k,$$

we have

$$\varphi_{\mathbf{t}} = f \odot \prod_{k=1}^{s} \mu_{t_k}(m_k) = \left(\prod_{k=1}^{s} \delta_{h_k, k}\right) f \odot \prod_{k=1}^{s} \mu_{0, t_k}(m_k)$$

$$= \int_{G_s} \prod_{k=1}^{s} K_{0,t_k}(x_k - z_k) \left(\left(\prod_{j=1}^{s} \delta_{h_j, j}\right) f(\mathbf{z})\right) d\mathbf{z}$$

by (6.9). Since

$$K_{l,t_k}^{(l)}(x) = K_{0,t_k}(x),$$

hence by partial integration $\rho$ times with respect to every variable $z_k$, we have

$$\varphi_{\mathbf{t}} = \int_{G_s} \left(\prod_{k=1}^{s} K_{\rho,t_k}(x_k - z_k)\right) \left(\left(\prod_{j=1}^{s} \delta_{h_j, j}\right) f(\mathbf{z})\right)^{(\rho, \cdots, \rho)} d\mathbf{z}$$

$$= \left(\left(\prod_{k=1}^{s} \delta_{h_k, k}\right) f\right)^{(\rho, \cdots, \rho)} \odot \prod_{j=1}^{s} \mu_{\rho, t_j}(m_j). \tag{6.12}$$

Since $f \in H_s^\alpha(C)$, therefore

$$\left\| \left(\left(\prod_{k=1}^{s} \delta_{h_k, k}\right) f\right)^{(\rho, \cdots, \rho)} \right\| \leqslant C \prod_{k=1}^{s} h_k^\beta \leqslant C 5^{-\beta s} 2^{-\beta t_0}. \tag{6.13}$$

Put

$$\psi(x) = ((2\pi i x 2^{t_k - 1})^\rho \sin (2\pi x 2^{t_k - 1} h_k))^{-1}.$$

Since

$$|\sin (2\pi x 2^{t_k - 1} h_k)| \geqslant \sin \frac{\pi}{10}$$

for $\frac{1}{2} \leqslant |x| \leqslant 2$, therefore

$$g(\psi) = \max_{\frac{1}{2}\leqslant|x|\leqslant 2} (|\psi|, |\psi'|, |\psi''|) \leqslant c(\alpha)2^{-\rho t_k}$$

and so

$$\|K_{\rho,t_k}(x)\|_1 \leqslant c(\alpha)2^{-\rho t_k} \tag{6.14}$$

by Lemma 6.5. Hence it follows by (6.12), (6.13), (6.14) and Lemma 6.3 that

$$\begin{aligned}\|\varphi_{\mathbf{t}}\| &\leqslant \left\|\left(\left(\prod_{k=1}^{s} \delta_{h_k,k}\right) f\right)^{(\rho,\cdots,\rho)}\right\| \prod_{j=1}^{s} \|K_{\rho,t_j}(x)\|_1 \\ &\leqslant Cc(\alpha)^s 2^{-\beta t_0 - \rho t_0} = Cc(\alpha)^s 2^{-\alpha t_0}\end{aligned}$$

and so $f \in Q_s^\alpha(Cc(\alpha)^s)$. The theorem is proved.

**Theorem 6.2.**

$$Q_s^\alpha(C) \subset E_s^\alpha(C2^s).$$

*Proof.* Suppose that $f \in Q_s^\alpha(C)$ and

$$\varphi_{\mathbf{t}} = f \odot \prod_{k=1}^{s} \mu_{t_k}(m_k) = \sum C_{\mathbf{t}}(\mathbf{m}) e^{2\pi i(\mathbf{m},\mathbf{x})}.$$

Since

$$C_{\mathbf{t}}(\mathbf{m}) = \int_{G_s} \varphi_{\mathbf{t}}(x) e^{-2\pi i(\mathbf{m},\mathbf{x})} d\mathbf{x},$$

hence

$$|C_{\mathbf{t}}(\mathbf{m})| \leqslant \int_{G_s} |\varphi_{\mathbf{t}}(\mathbf{x})| d\mathbf{x} \leqslant \|\varphi_{\mathbf{t}}\| \leqslant C2^{-\alpha t_0}.$$

Since $C_{\mathbf{t}}(\mathbf{m}) = 0$ for $|m_k| \geqslant 2^{t_k}$, therefore

$$|C_{\mathbf{t}}(\mathbf{m})| \leqslant C\|\mathbf{m}\|^{-\alpha}. \tag{6.15}$$

From (6.1),

$$C(\mathbf{m}) = C(\mathbf{m}) \sum_{t_1=0}^{\infty} \mu_{t_1}(m_1) \cdots \sum_{t_s=0}^{\infty} \mu_{t_s}(m_s) = \sum{}'' C_{\mathbf{t}}(\mathbf{m}), \tag{6.16}$$

where $\sum''$ denotes a sum of which $\mathbf{t}$ runs over all non-negative integral vectors. For any given $\mathbf{m}$, there are at most $2^s$ non-negative integral vectors $\mathbf{t}$ such that

$$2^{-1} < |2^{1-t_k} m_k| < 2, \quad k = 1, \cdots, s,$$

i.e.; $|C_{\mathbf{t}}(\mathbf{m})| \neq 0$. Hence

$$|C(\mathbf{m})| \leqslant C2^s\|\mathbf{m}\|^{-\alpha}$$

by (6.15) and (6.16) and so $f \in E_s^\alpha(C2^s)$. The theorem is proved.

**Theorem 6.3.** *Suppose that* $f \in Q_s^\alpha(C)$. *Then*

$$f(\mathbf{x}) = \sum{}'' \varphi_{\mathbf{t}}(\mathbf{x}).$$

*Proof.* Since $f \in Q_s^\alpha(C)$, therefore

$$\|\varphi_{\mathbf{t}}\| = \sup_{\mathbf{x} \in G_s} |\varphi_{\mathbf{t}}(\mathbf{x})| \leqslant C2^{-\alpha t_0}$$

and

$$\sum{}'' \|\varphi_{\mathbf{t}}\| \leqslant C \sum{}'' 2^{-\alpha t_0} = C\left(\sum_{t=0}^{\infty} 2^{-\alpha t}\right)^s = Cc(\alpha)^s.$$

Hence the series

$$\sum{}'' \varphi_{\mathbf{t}}(\mathbf{x})$$

is uniformly convergent on $G_s$ and so it represents a continuous function on $G_s$ and it is denoted by $f_0(\mathbf{x})$. For any given integral vector $\mathbf{n}$,

$$\begin{aligned}
&\int_{G_s} (f(\mathbf{x}) - f_0(\mathbf{x}))e^{-2\pi i(\mathbf{n},\mathbf{x})}d\mathbf{x} \\
&= \sum \int_{G_s} (C(\mathbf{m}) - \sum{}'' C_{\mathbf{t}}(\mathbf{m}))e^{2\pi i(\mathbf{m}-\mathbf{n},\mathbf{x})}d\mathbf{x} \\
&= C(\mathbf{n}) - \sum{}'' C_{\mathbf{t}}(\mathbf{n}) \\
&= C(\mathbf{n})(1 - \sum{}'' \mu_{t_1}(n_1)\cdots\mu_{t_s}(n_s)) = 0.
\end{aligned}$$

Hence $f(\mathbf{x})$ and $f_0(\mathbf{x})$ are equal almost everwhere on $G_s$. Since they are both continuous functions on $G_s$, we have $f(\mathbf{x}) = f(\mathbf{x}_0)$. The theorem is proved.

## § 6.4. Periodic functions

In these next two sections, we shall introduce methods of reducing the integral of a certain class of functions to the integral of a class of periodic functions.

Let $\mathbf{x}_\nu(x) = (x_1, \cdots, x_{\nu-1}, x, x_{\nu+1}, \cdots, x_s)$. Let $f(\mathbf{x})$ be a function on $s$-dimensional unit cube $G_s$. Let $\alpha > 0$ and $\alpha = \rho + \beta$, where $\rho$ is a non-negative integer and $0 < \beta \leqslant 1$. Define

$$\sigma_{h,k} f = f(\mathbf{x}_k(x_k + h)) - f(\mathbf{x})$$

for $\mathbf{x} \in G_s$ and $\mathbf{x}_k(x_k + h) \in G_s$. Suppose that the derivatives

$$\frac{\partial^{\tau_1+\cdots+\tau_s}}{\partial x_1^{\tau_1} \cdots \partial x_s^{\tau_s}} = f^{(\tau_1,\cdots,\tau_s)}, \quad 0 \leqslant \tau_1, \cdots, \tau_s \leqslant \rho$$

exist. Let

$$\|f^\alpha\| = \sup_{\substack{\mathbf{x} \in G_s \\ \mathbf{x}+\mathbf{h} \in G_s}} \left| \left( \left( \prod_{k=1}^{s} |h_k|^{-\beta} \sigma_{h_k,k} \right) f \right)^{(\rho,\cdots,\rho)} \right|$$

and $D_s^\alpha(C)$ be the class of functions on $G_s$ such that

$$\|f^\alpha\| \leqslant C$$

and the lower derivatives of $f$ are also bounded by $C$. For any $f \in D_s^\alpha(C)$, if there exists $\varphi(\mathbf{x}) \in D_s^\alpha(Cc(\alpha)^s)$ such that

$$\varphi(\mathbf{x}_\nu(1)) = \varphi(\mathbf{x}_\nu(0)), \quad \nu = 1, \cdots, s \tag{6.17}$$

and

$$\int_{G_s} f(\mathbf{x}) d\mathbf{x} = \int_{G_s} \varphi(\mathbf{x}) d\mathbf{x}, \tag{6.18}$$

then $\varphi(\mathbf{x})$ is called a simple periodic function of $f(\mathbf{x})$.

From the definition of $H_s^\alpha(C)$, we derive immediately

**Theorem 6.4.** *Suppose that* $\alpha \leqslant 1$. *If* $\varphi \in D_s^\alpha(C)$ *is a simple periodic function of* $f$, *then* $\varphi(\{x_1\}, \cdots, \{x_s\}) \in H_s^\alpha(C)$.

We know from Theorem 6.4 that if $\alpha \leqslant 1$, then the quadrature formula of the functions of $D_s^\alpha(C)$ may be deduced from the quadrature formula of the functions of $H_s^\alpha(C)$. Now we shall introduce several methods for constructing the simple periodic functions.

1. Let

$$\varphi_1(\mathbf{x}) = \frac{1}{2}(f(\mathbf{x}) + f(\mathbf{x}_1(1 - x_1))),$$

$$\varphi_2(\mathbf{x}) = \frac{1}{2}(\varphi_1(\mathbf{x}) + \varphi_1(\mathbf{x}_2(1 - x_2))),$$

$$\cdots\cdots$$

$$\varphi_s(\mathbf{x}) = \frac{1}{2}(\varphi_{s-1}(\mathbf{x}) + \varphi_{s-1}(\mathbf{x}_s(1 - x_s))) \tag{6.19}$$

and

$$\varphi(\mathbf{x}) = \varphi_s(\mathbf{x}).$$

Then $f \in D_s^\alpha(C)$ obviously implies that $\varphi \in D_s^\alpha(C)$. From (6.19),

$$\varphi_1(\mathbf{x}_1(1)) = \varphi_1(\mathbf{x}_1(0)) = \frac{f(\mathbf{x}_1(1)) + f(\mathbf{x}_1(0))}{2}.$$

Hence by the substitution $1 - x_1 = y_1$, we have

$$\int_{G_s} f(\mathbf{x})d\mathbf{x} = \int_{G_s} \varphi_1(\mathbf{x})d\mathbf{x}.$$

It is easily verified by induction that $\varphi$ satisfies (6.17) and (6.18). Hence $\varphi$ is a simple periodic function of $f$.

2. Let

$$\varphi_1(\mathbf{x}) = f(\mathbf{x}) + \left(x_1 - \frac{1}{2}\right)(f(\mathbf{x}_1(0)) - f(\mathbf{x}_1(1))),$$

$$\varphi_2(\mathbf{x}) = \varphi_1(\mathbf{x}) + \left(x_2 - \frac{1}{2}\right)(\varphi_1(\mathbf{x}_2(0)) - \varphi_1(\mathbf{x}_2(1))),$$

$$\cdots\cdots$$

$$\varphi_s(\mathbf{x}) = \varphi_{s-1}(\mathbf{x}) + \left(x_s - \frac{1}{2}\right)(\varphi_{s-1}(\mathbf{x}_s(0)) - \varphi_{s-1}(\mathbf{x}_s(1))) \quad (6.20)$$

and

$$\varphi(\mathbf{x}) = \varphi_s(\mathbf{x}).$$

Then $f \in D_s^\alpha(C)$ implies that $\varphi \in D_s^\alpha(C3^s)$ and from (6.20), we have

$$\varphi_1(\mathbf{x}_1(1)) = \varphi_1(\mathbf{x}_1(0)) = \frac{f(\mathbf{x}_1(1)) + f(\mathbf{x}_1(0))}{2}$$

and

$$\int_{G_s} f(\mathbf{x})d\mathbf{x} = \int_{G_s} \varphi_1(\mathbf{x})d\mathbf{x}.$$

It is easily verified by induction that $\varphi$ satisfies (6.17) and (6.18). Hence $\varphi$ is a simple periodic function of $f$.

3. Suppose that $\psi(x) \in D_1^{\alpha+1}(c(\alpha))$ and $\psi(x)$ is a non-decreasing function in $[0, 1]$ such that

$$\psi(0) = 0, \quad \psi(1) = 1, \quad \psi'(0) = \psi'(1) = 0. \quad (6.21)$$

Then it follows from (6.21) that if $f \in D_s^\alpha(C)$, then

$$\varphi(\mathbf{x}) = f(\psi(x_1), \cdots, \psi(x_s))\psi'(x_1)\cdots\psi'_s(x_s)$$

satisfies (6.17) and (6.18) and $\varphi(\mathbf{x}) \in D_s^\alpha(Cc(\alpha)^s)$. Hence $\varphi$ is a simple periodic function of $f$. For example, take $\psi(x) = \left(\sin\frac{\pi x}{2}\right)^2$. Then $\psi'(x) = \frac{\pi}{2}\sin\pi x$. Obviously $\psi(\mathbf{x})$ satisfies (6.21) and

$$\varphi(\mathbf{x}) = \left(\frac{\pi}{2}\right)^s f\left(\left(\sin\frac{\pi x_1}{2}\right)^2, \cdots, \left(\sin\frac{\pi x_s}{2}\right)^2\right)\sin\pi x_1\cdots\sin\pi x_s$$

is a simple periodic function of $f$ and $\varphi \in D_s^\alpha(C(2\pi)^{(\alpha+1)s})$.

## § 6.5. Continuation

Introduce the notation

$$\frac{\partial^l\varphi(\mathbf{x}_\nu(x))}{\partial x_\nu^l} = \frac{\partial^l\varphi(\mathbf{x})}{\partial x_\nu^l}\bigg|_{x_\nu=x}.$$

For any $f \in D_s^\alpha(C)$, if there exists $\varphi(\mathbf{x}) \in D_s^\alpha(Cc(\alpha)^s)$ such that

$$\frac{\partial^l\varphi(\mathbf{x}_\nu(1))}{\partial x_\nu^l} = \frac{\partial^l\varphi(\mathbf{x}_\nu(0))}{\partial x_\nu^l}, \quad l = 0, 1, \cdots, \rho; \nu = 1, \cdots, s \tag{6.22}$$

and

$$\int_{G_s} f(\mathbf{x})d\mathbf{x} = \int_{G_s}\varphi(\mathbf{x})d\mathbf{x}, \tag{6.23}$$

then $\varphi(\mathbf{x})$ is called a complete periodic function of $f(\mathbf{x})$.

From (6.22),

$$\varphi(\mathbf{x}_\nu(1))^{(l_1,\cdots,l_s)} = \varphi(\mathbf{x}_\nu(0))^{(l_1,\cdots,l_s)}, \quad 0 \leqslant l_1, \cdots, l_s \leqslant \rho$$

and so by the definitions of $H_s^\alpha(C)$ and $D_s^\alpha(C)$, we derive:

**Theorem 6.5.** *If $\varphi \in D_s^\alpha(C)$ and $\varphi$ is a complete periodic function of $f$, then $\varphi(\{x_1\}, \cdots, \{x_s\}) \in H_s^\alpha(C)$.*

Hence the quadrature formula for the class of functions $D_s^\alpha(C)$ can be deduced from the quadrature formula for the class of functions $H_s^\alpha(C)$. Now we shall introduce two methods of constructing the complete periodic functions.

1. Suppose that $\psi(x) \in D_1^{\alpha+1}(C)$ and $\psi(x)$ is a non-decreasing function in $[0, 1]$ such that

$$\psi(0)=0,\psi(1)=1,\quad \psi^{(l)}(0)=\psi^{(l)}(1)=0, l=1,\cdots,\rho+1. \quad (6.24)$$

Then if $f\in D_s^\alpha(C)$, the function

$$\varphi(\mathbf{x})=f(\psi(x_1),\cdots,\psi(x_s))\psi'(x_1)\cdots\psi'(x_s)$$

satisfies (6.22) and (6.23) and $\varphi\in D_s^\alpha(Cc(\alpha)^s)$. Hence $\varphi$ is a complete periodic function of $f$. For example, let $n$ be an integer $\geqslant 2$ and

$$\psi_n(x)=(2n-1)C_{n-1}^{2(n-1)}\int_0^x (t(1-t))^{n-1}dt.$$

Then $\psi_n(0)=0$ and

$$\int_0^1 t^{n-1}(1-t)^{n-1}dt=\frac{n-1}{n}\int_0^1 t^n(1-t)^{n-2}dt=\cdots$$

$$=\frac{(n-1)(n-2)\cdots 1}{n(n+1)\cdots(2n-2)}\int_0^1 t^{2n-2}dt$$

$$=\frac{1}{(2n-1)C_{n-1}^{2(n-1)}}$$

by integration by parts. Hence

$$\psi_n(1)=(2n-1)C_{n-1}^{2(n-1)}\int_0^1 t^{n-1}(1-t)^{n-1}dt=1.$$

Since

$$\psi_n'(x)=(2n-1)C_{n-1}^{2(n-1)}x^{n-1}(1-x)^{n-1},$$

therefore

$$\psi_n^{(l)}(0)=\psi_n^{(l)}(1)=0,\quad l=1,\cdots,n-1.$$

Hence $\psi_{\rho+2}(x)$ satisfies (6.24) and so if $f\in D_s^\alpha(C)$, then the function

$$\varphi(\mathbf{x})=f(\psi_{\rho+2}(x_1),\cdots,\psi_{\rho+2}(x_s))\psi'_{\rho+2}(x_1)\cdots\psi'_{\rho+2}(x_s)$$

is a complete periodic function of $f$ and $\varphi\in D_s^\alpha(Cc(\alpha)^s)$. In particular, we have

$$\psi_2(x)=6\int_0^x t(1-t)dt=3x^2-2x^3$$

and

$$\psi_3(x)=30\int_0^x t^2(1-t)^2dt=10x^3-15x^4+6x^5.$$

2. The rational numbers $B_n(n=0,1,\cdots)$ and the polynomials $B_n(x)$ $(n=0,1,\cdots)$ defined by the recurrence relations

$$B_0=1,\quad \sum_{k=0}^{n-1} C_k^n B_k=0,\quad n\geqslant 2 \quad (6.25)$$

and

$$B_0(x) = 1, \quad B_n(x) = \sum_{k=0}^{n} C_k^n B_k x^{n-k}, \quad n \geqslant 1 \tag{6.26}$$

are called the Bernoulli numbers and the Bernoulli polynomials respectively. For example,

$$C_0^2 B_0 + C_1^2 B_1 = 0, \quad B_1 = -\frac{1}{2},$$

$$C_0^3 B_0 + C_1^3 B_1 + C_2^3 B_2 = 0, \quad B_2 = \frac{1}{6}$$

and

$$B_1(x) = C_0^1 B_0 x + C_1^1 B_1 = x - \frac{1}{2},$$

$$B_2(x) = C_0^2 B_0 x^2 + C_1^2 B_1 x + C_2^2 B_2 = x^2 - x + \frac{1}{6}.$$

**Lemma 6.6.** *Bernoulli polynomials satisfy*

$$B_n(1) - B_n(0) = \begin{cases} 1, & \text{if } n = 1, \\ 0, & \text{if } n \neq 1, \end{cases} \tag{6.27}$$

$$B_n'(x) = nB_{n-1}(x), \quad n \geqslant 1 \tag{6.28}$$

*and*

$$\int_0^1 B_n(x)dx = 0, \quad n \geqslant 1.$$

*Proof.* Since

$$B_0(x) = 1, \quad B_1(x) = x - \frac{1}{2},$$

therefore (6.27) holds for $n = 0, 1$. Suppose that $n \geqslant 2$. Then

$$B_n(1) = \sum_{k=0}^{n} C_k^n B_k = B_n + \sum_{k=0}^{n-1} C_k^n B_k = B_n = B_n(0)$$

by (6.25) and (6.26). (6.28) may be derived immediately by the differentiation of (6.26)

$$B_n'(x) = \sum_{k=0}^{n-1} (n-k) C_k^n B_k x^{n-k-1}$$

$$= n \sum_{k=0}^{n-1} C_k^{n-1} B_k x^{n-k-1} = nB_{n-1}(x), \quad n \geqslant 1.$$

From (6.25) and (6.26), we have

$$\int_0^1 B_n(x)dx = \sum_{k=0}^{n} C_k^n B_k \frac{1}{n+1-k}$$

$$= \frac{1}{n+1} \sum_{k=0}^{n} C_k^{n+1} B_k = 0, \quad n \geqslant 1.$$

The lemma is proved.

**Lemma 6.7.** *Let*

$$P_n(x) = \frac{1}{(n+1)!} B_{n+1}(x).$$

*Then*

$$P_n^{(l)}(1) - P_n^{(l)}(0) = \begin{cases} 1, & \text{if } l = n, \\ 0, & \text{if } l \neq n. \end{cases}$$

*Proof.* The lemma follows for $l > n+1$, since $B_n(x)$ is a polynomial of degree $n$. Now suppose that $l \leqslant n+1$. Then it follows from Lemma 6.6 that

$$P'_n(x) = \frac{B'_{n+1}(x)}{(n+1)!} = \frac{B_n(x)}{n!},$$

$$\cdots\cdots$$

$$P_n^{(l)}(x) = \frac{B_{n+1-l}(x)}{(n+1-l)!}$$

and

$$P_n^{(l)}(1) - P_n^{(l)}(0) = \frac{B_{n+1-l}(1) - B_{n+1-l}(0)}{(n+1-l)!} = \begin{cases} 1, & \text{if } l = n, \\ 0, & \text{if } l \neq n. \end{cases}$$

The lemma is proved.

**Lemma 6.8.** *Suppose that $F(x) \in D_1^\alpha(C)$ and the function $\Phi(x)$ is defined by*

$$\Phi(x) = F(x) + \sum_{n=0}^{\rho} \sum_{\eta=0}^{1} (-1)^\eta P_n(x) F^{(n)}(\eta). \tag{6.29}$$

*Then*

$$\Phi^{(l)}(1) = \Phi^{(l)}(0), \quad l = 0, 1, \cdots, \rho.$$

*Proof*. Differentiating (6.29), we have

$$\Phi^{(l)}(x) = F^{(l)}(x) + \sum_{n=0}^{p} \sum_{\eta=0}^{1} (-1)^{\eta} P_n^{(l)}(x) F^{(n)}(\eta).$$

Hence by Lemma 6.7

$$\begin{aligned} \Phi^{(l)}(1) - \Phi^{(l)}(0) &= F^{(l)}(1) - F^{(l)}(0) \\ &\quad + \sum_{n=0}^{p} \sum_{\eta=0}^{1} (-1)^{\eta} (P_n^{(l)}(1) - P_n^{(l)}(0)) F^{(n)}(\eta) \\ &= F^{(l)}(1) - F^{(l)}(0) + \sum_{\eta=0}^{1} (-1)^{\eta} F^{(l)}(\eta) = 0. \end{aligned}$$

The lemma follows.

Let

$$\varphi_0(\mathbf{x}) = f(\mathbf{x})$$

and

$$\varphi_\nu(\mathbf{x}) = \varphi_{\nu-1}(\mathbf{x}) + \sum_{n_\nu=0}^{p} \sum_{\eta_\nu=0}^{1} (-1)^{\eta_\nu} P_{n_\nu}(x_\nu) \frac{\partial^{n_\nu} \varphi_{\nu-1}(\mathbf{x}_\nu(\eta_\nu))}{\partial x_\nu^{n_\nu}}. \quad (6.30)$$

for $\nu = 1, \cdots, s$. Let

$$\varphi(\mathbf{x}) = \varphi_s(\mathbf{x}).$$

Now we shall prove that if $f \in D_s^\alpha(C)$, then $\varphi$ is a complete periodic function of $f$.

By Lemma 6.6,

$$\int_0^1 P_n(x)dx = \frac{1}{(n+1)!} \int_0^1 B_{n+1}(x)dx = 0, \quad n \geqslant 0$$

and so by (6.30),

$$\int_{G_s} \varphi_\nu(\mathbf{x})d\mathbf{x} = \int_{G_s} \varphi_{\nu-1}(\mathbf{x})d\mathbf{x}, \quad \nu = 1, \cdots, s.$$

Hence

$$\begin{aligned} \int_{G_s} f(\mathbf{x})d\mathbf{x} &= \int_{G_s} \varphi_1(\mathbf{x})d\mathbf{x} = \cdots \\ &= \int_{G_s} \varphi_{s-1}(\mathbf{x})d\mathbf{x} = \int_{G_s} \varphi(\mathbf{x})d\mathbf{x} \end{aligned}$$

and (6.23) is proved.

Let

$$F(x_\nu) = \varphi_{\nu-1}(\mathbf{x}).$$

Then by (6.29) and (6.30),

$$\Phi(x_\nu) = \varphi_\nu(\mathbf{x}).$$

Hence it follows from Lemma 6.8 that

$$\frac{\partial^{l_\nu}\varphi_\nu(\mathbf{x}_\nu(1))}{\partial x_\nu^{l_\nu}} = \frac{\partial^{l_\nu}\varphi_\nu(\mathbf{x}_\nu(0))}{\partial x_\nu^{l_\nu}}, \quad \nu = 1, \cdots, s; \; l_\nu = 0, \cdots, \rho. \tag{6.31}$$

Now we proceed to prove that

$$\frac{\partial^{l_j}\varphi_\nu(\mathbf{x}_j(1))}{\partial x_j^{l_j}} = \frac{\partial^{l_j}\varphi_\nu(\mathbf{x}_j(0))}{\partial x_j^{l_j}}, \quad j = 1, \cdots, \nu; \; l_j = 0, \cdots, \rho. \tag{6.32}$$

holds for $1 \leqslant \nu \leqslant s$. For $\nu = 1$, (6.32) follows from (6.31). Suppose that $\nu > 1$ and (6.32) holds for $\nu - 1$, i.e.,

$$\frac{\partial^{l_j}\varphi_{\nu-1}(\mathbf{x}_j(1))}{\partial x_j^{l_j}} = \frac{\partial^{l_j}\varphi_{\nu-1}(\mathbf{x}_j(0))}{\partial x_j^{l_j}}, \quad j = 1, \cdots, \nu - 1; \; l_j = 0, \cdots, \rho.$$

It follows by differentiating $n_\nu (0 \leqslant n_\nu \leqslant \rho)$ times with respect to $x_\nu$ that

$$\frac{\partial^{l_j+n_\nu}\varphi_{\nu-1}(\mathbf{x}_j(1))}{\partial x_j^{l_j}\partial x_\nu^{n_\nu}} = \frac{\partial^{l_j+n_\nu}\varphi_{\nu-1}(\mathbf{x}_j(0))}{\partial x_j^{l_j}\partial x_\nu^{n_\nu}},$$

$$j = 1, \cdots, \nu - 1; \quad l_j = 0, \cdots, \rho.$$

and so by the differentiation of (6.30), we have

$$\frac{\partial^{l_j}\varphi_\nu(\mathbf{x}_j(1))}{\partial x_j^{l_j}} = \frac{\partial^{l_j}\varphi_\nu(\mathbf{x}_j(0))}{\partial x_j^{l_j}}, \quad j = 1, \cdots, \nu - 1; \quad l_j = 0, \cdots, \rho. \tag{6.33}$$

Hence (6.32) holds also for $\nu$ by (6.31) and (6.33). Consequently (6.32) holds for $1 \leqslant \nu \leqslant s$ by the induction. Especially, the case $\nu = s$ of (6.32) means that (6.22) holds.

$f \in D_s^\alpha(C)$ implies that $\varphi \in D_s^\alpha(Cc(\alpha)^s)$, since $B_n(x) \in H_1^\alpha(c(n))$ $(\alpha = 1, 2, \cdots)$.

Hence we have proved that $\varphi$ is a complete periodic function of $f$.

## Notes

The class of functions $E_s^\alpha(C)$ was introduced by N. M. Korobov [1,2,7] and the classes of functions $H_s^\alpha(C)$ and $Q_s^\alpha(C)$ were first introduced by N. S. Bahvalov [3,4] (with some modifications given by Hua Loo Keng and Wang Yuan [6,7]).

§ 2—§ 3: Cf. N. S. Bahvalov, [3,4].

§ 4—§ 5: Cf. N. M. Korobov [7] and I. F. Sarygin [1].

## Chapter 7

# Numerical Integration of Periodic Functions

## § 7.1. The set of equi-distribution and numerical integration

**Theorem 7.1.** *Suppose that* $\alpha > 1$. *Then*

$$\sup_{f\in E_s^\alpha(C)} \left| \int_{G_s} f(\mathbf{x})d\mathbf{x} - \frac{1}{n}\sum_{l_1=0}^{m-1}\cdots\sum_{l_s=0}^{m-1} f\left(\frac{l_1}{m},\cdots,\frac{l_s}{m}\right)\right|$$
$$\leqslant C(2\zeta(\alpha)+1)^s n^{-\alpha/s}, \tag{7.1}$$

*where* $n = m^s$ *and*

$$\zeta(\alpha) = \sum_{k=1}^{\infty}\frac{1}{k^\alpha}.$$

*Proof.* The function $f$ of $E_s^\alpha(C)$ has an absolutely convergent Fourier expansion for $\alpha > 1$ (Cf. §6.1)

$$f(\mathbf{x}) = \sum C(\mathbf{m})e^{2\pi i(\mathbf{m},\mathbf{x})}, \qquad |C(\mathbf{m})| \leqslant \frac{C}{\|\mathbf{m}\|^\alpha}. \tag{7.2}$$

Set $\mathbf{l} = (l_1,\cdots,l_s)$. Since

$$\sum_{k=0}^{m-1} e^{2\pi i n k/m} = \begin{cases} m, & \text{if } m\mid n,\\ 0, & \text{if } m\nmid n\end{cases} \tag{7.3}$$

(Cf. Lemma 3.6) and

$$C(\mathbf{0}) = \int_{G_s} f(\mathbf{x})d\mathbf{x}, \tag{7.4}$$

we have

$$\frac{1}{n}\sum_{l_1=0}^{m-1}\cdots\sum_{l_s=0}^{m-1} f\left(\frac{l_1}{m},\cdots,\frac{l_s}{m}\right) = \frac{1}{n}\sum_{l_1=0}^{m-1}\cdots\sum_{l_s=0}^{m-1}\sum C(\mathbf{m})e^{2\pi i(\mathbf{m},\mathbf{l})/m}$$
$$= \sum C(\mathbf{m})\prod_{j=1}^{s}\left(\frac{1}{m}\sum_{l_j=0}^{m-1} e^{2\pi i l_j m_j/m}\right)$$

$$= \sum_{\substack{m|m_j \\ 1\leqslant j\leqslant s}} C(\mathbf{m}) = C(\mathbf{0}) + \sum_{\substack{m|m_j \\ 1\leqslant j\leqslant s}}{}' C(\mathbf{m})$$

and

$$\left|\int_{G_s} f(\mathbf{x})d\mathbf{x} - \frac{1}{n}\sum_{l_1=0}^{m-1}\cdots\sum_{l_s=0}^{m-1} f\left(\frac{l_1}{m},\cdots,\frac{l_s}{m}\right)\right|$$

$$\leqslant \sum_{\substack{m|m_j \\ 1\leqslant j\leqslant s}}{}' |C(\mathbf{m})| = \Sigma'|C(m\mathbf{m})| \leqslant C\Sigma' \frac{1}{\|m\mathbf{m}\|^\alpha}$$

$$\leqslant C\left(\Sigma\frac{1}{\bar{k}^\alpha}\right)^s m^{-\alpha} = C(2\zeta(\alpha)+1)^s n^{-\alpha/s}.$$

The theorem is proved.

Take

$$f(\mathbf{x}) = C\frac{e^{2\pi i m x_1} + e^{-2\pi i m x_1}}{m^\alpha}.$$

Then $f \in E_s^\alpha(C)$ (of course $f \in H_s^\alpha(C(2\pi)^{\alpha+1})$ and

$$\left|\int_{G_s} f(\mathbf{x})d\mathbf{x} - \frac{1}{n}\sum_{l_1=0}^{m-1}\cdots\sum_{l_s=0}^{m-1} f\left(\frac{l_1}{m},\cdots,\frac{l_s}{m}\right)\right|$$

$$= \frac{2C}{m^\alpha} = 2Cn^{-\alpha/s},$$

i.e., there exists a function of $E_s^\alpha(C)(\alpha > 1)$ such that the error term in the quadrature formula (7.1) is not less than $2Cn^{-\alpha/s}$. Hence the term $n^{-\alpha/n}$ in (7.1) does not admit further essential improvement.

## § 7.2. The *p* set and numerical integration

**Theorem 7.2.** Suppose that $\alpha > 1$ and $n = p^2$. Then

$$\sup_{f\in E_s^\alpha(C)}\left|\int_{G_s} f(\mathbf{x})d\mathbf{x} - \frac{1}{n}\sum_{a=1}^{p}\sum_{k=1}^{p} f\left(\frac{a}{p},\frac{ak}{p},\cdots,\frac{ak^{s-1}}{p}\right)\right|$$

$$\leqslant Cs(2\zeta(\alpha)+1)^s n^{-\frac{1}{2}}. \tag{7.5}$$

*Proof.* Set $\mathbf{k} = (1, k, \cdots, k^{s-1})$. Then by (7.2), (7.3) and (7.4),

$$\sum_{a=1}^{p}\sum_{k=1}^{p} f\left(\frac{a}{p},\frac{ak}{p},\cdots,\frac{ak^{s-1}}{p}\right)=\sum C(\mathbf{m})\sum_{a=1}^{p}\sum_{k=1}^{p} e^{2\pi i(\mathbf{k},\mathbf{m})a/p}$$

$$=p^2C(\mathbf{0})+p\sum{}'C(\mathbf{m})\sum_{\substack{1\leqslant k\leqslant p\\(\mathbf{k},\mathbf{m})\equiv 0\,(\mathrm{mod}\,p)}} 1$$

and so

$$\sup_{f\in E_s^\alpha(C)}\left|\int_{G_s} f(\mathbf{x})d\mathbf{x}-\frac{1}{n}\sum_{a=1}^{p}\sum_{k=1}^{p} f\left(\frac{a}{p},\frac{ak}{p},\cdots,\frac{ak^{s-1}}{p}\right)\right|$$

$$\leqslant\frac{C}{p}\sum{}'\frac{1}{\|\mathbf{m}\|^\alpha}\sum_{\substack{1\leqslant k\leqslant p\\(\mathbf{k},\mathbf{m})\equiv 0\,(\mathrm{mod}\,p)}} 1.$$

Since the number of solutions of the congruence

$$(\mathbf{k},\mathbf{m})\equiv 0\,(\mathrm{mod}\,p),\quad 1\leqslant k\leqslant p$$

is at most $s-1$, if $m_1,\cdots,m_s$ are not all divisible by $p$, and is equal to $p$ otherwise (Cf. Lemma 4.5), we have

$$\sum{}'\frac{1}{\|\mathbf{m}\|}\sum_{\substack{1\leqslant k\leqslant p\\(\mathbf{k},\mathbf{m})\equiv 0\,(\mathrm{mod}\,p)}} 1\leqslant(s-1)\sum{}'\frac{1}{\|\mathbf{m}\|^\alpha}+p\sum{}'\frac{1}{\|p\mathbf{m}\|^\alpha}$$

$$\leqslant s\sum\frac{1}{\|\mathbf{m}\|^\alpha}=s\left(\sum\frac{1}{\overline{m}^\alpha}\right)^s=s(2\zeta(\alpha)+1)^s.$$

The theorem follows.

**Lemma 7.1.** *The number of non-negative integral solutions of*

$$n=r_1+\cdots+r_s$$

*is equal to* $C_{s-1}^{n+s-1}$.

*Proof.* For $|x|<1$,

$$\frac{1}{(1-x)^s}=\left(\sum_{r=0}^{\infty}x^r\right)^s=\sum_{\substack{r_1=0\\n=r_1+\cdots+r_s}}^{\infty}\cdots\sum_{r_s=0}^{\infty}x^n$$

and also

$$\frac{1}{(1-x)^s}=\sum_{n=0}^{\infty}C_{s-1}^{n+s-1}x^n.$$

The lemma follows by comparing the coefficients of $x^n$ in the above two formulas.

**Lemma 7.2.** *Suppose that $s$ is an integer $\geqslant 0$ and $\alpha > 0$. Then*

$$\sum_{t=n}^{\infty} t^s 2^{-\alpha t} \leqslant c(\alpha, s) n^s 2^{-\alpha n}.$$

*Proof.* Since

$$t^s 2^{-\alpha t} \leqslant \int_t^{t+1} u^s 2^{-\alpha(u-1)} du, \quad t = n, n+1, \cdots,$$

therefore

$$\begin{aligned}\sum_{t=n}^{\infty} t^s 2^{-\alpha t} &\leqslant \int_n^{\infty} u^s 2^{-\alpha(u-1)} du \\ &= \frac{n^s 2^{-\alpha(n-1)}}{\alpha \ln 2} + \frac{s}{\alpha \ln 2} \int_n^{\infty} u^{s-1} 2^{-\alpha(u-1)} du \\ &= \cdots \\ &= \sum_{k=0}^{s} \frac{k! C_k^s n^{s-k} 2^{-\alpha(n-1)}}{(\alpha \ln 2)^{k+1}} \leqslant c(\alpha, s) n^s 2^{-\alpha n}.\end{aligned}$$

The lemma is proved.

**Theorem 7.3.** *Suppose that $0 \leqslant \alpha < 1$. Then*

$$\sup_{f \in Q_s^{\alpha}(C)} \left| \int_{G_s} f(\mathbf{x}) d\mathbf{x} - \frac{1}{p} \sum_{k=1}^{p} f\left(\frac{k}{p}, \frac{k^2}{p}, \cdots, \frac{k^s}{p}\right) \right|$$

$$\leqslant \begin{cases} Cc(\alpha, s) p^{-\frac{1}{2}}, & \text{if } \frac{1}{2} < \alpha \leqslant 1, \\ Cc(\alpha, s) p^{-\alpha} (\ln p)^{s-1+\delta_{\frac{1}{2},\alpha}}, & \text{if } 0 < \alpha \leqslant \frac{1}{2}. \end{cases}$$

*Proof.* For $f \in Q_s^{\alpha}(C)$, let

$$S(f) = \int_{G_s} f(\mathbf{x}) d\mathbf{x} - \frac{1}{p} \sum_{k=1}^{p} f\left(\frac{k}{p}, \frac{k^2}{p}, \cdots, \frac{k^s}{p}\right).$$

Then by Theorem 6.3, we have

$$S(f) = \sum{}'' S(\varphi_{\mathbf{t}}),$$

where

$$S(\varphi_{\mathbf{t}}) = \int_{G_s} \varphi_{\mathbf{t}}(\mathbf{x}) d\mathbf{x} - \frac{1}{p} \sum_{k=1}^{p} \varphi_{\mathbf{t}}\left(\frac{k}{p}, \frac{k^2}{p}, \cdots, \frac{k^s}{p}\right).$$

Hence

$$\sup_{f\in Q_s^\alpha(C)} |S(f)| \leqslant \Sigma_1 + \Sigma_2, \tag{7.6}$$

where

$$\Sigma_1 = \sup_{f\in Q_s^\alpha(C)} \sum_{t_0 \geqslant \log_2 p}'' |S(\varphi_{\mathbf{t}})|$$

and

$$\Sigma_2 = \sup_{f\in Q_s^\alpha(C)} \sum_{t_0 < \log_2 p}'' |S(\varphi_{\mathbf{t}})|,$$

in which $t_0 = t_1 + \cdots + t_s$. Since $C_{s-1}^{t+s-1} \leqslant c(s)t^{s-1}$ and

$$|S(\varphi_{\mathbf{t}})| \leqslant 2\|\varphi_{\mathbf{t}}\| \leqslant C2^{1-\alpha t_0},$$

therefore

$$\Sigma_1 \leqslant Cc(s) \sum_{t_0 \geqslant \log_2 p}'' 2^{-\alpha t_0} \leqslant Cc(s) \sum_{t=[\log_2 p]}^{\infty} t^{s-1}2^{-\alpha t}$$

$$\leqslant Cc(\alpha, s)p^{-\alpha}(\ln p)^{s-1} \tag{7.7}$$

by Lemmas 7.1. and 7.2. Set $\mathbf{k} = (k, k^2, \cdots, k^s)$. Since

$$C_{\mathbf{t}}(0) = \int_{G_s} \varphi_{\mathbf{t}}(\mathbf{x})d\mathbf{x},$$

we have

$$\sum_{k=1}^{p} \varphi_{\mathbf{t}}\left(\frac{k}{p}, \frac{k^2}{p}, \cdots, \frac{k^s}{p}\right) = \sum_{k=1}^{p} \Sigma C_{\mathbf{t}}(\mathbf{m})e^{2\pi i(\mathbf{k},\mathbf{m})/p}$$

$$= pC_{\mathbf{t}}(\mathbf{0}) + \Sigma' C_{\mathbf{t}}(\mathbf{m}) \sum_{k=1}^{p} e^{2\pi i(\mathbf{k},\mathbf{m})/p}$$

and

$$|S(\varphi_{\mathbf{t}})| \leqslant \frac{1}{p} \Sigma' |C_{\mathbf{t}}(\mathbf{m})| \left|\sum_{k=1}^{p} e^{2\pi i(\mathbf{k},\mathbf{m})/p}\right|.$$

Suppose that $\mathbf{m} \neq \mathbf{0}$. If at least one of the relations $|m_i| < 2^{t_i} (1 \leqslant i \leqslant s)$ is not satisfied, then $C_{\mathbf{t}}(\mathbf{m}) = 0$. Otherwise if $|m_i| < 2^{t_i} (1 \leqslant i \leqslant s)$ and $t_0 < \log_2 p$, then $m_i (1 \leqslant i \leqslant s)$ are not all divisible by $p$. Hence

$$\Sigma_2 \leqslant (s-1)p^{-\frac{1}{2}} \sup_{f\in Q_s^\alpha(C)} \sum_{t_0 < \log_2 p}'' \sum_{\substack{|m_i| < 2^{t_i} \\ 1 \leqslant i \leqslant s}}' |C_{\mathbf{t}}(\mathbf{m})|$$

by Lemma 4.7. Let

$$\|F\|_2 = \left(\int_{G_s} |F(\mathbf{x})|^2 d\mathbf{x}\right)^{1/2}.$$

Then

$$\Sigma |C_{\mathbf{t}}(\mathbf{m})|^2 = \|\varphi_{\mathbf{t}}\|_2^2 \leqslant \|\varphi_{\mathbf{t}}\|^2 \leqslant C^2 c(\alpha, s)2^{-2\alpha t_0}$$

and so by Lemmas 7.1 and 7.2 and Schwarz's inequality,

$$\begin{aligned}\Sigma_2 &\leqslant (s-1)p^{-\frac{1}{2}} \sup_{f\in Q_s^\alpha(C)} \sum_{t_0<\log_2 p}'' \Big(\sum_{|m_i|<2^{t_i}}' 1\Big)^{1/2} \Big(\sum_{|m_i|<2^{t_i}}' |C_{\mathbf{t}}(\mathbf{m})|^2\Big)^{1/2} \\ &\leqslant (s-1)p^{-\frac{1}{2}} \sup_{f\in Q_s^\alpha(C)} \sum_{t_0<\log_2 p}'' 2^{\frac{s+t_0}{2}} \|\varphi_{\mathbf{t}}\|_2 \\ &\leqslant Cc(\alpha,s)p^{-\frac{1}{2}} \sum_{t_0<\log_2 p}'' 2^{-(\alpha-\frac{1}{2})t_0} \\ &\leqslant Cc(\alpha,s)p^{-\frac{1}{2}} \sum_{t=0}^{[\log_2 p]} (t+1)^{s-1}2^{-(\alpha-\frac{1}{2})t} \\ &\leqslant \begin{cases} Cc(\alpha,s)p^{-\frac{1}{2}}, & \text{if } \frac{1}{2}<\alpha\leqslant 1, \\ Cc(\alpha,s)p^{-\alpha}(\ln p)^{s-1+\delta_{\frac{1}{2},\alpha}}, & \text{if } 0<\alpha\leqslant\frac{1}{2}. \end{cases}\end{aligned} \tag{7.8}$$

The theorem follows from (7.6), (7.7) and (7.8).

For $\alpha>1$, we may also use the sets of points

$$\left(\left\{\frac{k}{p^2}\right\}, \left\{\frac{k^2}{p^2}\right\}, \cdots, \left\{\frac{k^s}{p^2}\right\}\right), \quad 1\leqslant k\leqslant p^2$$

and

$$\left(\left\{\frac{k}{p}\right\}, \left\{\frac{k^2}{p}\right\}, \cdots, \left\{\frac{k^s}{p}\right\}\right), \quad 1\leqslant k\leqslant p$$

(Cf. §4.3) to obtain the quadrature formulas for the class of functions $E_s^\alpha(C)$ which have the same precision as Theorem 7.2.

Now, we shall study the lower estimate of the error term of the quadrature formula given by Theorem 7.2. Set

$$f(\mathbf{x}) = C \sum_{m_1=-(p-1)}^{p-1}{}' \sum_{m_2=-(p-1)}^{p-1}{}' \frac{e^{2\pi i(m_1x_1+m_2x_2)}}{(\overline{m}_1\overline{m}_2)^\alpha}, \quad \alpha>0.$$

Then $f\in E_s^\alpha(C)$,

$$\int_{G_s} f(\mathbf{x})d\mathbf{x} = 0$$

and

$$\frac{1}{p^2}\sum_{a=1}^{p}\sum_{k=1}^{p} f\left(\frac{a}{p}, \frac{ak}{p}, \cdots, \frac{ak^{s-1}}{p}\right)$$

$$= \frac{C}{p^2}\sum_{a=1}^{p}\sum_{k=1}^{p} \sum_{m_1=-(p-1)}^{p-1}{}' \sum_{m_2=-(p-1)}^{p-1}{}' \frac{e^{2\pi i(m_1+m_2k)a/p}}{(\overline{m}_1\overline{m}_2)^\alpha}$$

$$= \frac{C}{p} \sum_{k=1}^{p} \sum_{\substack{m_1=-(p-1) \\ m_1+m_2k\equiv 0 \pmod p}}^{p-1}{}' \sum_{m_2=-(p-1)}^{p-1}{}' \frac{1}{(\overline{m}_1\overline{m}_2)^\alpha} \geqslant Cp^{-1}.$$

Hence there exists a function of $E_s^\alpha(C)$ such that the error term in quadrature formula (7.5) is not less than $Cn^{-1/2}$ and so the error term in (7.5) does not admit further essential improvement.

## § 7.3. the *gp* set and numerical integration

**Lemma 7.3.** *Suppose that* $\alpha \geqslant 1$ *and* $a_i \geqslant 0 (-\infty < i < \infty)$. *If*

$$\sum a_i < \infty,$$

*then*

$$\sum a_i^\alpha \leqslant (\sum a_i)^\alpha.$$

*Proof*. Clearly, the lemma is true if all $a_i = 0$. Now suppose that $\sum a_i > 0$. Then

$$\sum a_i^\alpha = \sum_i \left(\frac{a_i}{\sum_j a_j}\right)^\alpha \left(\sum_k a_k\right)^\alpha$$

$$\leqslant \left(\sum_k a_k\right)^\alpha \sum_i \frac{a_i}{\sum_j a_j} \leqslant (\sum a_k)^\alpha.$$

The lemma is proved.

Let $\alpha$ be a positive number and $l$ be the least integer $\geqslant \alpha$. Let $\mu_{n,l,k}$ be the set of integers defined by

$$\left(\sum_{k=-n}^{n} z^k\right)^l = \sum_{k=-nl}^{nl} \mu_{n,l,k} z^k.$$

**Theorem 7.4.** *Suppose that* $\alpha > 1$. *If* $\boldsymbol{\gamma}$ *is a real vector such that*

$$\langle(\mathbf{m}, \boldsymbol{\gamma})\rangle > b\|\mathbf{m}\|^{-a}$$

*holds for all integral vectors* $\mathbf{m} \neq \mathbf{0}$, *where* $a, b$ *are two constants satisfying* $s \geqslant a \geqslant 1$ *and* $b > 0$, *then*

$$\sup_{f\in E_s^\alpha(C)}\left|\int_{G_s} f(\mathbf{x})d\mathbf{x}-\frac{1}{(2n+1)^l}\sum_{k=-nl}^{nl}\mu_{n,l,k}f(k\boldsymbol{\gamma})\right|$$
$$\leqslant Cc(b,\alpha,s)n^{-\alpha+\frac{s\alpha^2(\alpha-1)}{\alpha-1}}(\ln n)^{\alpha+s\alpha\delta_{1,\alpha}}.$$

*Proof*. For $f\in E_s^\alpha(C)$, we have

$$\frac{1}{(2n+1)^l}\sum_{k=-nl}^{nl}\mu_{n,l,k}f(k\boldsymbol{\gamma})$$
$$=\frac{1}{(2n+1)^l}\sum C(\mathbf{m})\sum_{k=-nl}^{nl}\mu_{n,l,k}e^{2\pi i(\mathbf{m},\gamma)k}$$
$$=C(\mathbf{0})+\frac{1}{(2n+1)^l}\sum{}'C(\mathbf{m})\left(\sum_{k=-n}^{n}e^{2\pi i(\mathbf{m},\gamma)k}\right)^l$$

and so

$$\sup_{f\in E_s^\alpha(C)}\left|\int_{G_s} f(\mathbf{x})d\mathbf{x}-\frac{1}{(2n+1)^l}\sum_{k=-nl}^{nl}\mu_{n,l,k}f(k\boldsymbol{\gamma})\right|$$
$$\leqslant C\sum{}'\frac{1}{\|\mathbf{m}\|^\alpha}\left|\frac{1}{2n+1}\sum_{k=-n}^{n}e^{2\pi i(\mathbf{m},\gamma)k}\right|^l$$
$$=C(\Sigma_1+\Sigma_2),\tag{7.9}$$

where $\Sigma_1$ denotes a sum of $\mathbf{m}$ satisfying $|m_i|\leqslant n^{\frac{\alpha}{\alpha-1}}(1\leqslant i\leqslant s)$ and $\Sigma_2$ the remaining part. Since

$$\left|\sum_{k=-n}^{n}e^{2\pi i(\mathbf{m},\gamma)k}\right|\leqslant\min\left(2n+1,\frac{1}{2\langle(\mathbf{m},\boldsymbol{\gamma})\rangle}\right)$$

by Lemma 3.10, we have

$$\Sigma_1\leqslant\frac{1}{2^\alpha(2n+1)^\alpha}\sum_{|m_i|\leqslant n^{\frac{\alpha}{\alpha-1}}}{}'\frac{1}{\|\mathbf{m}\|^\alpha\langle(\mathbf{m},\boldsymbol{\gamma})\rangle^\alpha}$$
$$\leqslant c(\alpha)n^{-\alpha}\left(\sum_{|m_i|\leqslant n^{\frac{\alpha}{\alpha-1}}}{}'\frac{1}{\|\mathbf{m}\|\,\langle(\mathbf{m},\boldsymbol{\gamma})\rangle}\right)^\alpha$$
$$\leqslant c(b,\alpha,s)n^{-\alpha+\frac{s\alpha^2(\alpha-1)}{\alpha-1}}(\ln n)^{\alpha+s\alpha\delta_{1,\alpha}}\tag{7.10}$$

by Lemmas 3.14 and 7.3. Since

$$\sum_{k>n}\frac{1}{k^\alpha}\leqslant\int_n^\infty\frac{dt}{t^\alpha}=\frac{1}{\alpha-1}n^{-(\alpha-1)},$$

therefore

$$\Sigma_2 \leqslant \sum_{i=1}^{s} \sum_{|m_i| > n^{\frac{\alpha}{\alpha-1}}} \frac{1}{|m_i|^\alpha} \sum \frac{1}{\|\mathbf{m}\|^\alpha} \leqslant c(\alpha, s)n^{-\alpha}. \tag{7.11}$$

The theorem follows from (7.9), (7.10) and (7.11).

**Theorem 7.5.** *Suppose that* $0 < \alpha \leqslant 1$. *Then under the assumption of Theorem* 7.4,

$$\sup_{f \in Q_s^\alpha(C)} \left| \int_{G_s} f(\mathbf{x})d\mathbf{x} - \frac{1}{n} \sum_{k=1}^{n} f(k\boldsymbol{\gamma}) \right|$$
$$\leqslant Cc(b, \alpha, s)n^{-\alpha+s(\alpha-1)} (\ln n)^{s-1+s\delta_{1,\alpha}}.$$

*Proof.* For given $f \in Q_s^\alpha(C)$, let

$$S(f) = \int_{G_s} f(\mathbf{x})d\mathbf{x} - \frac{1}{n} \sum_{k=1}^{n} f(k\boldsymbol{\gamma}).$$

Then by Theorem 6.3,

$$S(f) = \sum{}'' S(\varphi_{\mathbf{t}}),$$

where

$$S(\varphi_{\mathbf{t}}) = \int_{G_s} \varphi_{\mathbf{t}}(\mathbf{x})d\mathbf{x} - \frac{1}{n} \sum_{k=1}^{n} \varphi_{\mathbf{t}}(k\boldsymbol{\gamma}).$$

Hence

$$\sup_{f \in Q_s^\alpha(C)} |S(f)| \leqslant \Sigma_1 + \Sigma_2, \tag{7.12}$$

where

$$\Sigma_1 = \sup_{f \in Q_s^\alpha(C)} \sum_{t_0 \geqslant \log_2 n}'' |S(\varphi_{\mathbf{t}})|$$

and

$$\Sigma_2 = \sup_{f \in Q_s^\alpha(C)} \sum_{t_0 < \log_2 n}'' |S(\varphi_{\mathbf{t}})|.$$

Similar to (7.7), we have

$$\Sigma_1 \leqslant Cc(s) \sum_{t_0 \geqslant \log_2 n}'' 2^{-\alpha t_0} \leqslant Cc(\alpha, s)n^{-\alpha} (\ln n)^{s-1}. \tag{7.13}$$

Since $C_{\mathbf{t}}(\mathbf{m}) = 0$ for $\|\mathbf{m}\| \geqslant 2^{t_0}$, therefore by Lamma 3.10,

$$\sum_{k=1}^{n} \varphi_{\mathbf{t}}(k\boldsymbol{\gamma}) = \sum_{k=1}^{n} \sum_{\|\mathbf{m}\| < 2^{t_0}} C_{\mathbf{t}}(\mathbf{m}) e^{2\pi i(\mathbf{m},\boldsymbol{\gamma})k}$$

$$= nC_{\mathbf{t}}(\mathbf{0}) + \sum_{\|\mathbf{m}\| < 2^{t_0}}' C_{\mathbf{t}}(\mathbf{m}) \sum_{k=1}^{n} e^{2\pi i(\mathbf{m},\boldsymbol{\gamma})k}$$

and so

$$|S(\varphi_t)| \leqslant n^{-1} \sum_{\|\mathbf{m}\|<2^{t_0}}' |C_t(\mathbf{m})| \frac{1}{2\langle(\mathbf{m}, \boldsymbol{\gamma})\rangle}.$$

Hence it follows by (6.16), Theorem 6.2 and Lemma 3.14 that

$$\begin{aligned}
\Sigma_2 &\leqslant \sup_{f\in Q_s^\alpha(C)} n^{-1} \sum_{\|\mathbf{m}\|<n}' \frac{1}{\langle(\mathbf{m}, \boldsymbol{\gamma})\rangle} \Sigma'' |C_t(\mathbf{m})| \\
&\leqslant \sup_{f\in Q_s^\alpha(C)} n^{-1} \sum_{\|\mathbf{m}\|<n}' \frac{|C(\mathbf{m})|}{\langle(\mathbf{m}, \boldsymbol{\gamma})\rangle} \\
&\leqslant Cc(\alpha, s) n^{-1} \sum_{\|\mathbf{m}\|<n}' \frac{1}{\|\mathbf{m}\|^\alpha \langle(\mathbf{m}, \boldsymbol{\gamma})\rangle} \\
&\leqslant Cc(\alpha, s) n^{-1} \sum_{\|\mathbf{m}\|<n}' \frac{n^{1-\alpha}}{\|\mathbf{m}\| \langle(\mathbf{m}, \boldsymbol{\gamma})\rangle} \\
&\leqslant Cc(b, \alpha, s) n^{-\alpha+s(\alpha-1)} (\ln n)^{1+s\delta_{1,\alpha}}. \qquad (7.14)
\end{aligned}$$

The theorem follows by (7.12), (7.13) and (7.14).

For $\alpha = 2$, the weight $\mu_{n,l,k}$ may be simplified. First, we shall state the following lemma.

**Lemma 7.4.** *Let $\delta$ be a real number. Then*

$$\left| \sum_{k=-(n-1)}^{n-1} (n - |k|) e^{2\pi i k\delta} \right| \leqslant \min\left(n^2, \frac{1}{4\langle\delta\rangle^2}\right).$$

*Proof.* Since

$$\sum_{k=-(n-1)}^{n-1} (n - |k|) = \sum_{j=0}^{n-1} \sum_{k=-j}^{j} 1 = n^2$$

and $\sin \pi\delta \geqslant 2\delta$ for $0 \leqslant \delta \leqslant \frac{1}{2}$, so if $\delta$ is not an integer, then

$$\sum_{k=-(n-1)}^{n-1} (n - |k|) e^{2\pi i k\delta} = \sum_{j=0}^{n-1} \sum_{k=-j}^{j} e^{2\pi i k\delta}$$

$$= \frac{1}{\sin \pi\delta} \sum_{j=0}^{n-1} \sin(2j+1)\pi\delta = \left(\frac{\sin n\pi\delta}{\sin \pi\delta}\right)^2 \leqslant \frac{1}{4\langle\delta\rangle^2}.$$

The lemma is proved.

**Theorem 7.6.** *Under the assumption of Theorem 7.4,*

$$\sup_{f\in E_s^2(C)}\left|\int_{G_s} f(\mathbf{x})d\mathbf{x} - \frac{1}{n}\sum_{k=-(n-1)}^{n-1}\left(1-\frac{|k|}{n}\right)f(k\boldsymbol{\gamma})\right|$$
$$\leqslant Cc(b,s)n^{-2+4s(a-1)}(\ln n)^{2+2s\delta_{1,a}}.$$

*Proof.* Since

$$\frac{1}{n}\sum_{k=-(n-1)}^{n-1}\left(1-\frac{|k|}{n}\right)f(k\boldsymbol{\gamma})$$
$$=\frac{1}{n^2}\sum_{k=-(n-1)}^{n-1}(n-|k|)\sum C(\mathbf{m})e^{2\pi i(\mathbf{m},\gamma)k}$$
$$=\frac{1}{n^2}\sum C(\mathbf{m})\sum_{k=-(n-1)}^{n-1}(n-|k|)e^{2\pi i(\mathbf{m},\gamma)k}$$
$$=C(\mathbf{0})+\sum{}'C(\mathbf{m})\frac{1}{n^2}\sum_{k=-(n-1)}^{n-1}(n-|k|)e^{2\pi i(\mathbf{m},\gamma)k},$$

so by Lemma 7.4,

$$\sup_{f\in E_s^2(C)}\left|\int_{G_s} f(\mathbf{x})d\mathbf{x} - \frac{1}{n}\sum_{k=-(n-1)}^{n-1}\left(1-\frac{|k|}{n}\right)f(k\boldsymbol{\gamma})\right|$$
$$\leqslant Cn^{-2}\sum{}'\frac{1}{\|\mathbf{m}\|^2}\min\left(\frac{1}{4\langle(\mathbf{m},\boldsymbol{\gamma})\rangle^2},n^2\right)$$

and so the theorem may be easily proved by a method similar to the proof of Theorem 7.4.

Using the notation of §4.5, it follows by Theorems 4.12, 4.13 and 7.6 that

**Theorem 7.7.** *We have*

$$\sup_{f\in E_s^2(C)}\left|\int_{G_s} f(\mathbf{x})d\mathbf{x} - \frac{1}{n}\sum_{k=-(n-1)}^{n-1}\left(1-\frac{|k|}{n}\right)f(k\boldsymbol{\alpha})\right|$$
$$\leqslant Cc(\boldsymbol{\alpha},\varepsilon)n^{-2+\varepsilon}$$

*and*

$$\sup_{f\in E_s^2(C)}\left|\int_{G_s} f(\mathbf{x})d\mathbf{x} - \frac{1}{n}\sum_{k=-(n-1)}^{n-1}\left(1-\frac{|k|}{n}\right)f(k\boldsymbol{\beta})\right|$$
$$\leqslant Cc(\boldsymbol{\beta},\varepsilon)n^{-2+\varepsilon}.$$

From Theorems 4.12, 4.13, 7.4 and 7.5 we may obtain similar quadrature formulas.

## § 7.4. The lower estimation of the error term for the quadrature formula

**Theorem 7.8.** *For any given sequence of* $G_s$

$$P(k) = (x_1(k), \cdots, x_s(k)), \quad k = 1, 2, \cdots,$$

*the error term of the quadrature formula*

$$\int_{G_s} f(\mathbf{x})d\mathbf{x} - \frac{1}{n}\sum_{k=1}^{n} f(P(k))$$

*for the class of analytic functions on* $G_s$ *can not be better than* $O(n^{-1})$, *where the constant implied by the symbol "O" depends on* $f$ *only.*

*Proof*. Suppose that the theorem is not true. Then there exists a sequence $P(k)(k = 1, 2, \cdots)$ such that

$$\int_{G_s} f(\mathbf{x})d\mathbf{x} - \frac{1}{n}\sum_{k=1}^{n} f(P(k)) = o(n^{-1})$$

holds for any analytic function $f$, where the constant implied by the symbol "$o$" depends on $f$ only. For example, if we take $f(\mathbf{x}) = g(x_1)$, then

$$\sum_{k=1}^{n} g(x_1(k)) = n\int_0^1 g(x)dx + o(1).$$

Hence

$$g(x_1(n)) = \sum_{k=1}^{n} g(x_1(k)) - \sum_{k=1}^{n-1} g(x_1(k)) = \int_0^1 g(x)dx + o(1).$$

In particular, we have

$$\sin(2\pi x_1(n)) = o(1)$$

and

$$\cos(2\pi x_1(n)) = o(1),$$

if we take $g(x) = \sin 2\pi x$ and $g(x) = \cos 2\pi x$ respectively. Hence

$$1 = (\sin(2\pi x_1(n)))^2 + (\cos(2\pi x_1(n)))^2 = o(1).$$

This leads to a contradiction. The theorem follows.

From Theorem 7.8 we know that for any given sequence of points $P(k)(k=1,2,\cdots)$ of $G_s$, if we use the simple sum

$$n^{-1}\sum_{k=1}^{n} f(P(k))$$

to approximate the definite integral on $G_s$, then the error term is comparatively large. The quadrature formula given by Theorem 7.4 is quite precise but its disadvantage is the complicated weight $\mu_{n,l,k}$ which depends on $\alpha$.

## § 7.5. The solutions of congruences and numerical integration

**Theorem 7.9.** *Suppose that $\alpha > 1$. If the congruence*

$$(\mathbf{a}, \mathbf{m}) \equiv 0 \pmod{n} \tag{7.15}$$

*has no solution in the domain*

$$\|\mathbf{m}\| \leqslant M, \quad \mathbf{m} \neq \mathbf{0},$$

*then*

$$\sup_{f\in E_s^\alpha(C)} \left| \int_{G_s} f(\mathbf{x})d\mathbf{x} - \frac{1}{n}\sum_{k=1}^{n} f\left(\frac{k\mathbf{a}}{n}\right)\right| \leqslant Cc(\alpha, \varepsilon)^s M^{-\alpha+\varepsilon}. \tag{7.16}$$

*Proof.* Obviously, we may suppose that $\varepsilon < \alpha - 1$. Since

$$\frac{1}{n}\sum_{k=1}^{n} f\left(\frac{k\mathbf{a}}{n}\right) = \frac{1}{n}\sum_{k=1}^{n}\sum C(\mathbf{m})e^{2\pi i(\mathbf{a},\mathbf{m})k/n}$$

$$= C(\mathbf{0}) + \sum{}' C(\mathbf{m})\frac{1}{n}\sum_{k=1}^{n} e^{2\pi i(\mathbf{a},\mathbf{m})k/n}$$

$$= C(\mathbf{0}) + \sum_{(\mathbf{a},\mathbf{m})\equiv 0 \pmod{n}}{}' C(\mathbf{m}),$$

we have

$$\sup_{f\in E_s^\alpha(C)} \left| \int_{G_s} f(\mathbf{x})d\mathbf{x} - \frac{1}{n}\sum_{k=1}^{n} f\left(\frac{k\mathbf{a}}{n}\right)\right| \leqslant C \sum_{(\mathbf{a},\mathbf{m})\equiv 0 \pmod{n}}{}' \frac{1}{\|\mathbf{m}\|^\alpha}.$$

Let $T_{l,M}$ be the number of solutions of (7.15) in the domain

$$\|\mathbf{m}\| < lM$$

where $l$ is a positive integer. Then

$$T_{l,M} \leqslant c(\varepsilon)^l l^{1+\varepsilon} M^{\varepsilon}$$

by Lemma 3.3. Since

$$l^{-\alpha} - (1+l)^{-\alpha} = \alpha \int_l^{l+1} x^{-\alpha-1} dx \leqslant \alpha l^{-(\alpha+1)},$$

therefore

$$\sum_{(\mathbf{a},\mathbf{m})\equiv 0 \pmod n}{}' \frac{1}{\|\mathbf{m}\|^\alpha} \leqslant \sum_{l=1}^{\infty} \frac{T_{l+1,M} - T_{l,M}}{(lM)^\alpha}$$

$$= \sum_{l=1}^{\infty} T_{l+1,M}(l^{-\alpha} - (l+1)^{-\alpha}) M^{-\alpha}$$

$$\leqslant c(\alpha, \varepsilon)^s M^{-\alpha+\varepsilon} \sum_{l=1}^{\infty} l^{-\alpha+\varepsilon} \leqslant c(\alpha, \varepsilon)^s M^{-\alpha+\varepsilon}.$$

The theorem follows.

**Theorem 7.10.** *Suppose that* $0 < \alpha \leqslant 1$. *Then under the assumption of Theorem* 7.9,

$$\sup_{f \in Q_s^\alpha(C)} \left| \int_{G_s} f(\mathbf{x}) d\mathbf{x} - \frac{1}{n} \sum_{k=1}^{n} f\left(\frac{k\mathbf{a}}{n}\right) \right| \leqslant C c(\alpha, \varepsilon)^s M^{-\alpha+\varepsilon}. \quad (7.17)$$

*Proof.* We may suppose that $\varepsilon < \alpha$. Let

$$S(f) = \int_{G_s} f(\mathbf{x}) d\mathbf{x} - \frac{1}{n} \sum_{k=1}^{n} f\left(\frac{k\mathbf{a}}{n}\right).$$

Then

$$S(f) = \sum{}'' S(\varphi_{\mathbf{t}}),$$

where

$$S(\varphi_{\mathbf{t}}) = \int_{G_s} \varphi_{\mathbf{t}}(\mathbf{x}) d\mathbf{x} - \frac{1}{n} \sum_{k=1}^{n} \varphi_{\mathbf{t}}\left(\frac{k\mathbf{a}}{n}\right).$$

Hence

$$\sup_{f \in Q_s^\alpha(C)} |S(f)| \leqslant \Sigma_1 + \Sigma_2,$$

where

$$\Sigma_1 = \sup_{f \in Q_s^\alpha(C)} \sum_{t_0 \geqslant \log_2 M}{}'' |S(\varphi_{\mathbf{t}})|$$

and

$$\Sigma_2 = \sup_{f \in Q_s^\alpha(C)} \sum_{t_0 < \log_2 M}{}'' |S(\varphi_{\mathbf{t}})|.$$

Similar to (7.7), we have

$$\Sigma_1 \leqslant 2C \sum_{t_0 \geqslant \log_2 M}'' 2^{-\alpha t_0} = 2C \sum_{t_0 \geqslant \log_2 M}'' 2^{-(\alpha-\varepsilon)t_0 - \varepsilon t_0}$$

$$\leqslant 2CM^{-\alpha+\varepsilon}\left(\sum_{t=0}^{\infty} 2^{-\varepsilon t}\right)^s = Cc(\alpha, \varepsilon)^s M^{-\alpha+\varepsilon}.$$

Since $C_{\mathbf{t}}(\mathbf{m}) = 0$ for $\|\mathbf{m}\| \geqslant 2^{t_0}$ and

$$|S(\varphi_{\mathbf{t}})| = \left|\sum{}' C_{\mathbf{t}}(\mathbf{m}) \frac{1}{n} \sum_{k=1}^{n} e^{2\pi i(\mathbf{a}, \mathbf{m})k/n}\right|$$

$$\leqslant \sum_{(\mathbf{a},\mathbf{m}) \equiv 0 \, (\mathrm{mod}\, n)}' |C_{\mathbf{t}}(\mathbf{m})|,$$

hence

$$\Sigma_2 \leqslant \sup_{f \in Q_s^\alpha(C)} \sum_{(\mathbf{a},\mathbf{m}) \equiv 0 \, (\mathrm{mod}\, n)}' \sum_{t_0 < \log_2 M}'' |C_{\mathbf{t}}(\mathbf{m})| = 0.$$

The theorem follows.

If we use Lemma 3.5 instead of Lemma 3.3 in the proof of Theorem 7.9 and the Lemmas 7.1 and 7.2 to evaluate $\Sigma_1$ in the proof of Theorem 7.10, then we have

**Theorem 7.11.** *Under the assumptions of Theorems* 7.9 *and* 7.10, *the right hand sides of* (7.16) *and* (7.17) *may be replaced by*

$$Cc(\alpha, s) M^{-\alpha} (\ln 3M)^{s-1}.$$

Using the notations of §4.6 and §4.7, we have the following two lemmas by (1.14), Theorems 4.12 and 2.8 and Lemma 3.9.

**Lemma 7.5.** *The congruence*

$$c_1 m_1 + c_2 m_2 + \cdots + c_s m_s \equiv 0 \, (\mathrm{mod}\, n), \qquad s = \frac{\varphi(m)}{2}$$

*has no solution in the domain*

$$\|\mathbf{m}\| \leqslant c(\mathscr{R}_s, \varepsilon) n^{\frac{1}{2} + \frac{1}{2(s-1)} - \varepsilon}, \quad \mathbf{m} \neq \mathbf{0}.$$

**Lemma 7.6.** *The congruence*

$$m_1 + F_n(2) m_2 + \cdots + F_n(s) m_s \equiv 0 \, (\mathrm{mod}\, F_n)$$

*has no solution in the domain*

$$\|\mathbf{m}\| \leqslant c(\eta) F_n^{\frac{1}{2}+\frac{1}{2^s+1}\frac{1}{\ln 2}+\frac{1}{2^{2s+3}}}, \quad \mathbf{m} \neq \mathbf{0}.$$

From Lemmas 7.5 and 7.6, and Theorem 7.9, we can derive

**Theorem 7.12.** *Suppose that* $\alpha > 1$. *Then*

$$\sup_{f \in E_s^\alpha(C)} \left| \int_{G_s} f(\mathbf{x}) d\mathbf{x} - \frac{1}{n} \sum_{k=1}^{n} f\left(\frac{c_1 k}{n}, \frac{c_2 k}{n}, \cdots, \frac{c_s k}{n}\right) \right|$$
$$\leqslant C c(\mathscr{R}_s, \alpha, \varepsilon) n^{-\frac{\alpha}{2}-\frac{\alpha}{2(s-1)}+\varepsilon}, \quad s = \frac{\varphi(m)}{2},$$

*and*

$$\sup_{f \in E_s^\alpha(C)} \left| \int_{G_s} f(\mathbf{x}) d\mathbf{x} - \frac{1}{F_n} \sum_{k=1}^{F_n} f\left(\frac{k}{F_n}, \frac{F_n(2) k}{F_n}, \cdots, \frac{F_n(s) k}{F_n}\right) \right|$$
$$\leqslant C c(\eta, \alpha) F_n^{-\frac{\alpha}{2}-\frac{\alpha}{2^s+1}\frac{1}{\ln 2}-\frac{\alpha}{2^{2s+4}}}. \tag{7.18}$$

Using Theorems 7.11 and 2.9 instead of Theorems 7.9 and 2.8 respectively, we have

**Theorem 7.13.** *The right hand side of* (7.18) *can be replaced by* $C c(\eta, \alpha) F_n^{-\alpha} \ln 3F_n$ *for* $s = 2$ *and* $C c(\eta, \alpha, \varepsilon) F_n^{-3/4+\varepsilon}$ for $s = 3$ *respectively.*

we may also obtain the corresponding quadrature formula for the class of functions $Q_s^\alpha(C)$ by the use of Theorem 7.10 instead of Theorem 7.9.

## § 7.6. The *glp* set and numerical integration

**Lemma 7.7.** *Suppose that* $\alpha > 1$ *and* $N \geqslant 1$. *Then*

$$\sum_{\|\mathbf{m}\| \leqslant N} 1 \leqslant 3^s N (\ln 3N)^{s-1} \tag{7.19}$$

*and*

$$\sum_{\|\mathbf{m}\| \geqslant N} \frac{1}{\|\mathbf{m}\|^\alpha} \leqslant (5\zeta(\alpha))^s N^{-\alpha+1} (\ln 3N)^{s-1}. \tag{7.20}$$

*Proof*. (7.19) is obviously true for $s = 1$. Now suppose that $k \geqslant 1$ and (7.19) holds for $s \leqslant k$. Then

$$\sum_{\bar{m}_1\cdots\bar{m}_{k+1}\leqslant N} 1 = \sum_{\bar{m}_1\leqslant N}\ \sum_{\bar{m}_2\cdots\bar{m}_{k+1}\leqslant N/\bar{m}_1} 1 \leqslant 3^k N(\ln 3N)^{k-1}\sum_{\bar{m}_1\leqslant N}\frac{1}{\bar{m}}$$
$$< 3^{k+1}N(\ln 3N)^k.$$

Hence (7.19) follows by mathematical induction.

For $s=1$,

$$\sum_{\bar{m}\geqslant N}\frac{1}{\bar{m}^\alpha} = 2\sum_{m\geqslant N}\frac{1}{m^\alpha} \leqslant \frac{2}{N^\alpha} + 2\int_N^\infty \frac{dt}{t^\alpha}$$
$$= \frac{2}{N^\alpha} + \frac{2}{(\alpha-1)N^{\alpha-1}} < 5\zeta(\alpha)N^{-\alpha+1}.$$

Now suppose that $k\geqslant 1$ and (7.20) holds for $s\leqslant k$. Then

$$\sum_{\bar{m}_1\cdots\bar{m}_{k+1}\geqslant N}\frac{1}{(\bar{m}_1\cdots\bar{m}_{k+1})^\alpha} = \sum_{\bar{m}_1\leqslant N}\frac{1}{\bar{m}_1^\alpha}\sum_{\bar{m}_2\cdots\bar{m}_{k+1}\geqslant N/\bar{m}_1}\frac{1}{(\bar{m}_2\cdots\bar{m}_{k+1})^\alpha}$$
$$+\sum_{\bar{m}_1>N}\frac{1}{\bar{m}_1^\alpha}\sum_{\bar{m}_2\cdots\bar{m}_{k+1}\geqslant 1}\frac{1}{(\bar{m}_2\cdots\bar{m}_{k+1})^\alpha}$$
$$< (5\zeta(\alpha))^k N^{-\alpha+1}(\ln 3N)^{k-1}\sum_{\bar{m}_1\leqslant N}\frac{1}{\bar{m}_1} + (3\zeta(\alpha))^k\sum_{\bar{m}_1>N}\frac{1}{\bar{m}_1^\alpha}$$
$$< 3(5\zeta(\alpha))^k N^{-\alpha+1}(\ln 3N)^k + (3\zeta(\alpha))^{k+1}N^{-\alpha+1}$$
$$< (5\zeta(\alpha))^{k+1}N^{-\alpha+1}(\ln 3N)^k.$$

The lemma follows.

**Lemma 7.8.** *Suppose that* $0<\delta<1$. *Then there exist no less than* $p-[\delta p]$ *integers in the interval* $1\leqslant a\leqslant p$ *such that the congruence*

$$(\mathbf{a},\mathbf{m}) = m_1 + m_2 a + \cdots + m_s a^{s-1} \equiv 0 \pmod{p} \tag{7.21}$$

*has no solution in the domain*

$$\|\mathbf{m}\| \leqslant \delta s^{-1}3^{-s}p(\ln 3p)^{-(s-1)}, \quad \mathbf{m}\neq \mathbf{0}. \tag{7.22}$$

*Proof.* We may suppose that $\delta s^{-1}3^{-s}p(\ln 3p)^{-(s-1)}\geqslant 1$. Otherwise the lemma evidently holds. Since the number of solutions of the congruence (7.21) is at most $s-1$ in the interval $1\leqslant a\leqslant p$ for any given $\mathbf{m}$ belonging to (7.22) (Cf. Lemma 4.5), the total number of solutions of the congruence (7.21) with $\mathbf{m}$ satisfying (7.22) is at most

$$\sum_{\|\mathbf{m}\|\leqslant\delta s^{-1}3^{-s}p(\ln 3p)^{-(s-1)}}\ \sum_{\substack{(\mathbf{a},\mathbf{m})\equiv 0\,(\mathrm{mod}\,p)\\ 1\leqslant a\leqslant p}} 1$$
$$\leqslant (s-1)3^s(\ln 3p)^{s-1}\delta s^{-1}3^{-s}p(\ln 3p)^{-(s-1)} < \delta p.$$

Hence there exist at least $p-[\delta p]$ integers in the interval $1 \leqslant a \leqslant p$ such that the congruence (7.21) has no solution in the domain (7.22). The lemma is proved.

**Theorem 7.14.** *There exists an integral vector* $\mathbf{a}=(a_1,\cdots,a_s)$ *depending only on* $p$ *such that*

$$\sup_{f\in E_s^\alpha(C)}\left|\int_{G_s} f(\mathbf{x})d\mathbf{x}-\frac{1}{p}\sum_{k=1}^{p} f\left(\frac{k\mathbf{a}}{p}\right)\right| < Cc(\alpha,s)p^{-\alpha}(\ln p)^{(\alpha+1)(s-1)},\quad \alpha>1$$

*and*

$$\sup_{f\in Q_s^\alpha(C)}\left|\int_{G_s} f(\mathbf{x})d\mathbf{x}-\frac{1}{p}\sum_{k=1}^{p} f\left(\frac{k\mathbf{a}}{p}\right)\right| < Cc(\alpha,s)p^{-\alpha}(\ln p)^{(\alpha+1)(s-1)},\quad 0<\alpha\leqslant 1.$$

*Proof.* Let $M=2^{-1}s^{-1}3^{-s}p(\ln 3p)^{-(s-1)}$. Let $A$ denote the set of integers in the interval $1\leqslant a\leqslant p$ such that the congruence (7.21) has no solution in the domain

$$\|\mathbf{m}\|\leqslant M,\quad \mathbf{m}\neq\mathbf{0}.$$

Then it follows from Lemma 7.8 that the number of elements of $A$ is no less than $\frac{p+1}{2}$. Let $a_1=1, a_2=a,\cdots,a_s=a^{s-1}$, where $a\in A$. Then the theorem follows from Theorem 7.11.

We shall give another proof of Theorem 7.14 with a slight modification of error term in the following.

**Theorem 7.15.** *Suppose that* $\alpha>1$. *Then there exists an integral vector* $\mathbf{a}(=\mathbf{a}(p))$ *such that*

$$\sup_{f\in E_s^\alpha(C)}\left|\int_{G_s} f(\mathbf{x})d\mathbf{x}-\frac{1}{p}\sum_{k=1}^{p} f\left(\frac{k\mathbf{a}}{p}\right)\right| < Cc(\alpha,s)p^{-\alpha}(\ln p)^{\alpha(s-1)}.$$

*Proof.* The notations introduced in Theorem 7.14 are also used here. Let $\mathbf{a}=(1,a,\cdots,a^{s-1})$ and

$$\Omega(a)=\sum_{(\mathbf{a},\mathbf{m})\equiv 0(\mathrm{mod}\,p)}\frac{1}{\|\mathbf{m}\|^\alpha}.$$

Then

$$\sup_{f\in E_s^\alpha(C)}\left|\int_{G_s} f(\mathbf{x})d\mathbf{x}-\frac{1}{p}\sum_{k=1}^{p} f\left(\frac{k\mathbf{a}}{p}\right)\right|\leqslant C\Omega(a) \tag{7.23}$$

and by Lemma 7.7,

$$\begin{aligned}
\sum_{a\in A}\Omega(a) &= \sum_{a\in A}\ \sum_{(\mathbf{a},\mathbf{m})\equiv 0\,(\mathrm{mod}\,p)}{}' \frac{1}{\|\mathbf{m}\|^\alpha}\\
&\leqslant \sum_{\|\mathbf{m}\|\geqslant M}\ \sum_{\substack{(\mathbf{a},\mathbf{m})\equiv 0\,(\mathrm{mod}\,p)\\ 1\leqslant a\leqslant p}} \frac{1}{\|\mathbf{m}\|^\alpha}\\
&\leqslant p\sum{}' \frac{1}{\|p\mathbf{m}\|^\alpha}+(s-1)\sum_{\|\mathbf{m}\|\geqslant M}\frac{1}{\|\mathbf{m}\|^\alpha}\\
&< s(2\zeta(\alpha)+1)^s p^{-\alpha+1}+s(5\zeta(\alpha))^s M^{-\alpha+1}(\ln 3p)^{s-1}\\
&< \frac{1}{3}(2s)^\alpha(3^\alpha 5\zeta(\alpha))^s p^{-\alpha+1}(\ln 3p)^{\alpha(s-1)}.
\end{aligned} \tag{7.24}$$

There are at most $[p/3]$ elements of $A$ such that the corresponding $\Omega(a)$ satisfies

$$\Omega(a)\geqslant(2s)^\alpha(3^\alpha 5\zeta(\alpha))^s p^{-\alpha}(\ln 3p)^{\alpha(s-1)}.$$

Otherwise we have

$$\begin{aligned}
\sum_{a\in A}\Omega(a) &\geqslant\left(\left[\frac{p}{3}\right]+1\right)(2s)^\alpha 3^{\alpha s}(5\zeta(\alpha))^s p^{-\alpha}(\ln 3p)^{\alpha(s-1)}\\
&>\frac{1}{3}(2s)^\alpha(3^\alpha 5\zeta(\alpha))^s p^{-\alpha+1}(\ln 3p)^{\alpha(s-1)}
\end{aligned}$$

which leads to a contradiction with (7.24). Since

$$\frac{p+1}{2}-\left[\frac{p}{3}\right]\geqslant 1,$$

so there exists at least an integer $a$ of $A$ such that

$$\Omega(a)<(2s)^\alpha(3^\alpha 5\zeta(\alpha))^s p^{-\alpha}(\ln 3p)^{\alpha(s-1)} \tag{7.25}$$

and the theorem follows from (7.23) and (7.25).

**Theorem 7.16.** *Suppose that* $0<\alpha\leqslant 1$. *Then there exists an integral vector* $\mathbf{a}(=\mathbf{a}(p))$ *such that*

$$\sup_{f\in Q_s^\alpha(C)}\left|\int_{G_s} f(\mathbf{x})d\mathbf{x}-\frac{1}{p}\sum_{k=1}^{p} f\left(\frac{k\mathbf{a}}{p}\right)\right|$$
$$< Cc(\alpha, s, \varepsilon)p^{-\alpha}(\ln p)^{s-1+\varepsilon}.$$

*Proof.* By (7.7), we have

$$\sup_{f\in Q_s^\alpha(C)}\left|\int_{G_s} f(\mathbf{x})d\mathbf{x}-\frac{1}{p}\sum_{k=1}^{p} f\left(\frac{k\mathbf{a}}{p}\right)\right|$$
$$\leqslant Cc(\alpha, s)p^{-\alpha}(\ln p)^{s-1}+\Lambda(a),$$

where $\mathbf{a}=(1, a, \cdots, a^{s-1})$ and

$$\Lambda(a)=\sup_{f\in Q_s^\alpha(C)}\sum_{(\mathbf{a},\mathbf{m})\equiv 0(\mathrm{mod}\, p)}{}' \sum_{t_0<\log_2 p}{}''|C_{\mathbf{t}}(\mathbf{m})|.$$

By Theorem 6.2,

$$\sum_{a\in A}\Lambda(a)=\sup_{f\in Q_s^\alpha(C)}\sum_{a\in A}\sum_{(\mathbf{a},\mathbf{m})\equiv 0(\mathrm{mod}\, p)}{}' \sum_{t_0<\log_2 p}{}''|C_{\mathbf{t}}(\mathbf{m})|$$
$$\leqslant \sup_{f\in Q_s^\alpha(C)}\sum_{M<\|\mathbf{m}\|<p}\sum_{\substack{(\mathbf{a},\mathbf{m})\equiv 0(\mathrm{mod}\, p)\\ 1\leqslant a\leqslant p}}\Sigma''|C_{\mathbf{t}}(\mathbf{m})|$$
$$\leqslant \sup_{f\in Q_s^\alpha(C)}(s-1)\sum_{M<\|\mathbf{m}\|<p}|C(\mathbf{m})|\leqslant Cc(\alpha, s)\sum_{M<\|\mathbf{m}\|<p}\frac{1}{\|\mathbf{m}\|^\alpha}$$
$$=Cc(\alpha, s)\sum_{M<\|\mathbf{m}\|<p}\frac{\|\mathbf{m}\|^{1-\alpha+\varepsilon/s}}{\|\mathbf{m}\|^{1+\varepsilon/s}}$$
$$\leqslant Cc(\alpha, s)p^{1-\alpha+\varepsilon/s}\sum_{\|\mathbf{m}\|>M}\frac{1}{\|\mathbf{m}\|^{1+\varepsilon/s}}.$$

Hence

$$\sum_{a\in A}\Lambda(a)\leqslant Cc(\alpha, s, \varepsilon)p^{1-\alpha+\varepsilon/s}M^{-\varepsilon/s}(\ln M)^{s-1}$$
$$\leqslant Cc(\alpha, s, \varepsilon)p^{1-\alpha}(\ln p)^{s-1+\varepsilon}$$

by Lemma 7.7 and so there exists an integer $a$ of $A$ such that

$$\Lambda(a)\leqslant\frac{2}{p+1}\sum_{a\in A}\Lambda(a)\leqslant Cc(\alpha, s, \varepsilon)p^{-\alpha}(\ln p)^{s-1+\varepsilon}.$$

The theorem follows.

## § 7.7. The Sarygin theorem

**Theorem 7.17.** *For any given $n$ points of $G_s$*

$$P(k) = (x_1^{(k)}, \cdots, x_s^{(k)}), \quad k = 1, \cdots, n,$$

*there exists $f \in E_s^\alpha(C)$ such that*

$$f(P(k)) = 0, \quad k = 1, \cdots, n$$

*and*

$$\int_{G_s} f(\mathbf{x})d\mathbf{x} \geqslant Cc(\alpha, s)n^{-\alpha}(\ln n)^{s-1}.$$

*Proof.* Let $t$ denote the integer satisfying $2^{t-1} \leqslant n < 2^t$. Then we add any $2^t - n$ points

$$P(k) = (x_1^{(k)}, \cdots, x_s^{(k)}), \quad k = n+1, \cdots, 2^t$$

and consider the integral vectors

$$\mathbf{r} = (r_1, \cdots, r_s),$$

where $r_1 + \cdots + r_s = t, r_i \geqslant 0 (1 \leqslant i \leqslant s)$. Let $M(\mathbf{r})$ denote the set of integral vectors $\mathbf{m}$ such that

$$\overline{m}_i \leqslant 2^{r_i}, \quad i = 1, \cdots, s.$$

Then the number of elements of $M(\mathbf{r})$ is no less than $2^{t+1} + 1$. Now we shall prove that for any given $\mathbf{r}$, there exists a trigonometrical polynomial

$$T_{\mathbf{r}}(\mathbf{x}) = \sum_{\mathbf{m} \in M(\mathbf{r})} C_{\mathbf{r}}(\mathbf{m}) e^{2\pi i(\mathbf{m}, \mathbf{x})} \not\equiv 0 \tag{7.26}$$

such that

$$T_{\mathbf{r}}(P(k)) = 0, \quad k = 1, \cdots, 2^t. \tag{7.27}$$

Indeed, since (7.27) is a system of linear equations of the Fourier coefficients $C_{\mathbf{r}}(\mathbf{m})$ and the number of unknowns is greater then the number of equations, there exist $C_{\mathbf{r}}(\mathbf{m})$'s not all zero such that $T_{\mathbf{r}}(\mathbf{x})$ satisfies (7.26) and (7.27).

Let

$$T_{\mathbf{r}}^{(0)}(\mathbf{x}) = \frac{T_{\mathbf{r}}(\mathbf{x}) e^{-2\pi i(\mathbf{m}', \mathbf{x})}}{C_{\mathbf{r}}(\mathbf{m}') 2^{\alpha(t+s)}},$$

where $\mathbf{m}' = (m_1', \cdots, m_s')$ and

$$|C_{\mathbf{r}}(\mathbf{m}')| = \max_{\mathbf{m} \in M(\mathbf{r})} |C_{\mathbf{r}}(\mathbf{m})|.$$

Now we proceed to prove that we may choose constant $\chi = c(\alpha, s)$ such that

$$f(\mathbf{x}) = C\chi \sum_{\mathbf{r}} T_{\mathbf{r}}^{(0)}(\mathbf{x}) \in E_s^\alpha(C).$$

Since $e^{2\pi i(\mathbf{m},\mathbf{x})}$ may appear only in those $T_{\mathbf{r}}^{(0)}(\mathbf{x})$ with

$$\overline{m}_1 \leqslant 2^{r_1+1}, \cdots, \quad \overline{m}_s \leqslant 2^{r_s+1},$$

so $r_1$ satisfies

$$\begin{aligned} \log_2 \overline{m}_1 - 1 \leqslant r_1 &= t - r_2 - \cdots - r_s \\ &\leqslant t - \log_2 \overline{m}_2 - \cdots - \log_2 \overline{m}_s + s - 1 \end{aligned}$$

and so the values that may be taken by $r_1$ and also the other $r_i$'s do not exceed

$$t - \log_2 \overline{m}_1 - \cdots - \log_2 \overline{m}_s + s + 1 = \log_2 \frac{2^t}{\|\mathbf{m}\|} + s + 1.$$

Consequently, the number of polynomials $T_{\mathbf{r}}^{(0)}(\mathbf{x})$ which contain the term $e^{2\pi i(\mathbf{m},\mathbf{x})}$ is at most

$$\left(\log_2 \frac{2^{t+s}}{\|\mathbf{m}\|} + 1\right)^s.$$

Take

$$\chi = \inf_{0<y\leqslant 1} y^{-\alpha} \left(\log_2 \frac{1}{y} + 1\right)^{-s}.$$

Then $\chi = c(\alpha, s)$ and

$$C\chi \frac{\left(\log_2 \dfrac{2^{t+s}}{\|\mathbf{m}\|} + 1\right)^s}{2^{\alpha(t+s)}} \leqslant C\|\mathbf{m}\|^{-\alpha}.$$

Hence $f \in E_s^\alpha(C)$.

By (7.27), we have

$$f(P(k)) = 0, \quad k = 1, 2, \cdots, 2^t$$

and by Lemma 7.1,

$$\begin{aligned} \int_{G_s} f(\mathbf{x}) d\mathbf{x} &= C\chi 2^{-\alpha(t+s)} \sum_{\mathbf{r}} 1 = C\chi 2^{-\alpha(t+s)} C_{s-1}^{t+s-1} \\ &\geqslant Cc(\alpha, s) n^{-\alpha} (\ln n)^{s-1}. \end{aligned}$$

The theorem is proved.

We know by Theorem 7.17 that the error terms for the quadrature formulas given by Theorems 7.4 and 7.15 are of the best possible kind in principal order and the error term given by Theorem 7.13 for $s=2$ is of the best possible kind apart from some possible improvement about the constant $c(\eta, \alpha)$.

## § 7.8. The mean error of the quadrature formula

Let

$$c(s, \varepsilon) = 2((2\zeta(1+\varepsilon)+1)^s - 1),$$

where $0 < \varepsilon < 1$. Let $\Omega(=\Omega(\varepsilon))$ denote the set of points $\boldsymbol{\gamma}$ of $G_s$ such that the inequality

$$\langle(\mathbf{m}, \boldsymbol{\gamma})\rangle \geqslant \varepsilon c(s, \varepsilon)^{-1} \|\mathbf{m}\|^{-1-\varepsilon}$$

holds for any integral vector $\mathbf{m} \neq \mathbf{0}$. Then by Theorem 4.10, the Lebesgue measure of $\Omega$ satisfies

$$\operatorname{mes} \Omega > 1 - \varepsilon.$$

Let

$$S(n, \Omega, f) = \int_{\Omega} |S(n, \boldsymbol{\gamma}, f)| \, d\boldsymbol{\gamma},$$

where $f \in Q_s^\alpha(C)$ $\left(\alpha > \frac{1}{2}\right)$ and

$$S(n, \boldsymbol{\gamma}, f) = \int_{G_s} f(\mathbf{x}) d\mathbf{x} - \frac{1}{(2n+1)^l} \sum_{k=-nl}^{nl} \mu_{n,l,k} f(k\boldsymbol{\gamma}),$$

in which $l$ is the least integer $\geqslant \alpha + 1$ and $\mu_{n,l,k}$ is defined in §7.3. $S(n, \Omega, f)$ is called the mean error of the quadrature formula given by good points.

**Theorem 7.18.** *Suppose that* $\alpha > \frac{1}{2}$. *Then*

$$\sup_{f \in Q_s^\alpha(C)} S(n, \Omega, f) \leqslant C c(\alpha, s, \varepsilon) n^{-\alpha - \frac{1}{2} + \varepsilon}.$$

To proof Theorem 7.18, we shall need

**Lemma 7.9.** *Suppose that* $\alpha > \frac{1}{2}$ *and* $f \in Q_s^\alpha(C)$. *Then*

$$\sum |C(\mathbf{m})|^2 \|\mathbf{m}\|^{2(\alpha-\varepsilon)} \leqslant C^2 c(s, \varepsilon),$$

*where* $C(\mathbf{m})$ *denotes the Fourier coefficient of* $f$.

*Proof.* Since

$$\sum_{\mathbf{m}} |C_{\mathbf{t}}(\mathbf{m})|^2 = \|\varphi_{\mathbf{t}}\|_2^2 \leqslant \|\varphi_{\mathbf{t}}\|^2 \leqslant C^2 2^{-2\alpha t_0}$$

and $C_{\mathbf{t}}(\mathbf{m}) = 0$ for $\|\mathbf{m}\| \geqslant 2^{t_0}$, therefore

$$\sum_{\mathbf{m}} |C_{\mathbf{t}}(\mathbf{m})|^2 \|\mathbf{m}\|^{2(\alpha-\varepsilon)} \leqslant 2^{2(\alpha-\varepsilon)t_0} \sum_{\mathbf{m}} |C_{\mathbf{t}}(\mathbf{m})|^2 \leqslant C^2 2^{-2\varepsilon t_0}$$

and so

$$\begin{aligned}
\sum_{\mathbf{m}} |C(\mathbf{m})|^2 \|\mathbf{m}\|^{2(\alpha-\varepsilon)} &= \sum_{\mathbf{m}} \left| \sum_{\mathbf{t}}'' C_{\mathbf{t}}(\mathbf{m}) \right|^2 \|\mathbf{m}\|^{2(\alpha-\varepsilon)} \\
&\leqslant \sum_{\mathbf{m}} \left( \sum_{\mathbf{t}}'' |C_{\mathbf{t}}(\mathbf{m})|^2 \, 2^{\varepsilon t_0} \right) \left( \sum_{\mathbf{t}}'' 2^{-\varepsilon t_0} \right) \|\mathbf{m}\|^{2(\alpha-\varepsilon)} \\
&\leqslant c(s, \varepsilon) \sum_{\mathbf{t}}'' 2^{\varepsilon t_0} \sum_{\mathbf{m}} |C_{\mathbf{t}}(\mathbf{m})|^2 \|\mathbf{m}\|^{2(\alpha-\varepsilon)} \\
&\leqslant C^2 c(s, \varepsilon) \sum_{\mathbf{t}}'' 2^{-\varepsilon t_0} = C^2 c(s, \varepsilon).
\end{aligned}$$

The lemma is proved.

The proof of Theorem 7.18. We may suppose that $\varepsilon < \alpha - \frac{1}{2}$. Then by the argument of §7.3,

$$|S(n, \boldsymbol{\gamma}, f)| \leqslant \sum{}' \frac{|C(\mathbf{m})|}{(2n+1)^l} \left| \sum_{k=-n}^{n} e^{2\pi i (\mathbf{m}, \gamma) k} \right|^l.$$

Hence it follows by Schwarz's inequality that

$$S(n, \Omega, f) \leqslant \int_{\Omega} S_1^{1/2} S_2^{1/2} d\boldsymbol{\gamma}, \tag{7.28}$$

where

$$S_1 = \sum{}' |C(\mathbf{m})|^2 \frac{\|\mathbf{m}\|^{2(\alpha-\varepsilon/2)}}{(2n+1)^2} \left| \sum_{k=-n}^{n} e^{2\pi i (\mathbf{m}, \gamma) k} \right|^2$$

and

$$S_2 = \sum{}' \frac{1}{(2n+1)^{2(l-1)} \|\mathbf{m}\|^{2(\alpha-\varepsilon/2)}} \left| \sum_{k=-n}^{n} e^{2\pi i (\mathbf{m}, \gamma) k} \right|^{2(l-1)}$$

By (6.9) and Lemma 7.9,

$$\int_{G_s} S_1 d\boldsymbol{\gamma} = \sum{}' |C(\mathbf{m})|^2 \frac{\|\mathbf{m}\|^{2(\alpha-\varepsilon/2)}}{(2n+1)^2} \int_{G_s} \left| \sum_{k=-n}^{n} e^{2\pi i(\mathbf{m},\boldsymbol{\gamma})k} \right|^2 d\boldsymbol{\gamma}$$

$$= \sum{}' |C(\mathbf{m})|^2 \frac{\|\mathbf{m}\|^{2(\alpha-\varepsilon/2)}}{(2n+1)^2} \int_{G_s} \left( \sum_{k=-n}^{n} e^{2\pi i(\mathbf{m},\boldsymbol{\gamma})k} \right) \left( \sum_{j=-n}^{n} e^{-2\pi i(\mathbf{m},\boldsymbol{\gamma})j} \right) d\boldsymbol{\gamma}$$

$$= \frac{1}{2n+1} \sum{}' |C(\mathbf{m})|^2 \|\mathbf{m}\|^{2(\alpha-\varepsilon/2)} \leqslant C^2 c(s,\varepsilon) n^{-1}.$$

Since $2(l-1) \geqslant 2\alpha > 1$ and $2\alpha - \varepsilon > 1$, hence in a way similar to the proof of Theorem 7.4,

$$S_2 \leqslant c(\alpha, s, \varepsilon) n^{-2\alpha+2\varepsilon}$$

and so by (7.28) and Schwarz's inequality, we have

$$S(n, \Omega, f) \leqslant c(\alpha, s, \varepsilon) n^{-\alpha+\varepsilon} \int_{G_s} S_1^{1/2} d\boldsymbol{\gamma}$$

$$\leqslant c(\alpha, s, \varepsilon) n^{-\alpha+\varepsilon} \left( \int_{G_s} S_1 d\boldsymbol{\gamma} \right)^{1/2}$$

$$\leqslant Cc(\alpha, s, \varepsilon) n^{-\alpha-\frac{1}{2}+\varepsilon}.$$

The theorem is proved.

We know from Theorem 7.18 that the mean error of the quadrature formula given by the good points is better by a factor $O(n^{-\frac{1}{2}})$ compared with the error term of the quadrature formula given by Theorem 7.4 for the class of functions $Q_s^\alpha(C)$ $\left(\alpha > \frac{1}{2}\right)$. Hence we may expect that perhaps the error term of the quadrature formula given by good points is $O(n^{-\alpha-\frac{1}{2}+\varepsilon})$.

## §7.9. Continuation

Suppose that $0 < \varepsilon < \frac{1}{2}$. Then it follows by Lemma 7.8 that for any prime $p$ in the interval $n/2 < p \leqslant n$, there exist no less than $p - [\varepsilon p] \geqslant (1-\varepsilon)p$ integral vectors $\mathbf{a} = (1, a, \cdots, a^{s-1})(1 \leqslant a \leqslant p)$ such that the non-trivial solutions $\mathbf{m}$ of the congruence

$$(\mathbf{a}, \mathbf{m}) \equiv 0 \pmod{p}$$

satisfy

$$\|\mathbf{m}\| \geqslant \varepsilon s^{-1} 3^{-s} p\,(\ln 3p)^{-(s-1)} \geqslant \varepsilon 2^{-1} s^{-1} 3^{-s} n\,(\ln 3n)^{-(s-1)} = M \text{ (say)}.$$

Let $\omega(=\omega(\varepsilon, n))$ be the set of these $(\mathbf{a}, p)$. Denote the number of elements of $\omega$ by $|\omega|$. Then by the prime number theorem (Cf. Hua Loo Keng [2], Chap. 9),

$$|\omega| \geqslant \sum_{\frac{n}{2}<p\leqslant n} (1-\varepsilon)p \geqslant \frac{cn^2}{\ln n}. \tag{7.29}$$

Let

$$S(n, \omega, f) = \frac{1}{|\omega|} \sum_{(\mathbf{a}, p)\in\omega} |S(p, \mathbf{a}, f)|,$$

where $f \in Q_s^\alpha(C)\left(\alpha > \frac{1}{2}\right)$ and

$$S(p, \mathbf{a}, f) = \int_{G_s} f(x)dx - \frac{1}{p}\sum_{k=1}^{p} f\left(\frac{k\mathbf{a}}{p}\right).$$

$S(n, \omega, f)$ is called the mean error of the quadrature formula given by good lattice points.

**Theorem 7.19.** *Suppose that* $\alpha > \frac{1}{2}$. *Then*

$$\sup_{f\in Q_s^\alpha(C)} S(n, \omega, f) \leqslant Cc(\alpha, s, \varepsilon) n^{-\alpha-\frac{1}{2}+\varepsilon}.$$

To prove Theorem 7.19, we shall need

**Lemma 7.10.**

$$A(\omega, \mathbf{m}) = \sum_{\substack{(\mathbf{a}, p)\in\omega \\ (\mathbf{a}, \mathbf{m})\equiv 0 \pmod p}} 1 \leqslant n \log_{\frac{n}{2}}(s\|\mathbf{m}\| n^{s-1}).$$

*Proof.* Let $\sigma_m(l)$ be the number of prime divisors of $l$ which are $\geqslant m$. Then

$$\sigma_m(l) \leqslant \log_m l.$$

Hence

$$A(\omega, \mathbf{m}) \leqslant \sum_{1\leqslant a\leqslant p} \sum_{\substack{n/2<p\leqslant n \\ (\mathbf{a}, \mathbf{m})\equiv 0 \pmod p}} 1 \leqslant \sum_{1\leqslant a\leqslant p} \sigma_{\frac{n}{2}}((\mathbf{a}, \mathbf{m}))$$
$$\leqslant n \log_{\frac{n}{2}}(s\|\mathbf{m}\| n^{s-1}).$$

The lemma is proved.

The proof of Theorem 7.19. We may suppose that $\varepsilon < \alpha - \frac{1}{2}$. Then by the argument of §7.5,

$$|S(p, \mathbf{a}, f)| = \Big| \sum_{(\mathbf{a},\mathbf{m}) \equiv 0 \pmod p}{}' C(\mathbf{m}) \Big|.$$

Hence by (7.29), Lemma 7.10 and Schwarz's inequality,

$$\begin{aligned}
S(n, \omega, f) &\leqslant \frac{1}{|\omega|} \sum_{(\mathbf{a}, p) \in \omega} \sum_{(\mathbf{a},\mathbf{m}) \equiv 0 \pmod p}{}' |C(\mathbf{m})| \\
&\leqslant \frac{1}{|\omega|} \sum_{\|\mathbf{m}\| \geqslant M} A(\omega, \mathbf{m}) |C(\mathbf{m})| \\
&\leqslant \frac{1}{|\omega|} \left(\sum |C(\mathbf{m})|^2 \|\mathbf{m}\|^{2(\alpha-\varepsilon/2)}\right)^{1/2} \left( \sum_{\|\mathbf{m}\| \geqslant M} A(\omega, \mathbf{m})^2 \|\mathbf{m}\|^{-2(\alpha-\varepsilon/2)} \right)^{1/2} \\
&\leqslant Cc(s, \varepsilon) n^{-2} (\ln n) \left( \sum_{\|\mathbf{m}\| \geqslant M} n^2 (1 + \log_{\frac{n}{2}} \|\mathbf{m}\|)^2 \|\mathbf{m}\|^{-2(\alpha-\varepsilon/2)} \right)^{1/2} \\
&\leqslant Cc(s, \varepsilon) n^{-1} (\ln n) \left( \sum_{\|\mathbf{m}\| \geqslant M} \frac{\|\mathbf{m}\|^{-2\alpha+1+3\varepsilon/2} (\ln \|\mathbf{m}\|)^2}{\|\mathbf{m}\|^{1+\varepsilon/2}} \right)^{1/2} \\
&\leqslant Cc(s, \varepsilon) n^{-1} (\ln n) M^{-\alpha+\frac{1}{2}+\frac{3}{4}\varepsilon} \left( \sum \frac{(\ln \|\mathbf{m}\|)^2}{\|\mathbf{m}\|^{1+\varepsilon/2}} \right)^{1/2} \\
&\leqslant Cc(s, \varepsilon) n^{-\alpha-\frac{1}{2}+\varepsilon}.
\end{aligned}$$

The theorem is proved.

We know from Theorem 7.19 that the mean error of the quadrature formula given by the good lattice points is better by a factor $O(n^{-1/2})$ compared with the error term of the quadrature formula given by Theorem 7.14 for the class of functions $Q_s^{\alpha}(C) \left(\alpha > \frac{1}{2}\right)$. Hence we may expect that perhaps the error term of the quadrature formula given by good lattice point is $O(n^{-\alpha-\frac{1}{2}+\varepsilon})$.

## Notes

N. M. Korobov [1] was the first to have proved a result with the same precision of Theorem 7.2 for the $p$ set $\left(\left\{\frac{k}{p^2}\right\}, \cdots, \left\{\frac{k^s}{p^2}\right\}\right) (1 \leqslant k \leqslant p^2)$ and Theorem 7.2 was proved by Hua Loo Keng and Wang Yuan [3]. Theorem 7.3 was proved by Yu. N. Sahov [4] (Cf. also V. M. Solodov [1]) for the case $\alpha > \frac{1}{2}$ and by Hua Loo Keng and Wang Yuan [6,7] for $\frac{1}{2} > \alpha \geqslant 0$.

R. D. Richtmyer [1] and L. Peck [1] proposed using the set $(\{r_1k\}, \cdots, \{r_sk\})$ $(k = 1, 2, \cdots)$ to evaluate the multiple integral. Theorem 7.4 was first proved by N. S. Bahvalov [1] and C. B. Haselgrove [1] (Cf. also Wang Yuan [2], N. M. Korobov [7], Hua Loo Keng and Wang Yuan [3,6,7], H. Niederreiter [2] and Wang Yuan [5]).

Theorem 7.8: Cf. N. M. Gelfand, A. S. Frolov and N. N. Cencov [1] and N. M. Korobov [7].

Theorem 7.11 was first proved by N. S. Bahvalov [1] (Cf. also Hua Loo Keng and Wang Yuan [3,6,7]). The case $s = 2$ of Theorem 7.13 was proved independently by Hua Loo Keng and Wang Yuan [1] and N. S. Bahvalov [1].

Theorem 7.15 was first proved by N. M. Korobov [2] with the error term $O(p^{-\alpha}(\ln p)^{\alpha s})$ and improved by N. S. Bahvalov [1] to $O(p^{-\alpha}(\ln p)^{-\alpha(s-1)})$ (Cf. also Wang Yuan [2]).

Theorem 7.17: Cf. I. F. Sarygin [2].

Theorems 7.18 and 7.19: Cf. N. S. Bahvalov [1,2] and Wang Yuan [2].

*Chapter 8*

# Numerical Error for Quadrature Formula

## § 8.1. The numerical error

We introduce the notations

$$S(n, \boldsymbol{\gamma}, f) = \int_{G_s} f(\mathbf{x})d\mathbf{x} - \frac{1}{n} \sum_{k=-(n-1)}^{n-1} \left(1 - \frac{|k|}{n}\right) f(k\boldsymbol{\gamma})$$

and

$$S(n, \mathbf{h}, f) = \int_{G_s} f(\mathbf{x})d\mathbf{x} - \frac{1}{n} \sum_{k=1}^{n} f\left(\frac{k\mathbf{h}}{n}\right),$$

where $\boldsymbol{\gamma}$ and $h$ denote real and integral vectors respectively.

**Theorem 8.1.**

$$\sup_{f \in E_s^2(C)} |S(n, \boldsymbol{\gamma}, f)| \leqslant CW_2(n, \boldsymbol{\gamma}),$$

*where*

$$W_2(n, \boldsymbol{\gamma}) = \frac{1}{n}\left(1 + \frac{\pi^2}{3}\right)^s + \frac{2}{n} \sum_{k=1}^{n} \left(1 - \frac{k}{n}\right) \prod_{\nu=1}^{s} (1 + 2\pi^2 B_2(\{k\gamma_\nu\})) - 1,$$

in which $B_2(x) = x^2 - x + \dfrac{1}{6}$ denotes the Bernoulli polynomial (Cf. §6.5).

**Theorem 8.2.**

$$\sup_{f \in E_s^2(C)} |S(n, \mathbf{h}, f)| \leqslant CW_2(n, \mathbf{h}),$$

*where*

$$W_2(n, \mathbf{h}) = \begin{cases} \dfrac{1}{n}\left(1+\dfrac{\pi^2}{3}\right)^s + \dfrac{2}{n}\displaystyle\sum_{k=1}^{\frac{n-1}{2}}\prod_{\nu=1}^{s}\left(1+2\pi^2 B_2\left(\left\{\frac{kh_\nu}{n}\right\}\right)\right), & \text{if } 2\nmid n, \\ \dfrac{1}{n}\left(1+\dfrac{\pi^2}{3}\right)^s + \dfrac{1}{n}\left(1-\dfrac{\pi^2}{6}\right)^\mu \left(1+\dfrac{\pi^2}{3}\right)^{s-\mu} & \\ \quad + \dfrac{2}{n}\displaystyle\sum_{k=1}^{\frac{n}{2}-1}\prod_{\nu=1}^{s}\left(1+2\pi^2 B_2\left(\left\{\frac{kh_\nu}{n}\right\}\right)\right) - 1, & \text{if } 2\mid n, \end{cases}$$

*in which* $\mu$ *denotes the number of odd integers of* $h_\nu (1 \leqslant \nu \leqslant s)$.

**Lemma 8.1.**

$$\sum_{m=-\infty}^{\infty} \frac{e^{2\pi i m x}}{\overline{m}^2} = 1 + 2\pi^2 B_2(\{x\}).$$

*Proof.* Since

$$\int_0^1 (1 + 2\pi^2 B_2(x)) e^{-2\pi i m x} dx$$
$$= \int_0^1 \left(1 - \frac{\pi^2}{6} + \frac{\pi^2}{2}(1-2x)^2\right) e^{-2\pi i m x} dx = \overline{m}^{-2},$$

the lemma follows.

The proof of Theorem 8.1. By Lemma 7.4,

$$W_2(n, \boldsymbol{\gamma}) = n^{-1}\sum{}' \frac{1}{\|\mathbf{m}\|^2}\left|\sum_{k=-(n-1)}^{n-1}(n - |k|) e^{2\pi i(\mathbf{m}, \boldsymbol{\gamma})k}\right|$$
$$= n^{-2}\sum{}' \frac{1}{\|\mathbf{m}\|^2}\sum_{j=0}^{n-1}\sum_{k=-j}^{j} e^{2\pi i(\mathbf{m}, \boldsymbol{\gamma})k}.$$

Hence by Lemma 8.1

$$W_2(n, \boldsymbol{\gamma}) = n^{-2}\sum_{j=0}^{n-1}\sum_{k=-j}^{j}\left(\sum \frac{1}{\|\mathbf{m}\|^2} e^{2\pi i(\mathbf{m}, \boldsymbol{\gamma})k} - 1\right)$$
$$= n^{-2}\sum_{j=0}^{n-1}\sum_{k=-j}^{j}\prod_{\nu=1}^{s}\left(\sum_{m_\nu=-\infty}^{\infty}\frac{e^{2\pi i m_\nu \gamma_\nu k}}{\overline{m}_\nu^2}\right) - 1$$

$$= n^{-2} \sum_{j=0}^{n-1} \sum_{k=-j}^{j} \prod_{\nu=1}^{s} (1 + 2\pi^2 B_2(\{k\gamma_\nu\})) - 1$$

$$= \frac{1}{n} \sum_{k=-n}^{n} \left(\frac{n-|k|}{n}\right) \prod_{\nu=1}^{s} (1 + 2\pi^2 B_2(\{k\gamma_\nu\})) - 1.$$

Since $B_2(\{x\})$ is an even function of $x$, the theorem follows.

The proof of Theorem 8.2. Since

$$W_2(n, \mathbf{h}) = \frac{1}{n} \sum_{k=1}^{n} \sum{}' \frac{e^{2\pi i(\mathbf{h},\mathbf{m})k/n}}{\|\mathbf{m}\|^2}$$

$$= \frac{1}{n} \sum_{k=1}^{n} \prod_{\nu=1}^{s} \left( \sum_{m_\nu=-\infty}^{\infty} \frac{e^{2\pi i h_\nu m_\nu k/n}}{\overline{m}_\nu^2} \right) - 1$$

$$= \frac{1}{n} \sum_{k=1}^{n} \prod_{\nu=1}^{s} \left(1 + 2\pi^2 B_2\left(\left\{\frac{kh_\nu}{n}\right\}\right)\right) - 1.$$

The theorem follows.

Let

$$\sup_{f \in E_s^4(C)} |S(n, \mathbf{h}, f)| \leqslant C W_4(n, \mathbf{h}).$$

Then in a way similar to Theorem 8.2, we may derive

**Theorem 8.3.**

$$W_4(n, \mathbf{h}) = \begin{cases} \dfrac{1}{n}\left(1 + \dfrac{\pi^4}{45}\right)^s + \dfrac{2}{n} \displaystyle\sum_{k=1}^{\frac{n-1}{2}} \prod_{\nu=1}^{s} \left(1 - \dfrac{2\pi^4}{3} B_4\left(\left\{\dfrac{kh_\nu}{n}\right\}\right)\right) - 1, \\ \qquad\qquad\qquad\qquad\qquad\qquad\qquad\qquad \text{if } 2 \nmid n, \\ \dfrac{1}{n}\left(1 + \dfrac{\pi^4}{45}\right)^s + \dfrac{1}{n}\left(1 - \dfrac{7\pi^4}{360}\right)^\mu \left(1 + \dfrac{\pi^4}{45}\right)^{s-\mu} \\ \quad + \dfrac{2}{n} \displaystyle\sum_{k=1}^{\frac{n}{2}-1} \prod_{\nu=1}^{s} \left(1 - \dfrac{2\pi^4}{3} B_4\left(\left\{\dfrac{kh_\nu}{n}\right\}\right)\right) - 1, \quad \text{if } 2 \mid n, \end{cases}$$

*where $\mu$ denotes the number of odd integers of $h_\nu (1 \leqslant \nu \leqslant s)$.*

We may also obtain the expressions for the

$$\sup_{f \in E_s^{2l}(C)} |S(n, \mathbf{h}, f)|, \quad l = 3, 4, \cdots.$$

## § 8.2. The comparison of good points

We shall construct the vectors $\boldsymbol{\gamma}$ by the use of the cyclotomic field, the Dirichlet field and Theorem 4.13 and then we shall compare their corresponding $W_2(n, \boldsymbol{\gamma})$ as follows.

**Table 1.** $s = 3$

| $n$ | $W_2(n, (e, e^2, e^3))$ | $W_2\left(n, \left(\frac{\sqrt{5}-1}{2}, \sqrt{2}, \sqrt{10}\right)\right)$ |
|---|---|---|
| $10^2$ | $3.7687 \times 10^{-1}$ | $3.1740 \times 10^{-1}$ |
| $5 \times 10^2$ | $3.4290 \times 10^{-2}$ | $5.3640 \times 10^{-2}$ |
| $10^3$ | $1.1180 \times 10^{-2}$ | $2.2850 \times 10^{-2}$ |

**Table 2.** $s = 4$

| $n$ | $W_2\left(n, \left(2\cos\frac{2\pi}{11}, \cdots, 2\cos\frac{8\pi}{11}\right)\right)$ | $W_2(n, (e, \cdots, e^4))$ |
|---|---|---|
| $10^3$ | $1.5683 \times 10^{-1}$ | $6.4835 \times 10^{-1}$ |
| $1.5 \times 10^3$ | $1.0070 \times 10^{-1}$ | $5.5341 \times 10^{-1}$ |
| $3 \times 10^3$ | $4.0200 \times 10^{-2}$ | $3.7821 \times 10^{-1}$ |

**Table 3.** $s = 5$

| $n$ | $W_2\left(n, \left(2\cos\frac{2\pi}{13}, \cdots, 2\cos\frac{10\pi}{13}\right)\right)$ | $W_2(n, (e, \cdots, e^5))$ |
|---|---|---|
| $10^4$ | $6.7819 \times 10^{-2}$ | $7.4518 \times 10^{-2}$ |
| $10^5$ | $3.6400 \times 10^{-3}$ | $3.6966 \times 10^{-3}$ |

**Table 4.** $s = 7$

| $n$ | $W_2\left(n, \left(2\cos\frac{2\pi}{17}, \cdots, 2\cos\frac{14\pi}{17}\right)\right)$ | $W_2(n, (e, \cdots, e^7))$ | $W_2\left(n, \left(\frac{\sqrt{5}-1}{2}, \sqrt{2}, \cdots, \sqrt{30}\right)\right)$ |
|---|---|---|---|
| $10^3$ | $2.6730 \times 10$ | $3.1621 \times 10$ | $2.8308 \times 10$ |
| $10^4$ | $2.1492$ | $2.2513$ | $3.9925$ |
| $10^5$ | $1.9503 \times 10^{-1}$ | $1.9592 \times 10^{-1}$ | $3.0794 \times 10^{-1}$ |

*Remark.* It is suggested by the above tables that the $\boldsymbol{\gamma}$ given by $\mathscr{R}_s$ or $\boldsymbol{\gamma} = (e, e^2, \cdots, e^s)$ is more advantageous compared to the $\boldsymbol{\gamma}$ given by $\mathscr{D}_s$ (Cf. P. J. Davis and P. Robinowitz [1]).

## § 8.3. The computation of the $\eta$ set

Let $F_n (\equiv F_{s,n})$ be the generalized Fibonacci sequence of dimension $s$, i.e., the sequence of integers defined by the recurrent formula

$$F_0 = F_1 = \cdots = F_{s-2} = 0, \quad F_{s-1} = 1,$$

$$F_{n+s} = F_{n+s-1} + \cdots + F_{n+1} + F_n, \quad n \geqslant 0$$

(Cf. §2.8). Let $n = F_m, h_1 = 1, h_2 = F_{m+1} - F_m, \cdots, h_s = F_{m+s-1} - F_{m+s-2} - \cdots - F_{m+1} - F_m$. Then we have the $\eta$ set (Cf. §4.7) and the following examples.

1. $F_{2,m} = 0, 1, 1, 2, 3, 5, 8, 13, 21, 34, 55, 89, 144, 233, 377$ for $0 \leqslant m \leqslant 15$ and

$$W_2(55, (1, 34)) \leqslant 3.8148 \times 10^{-2}.$$

2. Take $n$ to be $F_{13} = 401, F_{16} = 2{,}872$ and $F_{18} = 10{,}671$ for $s = 4$. Then we have

**Table 1** ($h_1 = 1$)

| $n$ | $h_2$ | $h_3$ | $h_4$ | $W_2(n, \mathbf{h})$ |
|---|---|---|---|---|
| 401 | 372 | 316 | 208 | $4.5260 \times 10^{-1}$ |
| 2,872 | 2,664 | 2,263 | 1,490 | $2.3845 \times 10^{-2}$ |
| 10,671 | 9,898 | 8,408 | 5,536 | $3.8520 \times 10^{-3}$ |

3. Take $n = F_{19} = 13{,}624, h_1 = 1, h_2 = 13{,}160, h_3 = 12{,}248, h_4 = 10{,}455, h_5 = 6{,}930$ for $s = 5$. Then

$$W_2(n, \mathbf{h}) \leqslant 3.0738 \times 10^{-2}.$$

4. Take $n = F_{21} = 29{,}970, h_1 = 1, h_2 = 29{,}478, h_3 = 28{,}502, h_4 = 26{,}566, h_5 = 22{,}726, h_6 = 15{,}109$ for $s = 6$. Then

$$W_2(n, \mathbf{h}) \leqslant 1.2002 \times 10^{-1}.$$

We may also obtain some data for $s \geqslant 7$ but the precision of the corresponding $W_2(n, \mathbf{h})$ is not so good (Cf. §8.7)

For $s=3$, we suggest using the sequence of integers defined by the recurrent formula

$$G_0 = G_1 = 0,\ G_2 = 1,\ G_{m+3} = G_{m+1} + G_m, \quad m \geqslant 0$$

(Cf. §1.7). Let $n = G_m, h_1 = 1, h_2 = G_{m+1} - G_m, h_3 = G_{m+2} - G_m$. Then we have

**Table 2** ($h_1 = 1$)

| $n$ | $h_2$ | $h_3$ | $W_2(n,\mathbf{h})$ |
|---|---|---|---|
| 151 | 49 | 114 | $1.5035 \times 10^{-1}$ |
| 1,081 | 351 | 816 | $1.1229 \times 10^{-2}$ |
| 1,897 | 616 | 1,432 | $3.8875 \times 10^{-3}$ |
| 13,581 | 4,410 | 10,252 | $1.2459 \times 10^{-4}$ |

## § 8.4. The computation of the $\mathscr{R}_s$ set

Let $s = \dfrac{p-1}{2}$. Then

$$\rho_l = \frac{\sin \dfrac{\pi}{p} g^{l+1}}{\sin \dfrac{\pi}{p} g^l}, \quad 1 \leqslant l \leqslant s-1$$

is a set of independent units of $\mathscr{R}_s = Q\left(\cos \dfrac{2\pi}{p}\right)$, where $g$ denotes a primitive root mod $p$. In particular, if $g = 2$, then

$$\rho_l = 2\cos \frac{2\pi}{p} 2^{l-1}, \quad 1 \leqslant l \leqslant s-1.$$

Let

$$\xi^{(i)} = \rho_1^{(i)x_1} \cdots \rho_{s-1}^{(i)x_{s-1}}, \quad 2 \leqslant i \leqslant s,$$

where $\rho_j^{(i)} (2 \leqslant i \leqslant s)$ denote the conjugates of $\rho_j$. Solving the system of linear equations

$$\ln |\xi^{(2)}| = \cdots = \ln |\xi^{(s)}|, \tag{8.1}$$

we obtain a unique set of ratios

$$\frac{x_1}{x_{s-1}}, \cdots, \frac{x_{s-2}}{x_{s-1}}.$$

Let $l$ be $a$ given real number and $l_i$ and $l_{s-1}$ be the integers nearest to $lx_i/l_{s-1} (1 \leqslant i \leqslant s-2)$ and $l$ respectively. Then

$$\eta (= \eta_l) = |\rho_1^{l_1} \cdots \rho_{s-1}^{l_{s-1}}|$$

is an algebraic integer and so

$$n (= n_l) = \eta + \eta^{(2)} + \cdots + \eta^{(s)}$$

is a rational integer. For practical use, we may take $n$ to be the integer nearest to $\eta$ which is denoted by $n \doteq \eta$. Set

$$h_1 = 1, \quad h_i = n \left| \left\{ 2 \cos \frac{2\pi i}{p} \right\} \right|, \quad 2 \leqslant i \leqslant s.$$

Then we have a $\mathscr{R}_s$ set (Cf. §4.6) and the following examples.

1. Take $\mathscr{R}_2 = Q\left(\cos \dfrac{2\pi}{5}\right)$. Then $n = F_{m+1}$, $h_1 = 1$, $h_2 = F_m$ is a $\eta$ set where $F_m = F_{2,m}$ (Cf. example 1 of §8.3).

2. Take $\mathscr{R}_3 = Q\left(\cos \dfrac{2\pi}{7}\right)$ and units

$$\varepsilon_1 = 2 \cos \frac{6\pi}{7} = -1.8019 \cdots,$$

$$\varepsilon_2 = 2 \cos \frac{2\pi}{7} = 1.2473 \cdots,$$

$$\varepsilon_3 = 2 \cos \frac{4\pi}{7} = -0.4447 \cdots.$$

Any two among these three units form a set of independent units of $\mathscr{R}_3$.

Solving the equation

$$|\varepsilon_2^\alpha \varepsilon_2^\beta| = |\varepsilon_3^\alpha \varepsilon_1^\beta|,$$

we have

$$\frac{\alpha}{\beta} = 1.357\cdots \simeq \frac{4}{3},$$

where "$a \simeq b$" means that $a$ and $b$ are "approximately equal". Hence we have, for example

$$n = 418 \doteqdot \eta = \varepsilon_1^8 \varepsilon_2^6, \quad h_1 = 1,$$

$$h_2 = 335 \doteqdot n\left\{\left|2\cos\frac{6\pi}{7}\right|\right\},$$

$$h_3 = 103 \doteqdot n\left\{\left|2\cos\frac{2\pi}{7}\right|\right\}$$

and also the following data:

| $n$ | $h_2$ | $h_3$ | $W_2(n,\mathbf{h})$ |
|---|---|---|---|
| 20 | 17 | 6 | 2.66 |
| 83 | 66 | 20 | $5.52\times 10^{-1}$ |
| 418 | 335 | 103 | $3.71\times 10^{-2}$ |
| 1,692 | 1,357 | 418 | $4.88\times 10^{-3}$ |

3. Take $\mathscr{R}_5 = Q\left(\cos\frac{2\pi}{11}\right)$ and units

$$\varepsilon_1 = 2\cos\frac{10\pi}{11}, \quad \varepsilon_2 = 2\cos\frac{2\pi}{11}, \quad \varepsilon_3 = 2\cos\frac{8\pi}{11},$$

$$\varepsilon_4 = 2\cos\frac{4\pi}{11}, \quad \varepsilon_5 = 2\cos\frac{6\pi}{11}.$$

Solving

$$|\varepsilon_2^\alpha \varepsilon_4^\beta \varepsilon_5^\gamma \varepsilon_3^\delta| = |\varepsilon_3^\alpha \varepsilon_5^\beta \varepsilon_2^\gamma \varepsilon_1^\delta| = |\varepsilon_4^\alpha \varepsilon_3^\beta \varepsilon_1^\gamma \varepsilon_5^\delta| = |\varepsilon_5^\alpha \varepsilon_1^\beta \varepsilon_4^\gamma \varepsilon_2^\delta|,$$

we have

$$\frac{\alpha}{\delta} = 1.412\cdots \simeq \frac{7}{5}, \quad \frac{\beta}{\delta} = 1.584\cdots \simeq \frac{8}{5}, \quad \frac{\gamma}{\delta} = 0.944\cdots \simeq \frac{5}{5}$$

and so

$$n = 9{,}389 \doteq \eta = |\varepsilon_1^7\varepsilon_2^8\varepsilon_3^5\varepsilon_4^5|, \quad h_1 = 1,$$

$$h_2 = 8{,}628 \doteq n\left\{\left|2\cos\frac{10\pi}{11}\right|\right\},$$

$$h_3 = 6{,}408 \doteq n\left\{\left|2\cos\frac{2\pi}{11}\right|\right\},$$

$$h_4 = 2{,}908 \doteq n\left\{\left|2\cos\frac{8\pi}{11}\right|\right\},$$

$$h_5 = 7{,}800 \doteq n\left\{\left|2\cos\frac{4\pi}{11}\right|\right\},$$

$$W_2(n, \mathbf{h}) \leqslant 6.69 \times 10^{-2}.$$

*Remark.* Suppose that $(n, \mathbf{h}(n))$ is an $\mathscr{R}_s$ set. Then we may obtain an $\mathscr{R}_{s'}$ set $(n, \mathbf{h}^*(n))$, where $\mathbf{h}^*(n)$ is obtained from the vector $\mathbf{h}(n)$ by neglecting any of its $s - s'$ components. The precision of $W_2(n, \mathbf{h}^*)$ is still good if $s/s'$ is close to 1. For example, suppose that $n = 462{,}891, h_1 = 1$, $h_2 = 450{,}265$, $h_3 = 412{,}730$, $h_4 = 351{,}310$, $h_5 = 267{,}681$, $h_6 = 164{,}124$, $h_7 = 43{,}464, h_8 = 371{,}882, h_9 = 277{,}266$. Then we obtain an $\mathscr{R}_7$ set $(n, \mathbf{h}_7^*)$ and an $\mathscr{R}_8$ set $(n, \mathbf{h}_8^*)$, where $\mathbf{h}_7^*$ and $\mathbf{h}_8^*$ are obtained from $\mathbf{h}$ by neglecting $h_8, h_9$ and $h_9$ respectively. We also obtain

$$W_2(n, \mathbf{h}_7^*) \leqslant 1.9397 \times 10^{-2}$$

and

$$W_2(n, \mathbf{h}_8^*) \leqslant 1.6240 \times 10^{-1}.$$

Hence we may obtain the $\mathscr{R}_s$ set of any dimension $s$, although the cyclotomic fields used here are confined to the fields with dimension $\frac{p-1}{2}$.

## § 8.5. Examples of other $\mathscr{F}_s$ sets

1. Take $\mathscr{D}_4 = Q(\sqrt{5}, \sqrt{2})$ and the units

$$\varepsilon_1 = \frac{1+\sqrt{5}}{2}, \quad \varepsilon_2 = 1 + \sqrt{2}, \quad \varepsilon_3 = 3 + \sqrt{10}.$$

Then from the unit

$$\varepsilon_1^6\varepsilon_2^4\varepsilon_3^2,$$

this gives

$$n = 11{,}574, \quad h_1 = 1, \quad h_2 = 7{,}153, \quad h_3 = 4{,}794, \quad h_4 = 1{,}878$$

and

$$W_2(n, \mathbf{h}) = 8.81 \times 10^{-3}.$$

2. Take $Q(\omega)$ and units

$$\varepsilon_1 = 2 + \omega^2, \quad \varepsilon_2 = 3 + 2\omega,$$

where $\omega = \sqrt[4]{5}$ (Cf. L. Bernstein [1]). Solving the equation

$$\ln|2 + \omega^2|^{x_1}|3 - 2\omega|^{x_2} = \ln|2 - \omega^2|^{x_1}|3 - 2\omega i|^{x_2},$$

we have

$$\frac{x_1}{x_2} = 2.1200\cdots.$$

From the unit

$$\varepsilon_1^4\varepsilon_2^2,$$

we have

$$n = 2{,}889, \quad h_1 = 1, \quad h_2 = 1{,}431, \quad h_3 = 862, \quad h_4 = 993$$

and

$$W_2(n, \mathbf{h}) = 2.8626 \times 10^{-2}.$$

And from the unit

$$\varepsilon_1^6\varepsilon_2^3,$$

we have

$$n = 310{,}563, \quad h_1 = 1, \quad h_2 = 153{,}837, \quad h_3 = 73{,}314, \quad h_4 = 106{,}741$$

and

$$W_2(n, \mathbf{h}) \leqslant 2.0948 \times 10^{-4}.$$

## § 8.6. The computation of a *glp* set

We introduce the notations

$$H_1(z)=\frac{3^s}{p_1}\left(1+2\sum_{k=1}^{\frac{p_1-1}{2}}\prod_{\nu=0}^{s-1}\left(1-2\left\{\frac{kz^\nu}{p_1}\right\}\right)^2\right),$$

$$H_2(z)=\frac{3^s}{p_1p_2}\left(1+2\sum_{k=1}^{\frac{p_1p_2-1}{2}}\prod_{\nu=0}^{s-1}\left(1-2\left\{\frac{k(p_2b_1^\nu+p_1b_2^\nu)}{p_1p_2}\right\}\right)^2\right),$$

$$\cdots\cdots$$

$$H_t(z)=\frac{3^s}{q}\left(1+2\sum_{k=1}^{\frac{q-1}{2}}\prod_{\nu=0}^{s-1}\left(1-2\left\{\frac{k(q_1b_1^\nu+\cdots+q_{t-1}b_{t-1}^\nu+q_tz^\nu)}{q}\right\}\right)^2\right),$$

where $p_1,\cdots,p_t$ denote the different primes, $q=p_1\cdots p_t$, $q_i=q/p_i (1\leqslant i\leqslant t)$ and where $b_i$ denotes the integer such that $H_i(z)$ takes the minimum for $z=1,\cdots,\frac{p_i-1}{2}(1\leqslant i\leqslant t)$.

**Lemma 8.2.** *Let $\alpha>1$. Let $q$ be a positive integer and $\mathbf{a}=(a_1,\cdots,a_s)$ be an integral vector. If $(a_i,q)=1(1\leqslant i\leqslant s)$, then*

$$\sum_{(\mathbf{a},\mathbf{m})\equiv 0 \pmod q}{}' \frac{1}{\|\mathbf{m}\|^\alpha}-\sum_{\substack{(\mathbf{a},\mathbf{m}^{(0)})\equiv 0 \pmod q\\ -\frac{q}{2}<m_i^{(0)}\leqslant\frac{q}{2}}}\frac{1}{\|\mathbf{m}^{(0)}\|^\alpha}\leqslant s2^\alpha(2\zeta(\alpha)+1)^s q^{-\alpha}.$$

*Proof.* Since

$$\sum_{-\frac{q}{2}<m\leqslant\frac{q}{2}}\sum_{l=-\infty}^{\infty}\frac{1}{\overline{(m+lq)}^\alpha}\leqslant 1+2\sum_{n=1}^{\infty}\frac{1}{n^\alpha}=1+2\zeta(\alpha)$$

and

$$\sum_{l=-\infty}^{\infty}{}'\frac{1}{\overline{(m+lq)}^\alpha}\leqslant 2\sum_{l=1}^{\infty}\frac{1}{\overline{\left(q\left(l-\frac{1}{2}\right)\right)}^\alpha}\leqslant 2\zeta(\alpha)\left(\frac{2}{q}\right)^\alpha$$

for $-q/2<m\leqslant q/2$, then in a way similar to the proof of Lemma 4.10, we have

$$\sum_{(\mathbf{a},\mathbf{m})\equiv 0 \pmod q}{}' \frac{1}{\|\mathbf{m}\|^\alpha}-\sum_{\substack{(\mathbf{a},\mathbf{m}^{(0)})\equiv 0 \pmod q\\ -\frac{q}{2}<m_i^{(0)}\leqslant\frac{q}{2}}}{}'\frac{1}{\|\mathbf{m}^{(0)}\|^\alpha}$$

$$\leqslant \sum_{\nu=1}^{s} \sum_{\substack{(\mathbf{a},\mathbf{m}^{(0)})\equiv 0 \pmod q \\ -\frac{q}{2}<m_i^{(0)}\leqslant\frac{q}{2}}} \left(\sum_{l_\nu=-\infty}^{\infty}{}' \frac{1}{(m_\nu^{(0)}+l_\nu q)^\alpha}\right) \prod_{\substack{\mu=1\\ \mu\neq\nu}}^{s} \left(\sum_{l_\mu=-\infty}^{\infty} \frac{1}{(m_\mu^{(0)}+l_\mu q)^\alpha}\right)$$

$$\leqslant s2^\alpha(2\zeta(\alpha)+1)^s q^{-\alpha}.$$

The lemma is proved.

**Lemma 8.3.** *Suppose that* $r$ *is a positive integer. If the congruence*

$$(\mathbf{a},\mathbf{m}) = \sum_{i=1}^{s} a_i m_i \equiv 0 \pmod n \tag{8.2}$$

*has no solution in the domain*

$$\|\mathbf{m}\| \leqslant M, \quad \mathbf{m} \neq \mathbf{0},$$

*then*

$$\sum_{\substack{(\mathbf{a},\mathbf{m})\equiv 0 \pmod n\\ |m_i|\leqslant rn}}{}' \frac{1}{\|\mathbf{m}\|} \leqslant c(s)M^{-1}(\ln rn)^s.$$

*Proof.* Let $T_{l,M}(l \geqslant 1)$ be the number of solutions of (8.2) in the domain

$$\|\mathbf{m}\| < lM, \quad \mathbf{m} \neq \mathbf{0}.$$

Then $T_{l,M} \leqslant c(s)l(\ln 3lM)^{s-1}$ (Cf. Lemma 3.5). Hence

$$\sum_{\substack{(\mathbf{a},\mathbf{m})\equiv 0 \pmod n\\ |m_i|\leqslant rn}}{}' \frac{1}{\|\mathbf{m}\|} \leqslant \sum_{l=1}^{(rn)^s} \frac{T_{l+1,M}-T_{l,M}}{lM}$$

$$= M^{-1}\sum_{l=1}^{(rn)^s} T_{l+1,M}\left(\frac{1}{l}-\frac{1}{l+1}\right) + \frac{T_{(rn)^s+1,M}}{((rn)^s+1)M}$$

$$\leqslant c(s)M^{-1}\sum_{l=1}^{(rn)^s}\frac{(\ln 3lM)^{s-1}}{l} + c(s)M^{-1}(\ln 3rn)^{s-1}$$

$$\leqslant c(s)M^{-1}(\ln rn)^s.$$

The lemma is proved.

**Theorem 8.4.** $\mathbf{b}_1 = (1, b_1, \cdots, b_1^{s-1})$ *is a good lattice point* mod $p_1$.

*Proof.* By Lemma 8.1,

$$B_2(\{x\}) = \{x\}^2 - \{x\} + \frac{1}{6} = \frac{1}{2\pi^2} \sum_{m=-\infty}^{\infty}{}' \frac{e^{2\pi imx}}{m^2}$$

and so

$$3(1-2\{x\})^2 = \frac{6}{\pi^2} \sum_{m=-\infty}^{\infty}{}' \frac{e^{2\pi imx}}{m^2} + 1 = \sum_{m=-\infty}^{\infty} \frac{e^{2\pi imx}}{\phi(m)}, \tag{8.3}$$

where

$$\phi(m) = \begin{cases} \frac{\pi^2}{6} m^2, & \text{if } m \neq 0, \\ 0, & \text{if } m = 0. \end{cases}$$

Since

$$\left(1 - 2\left\{\frac{(p_1-k)z^\nu}{p_1}\right\}\right)^2 = \left(1 - 2\left\{\frac{-kz^\nu}{p_1}\right\}\right)^2 = \left(1 - 2\left\{\frac{kz^\nu}{p_1}\right\}\right)^2$$

by (8.3), so

$$\begin{aligned} H_1(z) &= \frac{3^s}{p_1}\left(1 + \sum_{k=1}^{\frac{p_1-1}{2}} \prod_{\nu=0}^{s-1} \left(1-2\left\{\frac{kz^\nu}{p_1}\right\}\right)^2 \right. \\ &\quad \left. + \sum_{k=1}^{\frac{p_1-1}{2}} \prod_{\nu=0}^{s-1} \left(1-2\left\{\frac{(p_1-k)z^\nu}{p_1}\right\}\right)^2\right) \\ &= \frac{3^s}{p_1} \sum_{k=1}^{p_1} \prod_{\nu=0}^{s-1} \left(1-2\left\{\frac{kz^\nu}{p_1}\right\}\right)^2 \\ &= \frac{1}{p_1} \sum_{k=1}^{p_1} \sum \frac{e^{2\pi i(\mathbf{z},\mathbf{m})k/p_1}}{\prod_{\nu=1}^{s} \phi(m_\nu)} \\ &= 1 + \sum_{(\mathbf{z},\mathbf{m})\equiv 0 \pmod{p_1}}{}' \frac{1}{\prod_{\nu=1}^{s} \phi(m_\nu)} \\ &< 1 + \sum_{(\mathbf{z},\mathbf{m})\equiv 0 \pmod{p_1}}{}' \frac{1}{\|\mathbf{m}\|^2} \end{aligned}$$

and so

$$\begin{aligned} H_1(b_1) - 1 &\leq \min_{1\leq z\leq p_1-1} \sum_{(\mathbf{z},\mathbf{m})\equiv 0 \pmod{p_1}}{}' \frac{1}{\|\mathbf{m}\|^2} \\ &\leq c(s)p_1^{-2}(\ln p_1)^{2(s-1)} \end{aligned} \tag{8.4}$$

(Cf. Theorem 7.15). On the other hand

$$H_1(b_1) - 1 \geq \left(\frac{6}{\pi}\right)^s \sum_{(\mathbf{b}_1,\mathbf{m})\equiv 0 \pmod{p_1}}{}' \frac{1}{\|\mathbf{m}\|^2}.$$

Therefore it follows that there exists $c(s)$ such that the congruence

$$(\mathbf{b}_1, \mathbf{m}) \equiv 0 \pmod{p_1}$$

has no solution in the domain

$$\|\mathbf{m}\| < c(s)p_1 (\ln p_1)^{-(s-1)}, \quad \mathbf{m} \neq \mathbf{0}$$

and so the set

$$\left(\left\{\frac{k}{p_1}\right\}, \left\{\frac{kb_1}{p_1}\right\}, \cdots, \left\{\frac{kb_1^{s-1}}{p_1}\right\}\right), \quad 1 \leqslant k \leqslant p_1$$

has discrepancy

$$D(p_1) \leqslant c(s)p^{-1} (\ln p_1)^{2s-1}$$

by Theorem 3.2, i.e., $\mathbf{b}_1$ is a good lattice point mod $p_1$. The theorem is proved.

**Theorem 8.5.** $(p_1 + p_2, p_1b_2 + p_2b_1, \cdots, p_1b_2^{s-1} + p_2b_1^{s-1})$ *is a good lattice point* mod $p_1p_2$.

*Proof*. By (8.3), we have

$$H_2(z) = \frac{3^s}{p_1p_2} \sum_{k=1}^{p_1p_2} \prod_{\nu=0}^{s-1} \left(1 - 2\left\{\frac{(p_1z^\nu + p_2b_1^\nu)k}{p_1p_2}\right\}\right)^2$$

and

$$\begin{aligned} H_2(z) - 1 &= \frac{1}{p_1p_2} \sum_{k=1}^{p_1p_2} \sum{}' \frac{e^{2\pi i(p_1(\mathbf{z},\mathbf{m})+p_2(\mathbf{b}_1,\mathbf{m}))k/p_1p_2}}{\prod_{\nu=1}^{s} \phi(m_\nu)} \\ &= \sum_{p_1(\mathbf{z},\mathbf{m})+p_2(\mathbf{b}_1,\mathbf{m}) \equiv 0 \pmod{p_1p_2}}{}' \frac{1}{\prod_{\nu=1}^{s} \phi(m_\nu)} \\ &= \sum_{\substack{(\mathbf{b}_1,\mathbf{m}) \equiv 0 \pmod{p_1} \\ (\mathbf{z},\mathbf{m}) \equiv 0 \pmod{p_2}}}{}' \frac{1}{\prod_{\nu=1}^{s} \phi(m_\nu)}. \end{aligned}$$

Divide the last sum into two parts $\sum_1$ and $\sum_2$, where $\sum_1$ contains those $\mathbf{m}$ such that $p_2 | m_\nu$ for $1 \leqslant \nu \leqslant s$ and $\sum_2$ contains the remaining terms. By (8.4) and Lemma 8.2, we have

$$\sum_1 = \sum_{p_2(\mathbf{b}_1,\mathbf{m})\equiv 0 \pmod{p_1}}' \frac{1}{\prod_{\nu=1}^{s} \psi(p_2 m_\nu)}$$

$$\leqslant p_2^{-2} \sum_{(\mathbf{b}_1,\mathbf{m})\equiv 0 \pmod{p_1}}' \frac{1}{\prod_{\nu=1}^{s} \psi(m_\nu)}$$

$$= p_2^{-2}(H_1(b_1) - 1) \leqslant c(s)(p_1p_2)^{-2} (\ln p_1p_2)^{2(s-1)}$$

and

$$\sum_2 \leqslant \sum_{\substack{(\mathbf{b}_1,\mathbf{m})\equiv 0 \pmod{p_1} \\ (\mathbf{z},\mathbf{m})\equiv 0 \pmod{p_2}}}{}_2 \frac{1}{\|\mathbf{m}\|^2}$$

$$\leqslant \sum_{-\frac{p_1p_2}{2}<m_i\leqslant\frac{p_1p_2}{2}}{}_2 \quad \sum_{\substack{(\mathbf{b}_1,\mathbf{m})\equiv 0 \pmod{p_1} \\ (\mathbf{z},\mathbf{m})\equiv 0 \pmod{p_2}}} \frac{1}{\|\mathbf{m}\|^2} + c(s)(p_1p_2)^{-2}$$

$$\leqslant \Big( \sum_{-\frac{p_1p_2}{2}<m_i\leqslant\frac{p_1p_2}{2}}{}_2 \quad \sum_{\substack{(\mathbf{b}_1,\mathbf{m})\equiv 0 \pmod{p_1} \\ (\mathbf{z},\mathbf{m})\equiv 0 \pmod{p_2}}} \frac{1}{\|\mathbf{m}\|} \Big)^2 + c(s)(p_1p_2)^{-2}.$$

Hence it follows by Theorem 8.4 and Lemmas 4.5 and 8.3 that

$$\min_{1\leqslant z\leqslant\frac{p_2-1}{2}} \sum_2 \leqslant \Big( \min_{1\leqslant z\leqslant\frac{p_2-1}{2}} \sum_{\frac{-p_1p_2}{2}<m_i\leqslant\frac{p_1p_2}{2}}{}_2 \quad \sum_{\substack{(\mathbf{b}_1,\mathbf{m})\equiv 0 \pmod{p_1} \\ (\mathbf{z},\mathbf{m})\equiv 0 \pmod{p_2}}} \frac{1}{\|\mathbf{m}\|} \Big)^2 + c(s)(p_1p_2)^{-2}$$

$$\leqslant \Big( \frac{2}{p_2-1} \sum_{\substack{\frac{-p_1p_2}{2}<m_i\leqslant\frac{p_1p_2}{2} \\ (\mathbf{b}_1,\mathbf{m})\equiv 0 \pmod{p_1}}}{}_2 \frac{1}{\|\mathbf{m}\|} \sum_{\substack{1\leqslant z\leqslant\frac{p_2-1}{2} \\ (\mathbf{z},\mathbf{m})\equiv 0 \pmod{p_2}}} 1 \Big)^2 + c(s)(p_1p_2)^{-2}$$

$$\leqslant \Big( \frac{2(s-1)}{p_2-1} \sum_{\substack{\frac{-p_1p_2}{2}<m_i\leqslant\frac{p_1p_2}{2} \\ (\mathbf{b}_1,\mathbf{m})\equiv 0 \pmod{p_1}}}' \frac{1}{\|\mathbf{m}\|} \Big)^2 + c(s)(p_1p_2)^{-2}$$

$$\leqslant c(s)(p_1p_2)^{-2} (\ln p_1p_2)^{4s-2}.$$

Consequently.

$$H_2(b_2) - 1 \leqslant \sum_1 + \min_{1\leqslant z\leqslant\frac{p_2-1}{2}} \sum_2 \leqslant c(s)(p_1p_2)^{-2} (\ln p_1p_2)^{4s-2}.$$

On the other hand,

$$H_2(b_2) - 1 \geqslant \Big(\frac{b}{\pi^2}\Big)^s \sum_{p_1(\mathbf{b}_2,\mathbf{m})+p_2(\mathbf{b}_1,\mathbf{m})\equiv 0 \pmod{p_1p_2}}' \frac{1}{\prod_{\nu=1}^{s} \psi(m_\nu)}.$$

It follows that there exists $c(s)$ such that the congruence

$$p_1(\mathbf{b}_2, \mathbf{m}) + p_2(\mathbf{b}_1, \mathbf{m}) \equiv 0 \pmod{p_1 p_2}$$

has no solution in the domain

$$\|\mathbf{m}\| \leqslant c(s) p_1 p_2 (\ln p_1 p_2)^{-(2s-1)}, \quad \mathbf{m} \neq \mathbf{0}.$$

Hence the theorem follows by Theorem 3.2.

Similarly, we may prove

**Theorem 8.6.** *Let* $e_l = q_1 b_1^{l-1} + \cdots + q_t b_t^{l-1} (1 \leqslant l \leqslant s)$. *Then* $(e_1, \cdots, e_s)$ *is a good lattice point* mod $q$.

Let $x$ denote the solution of the congruence

$$(q_1 + \cdots + q_t)x \equiv 1 \pmod{q}, \quad 1 \leqslant x \leqslant q.$$

Let

$$h_{\nu+1} \equiv (q_1 b_1^{\nu} + \cdots + q_t b_t^{\nu})x \pmod{q},$$

$$1 \leqslant h_{\nu+1} \leqslant q, \quad 1 \leqslant \nu \leqslant s-1.$$

Then the good lattice point given by theorem 8.6 may be written as

$$(1, h_1, \cdots, h_s).$$

*Remark.* P. Keast [1, 2] has pointed out that the error term of the quadrature formula corresponding to a good lattice point given by Theorem 8.6 is comparatively small, if we take $p_1, p_2, \cdots$ and $p_t$ almost equal, when $t$ is confined by $t \leqslant 4$. It is also suggested by Theorem 8.5 that if we have a good lattice point mod $n$ and a prime $p$ which is not a divisor of $n$, then we may obtain a good lattice point mod $pn$.

## § 8.7. Several remarks

1. We have seen from §2—§4 that the precision of the error terms of the quadrature formulas given by $\mathscr{R}_s$ set and *glp* set are better than other quadrature formulas and another advantage is that the so obtained quadrature formula has a very simple form, i.e., the arithmatic mean of the values of the integrand at the given set of points is used to approximate the definite integral.

2. It follows from Theorem 8.4 that

$$c(s)n^2$$

elementary operations are required to obtain a good lattice point mod $n$, where $n=p_1$. Let $p_2 \doteq \sqrt{p_1}$ and $n=p_1p_2$. Then by Theorem 8.5, the number of elementary operations for obtaining a good lattice point mod $n$ is

$$c(s)(p_1^2+p_1p_2^2)=c(s)(p_1p_2)^{4/3}=c(s)n^{4/3}.$$

In general, we may take $p_{\nu+1}\doteq\sqrt{p_\nu}(1\leqslant\nu\leqslant t-1)$ and $n=p_1\cdots p_t$. Then the number of elementary operations for obtaining a good lattice point mod $n$ are

$$c(s)(p_1^2+p_1p_2^2+\cdots+p_1\cdots p_{t-1}p_t^2)=c(s)n^{\overline{2-2^{-(t-1)}}^{\,2}}. \tag{8.5}$$

If it is required that the error term of the corresponding quadrature formula should be comparatively small, then it is better to take the $p_i's$ almost equal according to the opinion of *P*. Keast [1, 2] and so (8.5) should be replaced by

$$c(s)n^{1+t^{-1}}.$$

However it requires only

$$O(\ln n)$$

elementary operations for obtaining the $\eta$ set or $\mathscr{R}_s$ set, where the constant implied by the symbol "$O$" depends on $\eta$ or $\mathscr{R}_s$ only. Of course, $O(n)$ elementary operations are still needed for obtaining the corrcsponding $W_2(n,\mathbf{h})$ of $\eta$ set or $\mathscr{R}_s$ set.

As for the comparison of *glp* set and $\mathscr{R}_s$ set, we quote the opinion expressed by S. Haber [2] as follows.

"The second method takes $AN^2s$ seconds (to get $a^*(N,s')$ and $B^*(N,s')$ for all $s'\leqslant s$), where now $A$ was .00001 for the calculations reported here (on a UNIVAC 1108). This is very much better, and the method can be carried out at reasonable expense for $N$ up to 10,000 or so and $s$ up to 10 or 20. However, there is reason to suspect that practical formulas for $s$ as high as 10 will reguire $N's$ of order 100,000 or more, and again the calculation becomes excessively long.

The third method requires only $As^3$ seconds, the only calculation of any significant length that is necessary is the solution of the linear system (12). $A$ is apt to be about $10^{-3}$ or lower. The length of calculation is thus no obstacle for $s$ up to 100, at least, and $N$ abitrary large. While it is not known that this method actually produces g.l.p. sequences, the numerical evidence indicates that it does. However, the evidence also indicates that the quadrature formulas produced have error bounds much higher than do those produced by the first two methods."

Here the second method refers to the method given by Theorem 8.4 of which $a^*(N, s')$ and $B^*(N, s')$ are $(1, b_1, \cdots, b_1^{s-1})$ mod $N$ and $W_2(N, \mathbf{b}_1)$ respectively. The third method means the method stated in §8.4 and the linear system (12) is (8.1).

The calculation of Y. S. Moon [1] given by an IBM 370/165 leads to a similar conclusion.

3. As for the comparison of the $W_2(n, \mathbf{h})$ given by *glp* set and $\mathscr{R}_s$ set, we may see *M*. Maisonneuve [1] and S. Haber [2] for the cases $s = 3, 5, 6$. Now we give some data for the cases $s = 7, 8, 9$ as follows.

**Table 1** ($s = 7$)

| $\mathscr{R}_s$ set | | *glp* set | |
|---|---|---|---|
| $n$ | $W_2(n, \mathbf{h})$ | $n$ | $W_2(n, \mathbf{h})$ |
| 11,215 | 1.9416 | 15,019 | 1.2 |
| 84,523 | $2.0407 \times 10^{-1}$ | 71,053 | $2.1 \times 10^{-1}$ |

**Table 2** ($s = 8$)

| $\mathscr{R}_s$ set | | *glp* set | |
|---|---|---|---|
| $n$ | $W_2(n, \mathbf{h})$ | $n$ | $W_2(n, \mathbf{h})$ |
| 28,832 | $3.4501 \times 10^{-1}$ | 24,041 | 3.9 |
| 84,523 | $9.8761 \times 10^{-1}$ | 100,063 | $7.6 \times 10^{-1}$ |

**Table 3** ($s = 9$)

| set | | *glp* set | |
|---|---|---|---|
| $n$ | $W_2(n, \mathbf{h})$ | $n$ | $W_2(n, \mathbf{h})$ |
| 42,570 | $1.0496 \times 10$ | 46,213 | 9.5 |
| 172,155 | 2.3708 | 159,053 | 2.5 |

4. The $W_2(n, \mathbf{h})$ given by $\mathscr{R}_s$ set is much better than that given by $\eta$ set, especially when $s$ is comparatively large. For example, we have

$$W_2(957, 833, \mathbf{h}) \leqslant 4.1494 \times 10^{-1}$$

from the $\mathscr{R}_s$ set and

$$W_2(1,035,269,\mathbf{h}) \leqslant 4.6013 \times 10^{-1}$$

from $\eta$ set for $s = 9$, and

$$W_2(7,494,007,\mathbf{h}) \leqslant 6.3956 \times 10^{-1}$$

from $\mathscr{R}_s$ set and

$$W_2(8,359,937,\mathbf{h}) \leqslant 1.0401$$

from $\eta$ set for $s = 11$.

5. So far as we know, the cyclotomic field often gives the most precise results in applications to the problems of mumerical analysis among the real algebraic number fields of the same degree and its other advantage is the convenience for computation.

## § 8.8. Tables

The tables given in the Appendix contain the $n$, $\mathbf{h}(n)$, $\rho(n,\mathbf{h})$, $W_2(n,\mathbf{h})$ and $W_4(n,\mathbf{h})$ of various dimensions. It is not only necessary for the approximate evaluation of multiple integral but also important for theoretical work in numerical analysis. The most useful table of A. I. Saltykov [1] was made according to Korobov's method (Cf. §8.6). It is confined only to $3 \leqslant s \leqslant 10$ and $100 \leqslant n \leqslant 155{,}093$, since it requires long calculations for obtaining good lattice points. He also gave the upper estimate of $\sup_{f \in E_s^2(C)} |S(n, \mathbf{h}, f)|$ which is rougher than the function of $W_2(n, \mathbf{h})$. Saltykov's table is contained in many monographs (Cf. N. M. Korobov [7], Hua Loo Kang and Wang Yuan [3] and A. H. Stroud [1]). By the use of a CDC 6400, M. Maisonneuve [1] published the calculation of $W_2(n,\mathbf{h})$ of the pairs $(n,\mathbf{h})$ contained in Saltykov's table. H. Conroy [1] gave several tables of good lattice points of $s \leqslant 12$ and P. Keast [1, 2] gave some data obtained by the use of Theorem 8.6. By a more complicated but precise method, M. Maisonueuve [1], G. Kedem and S. K. Zaremba [1] gave some data for $(n,\mathbf{h})$, $W_2(n,\mathbf{h})$ and $W_4(n,\mathbf{h})$ for $s = 3.4$ and $n \leqslant 6.606$. Recently,it was pointed out by R. Cranley and T. N. L. Patterson [1]:

"The number of rules available, particular with large number of points, is also very limited, the most extensive table been Saltykov and Conroy, Further rule would have to be computed although this task would be easier by the work of Hua and Wang."

The first $\mathscr{R}_s$ sets were given by Hua Loo Keng and Wang Yuan [4, 5] of which one is $s=11$ and $n=698{,}047$. Later, S. Haber [2] and Y. S. Moon [1] gave some tables of $\mathscr{R}_s$ sets of $s \leqslant 14$ and $n \leqslant 10^6$ by the use of computers UNIVAC 1108 and IBM 370/165 respectively. Recently Wang Yuan, Xu Guang Shan and Zhang Rong Xiao [1] published a more extended table of $\mathscr{R}_s$ set of $s \leqslant 18$ by the use of Djs-013.

Table 1 and tables 10—12 in the Appendix are given by $\mathscr{R}_s$. In tables 2—9, the data marked with star are given by $\eta$ or $\mathscr{R}_s$ and the others are good lattice points given by Korobov's method.

## § 8.9. Some examples

Suppose that $D$ is a bounded domain with piecewise smooth boundary. If $\mathscr{D} = G_s$, then we may use a simple sum constructed by the values of integrand at the given uniformly distributed set of points to approximate the definite integral

$$\int_{\mathscr{D}} f(\mathbf{x})d\mathbf{x}$$

(Cf. §5.4). If $D$ is not $G_s$, then without loss of generality, we may suppose that $\mathscr{D} \subset G_s$. Let

$$F(\mathbf{x}) = \begin{cases} f(\mathbf{x}), & \text{if } \mathbf{x} \in \mathscr{D}, \\ 0, & \text{if } \mathbf{x} \bar{\in} \mathscr{D}. \end{cases}$$

Then

$$\int_{\mathscr{D}} f(\mathbf{x})d\mathbf{x} = \int_{G_s} f(\mathbf{x})d\mathbf{x}.$$

Now we give some examples as follows.

1. Denote

$$I = 4\int_0^1\int_0^1 x_1x_2dx_1dx_2 = 1.$$

Let $I_1$ be the approximate value of $I$ given by the quadrature formula constructed by $\eta$ set of dimension 2 (Cf. example 3 of §8.3). Further let $I_2, I_3$ and $I_4$ denote the approximate values of $I$ given also by the quadrature formula of which the integrand are obtained by changing variables

$$x_i = y_i^2(1-y_i),$$

$$x_i = y_i^3(10 - 15y_i + 6y_i^2)$$

and

$$x_i = y_i^4(35 - 84y_i + 70y_i^2 - 20y_i^3), \quad i = 1, 2$$

respectively (§6.5). Then we have

**Table 1**

| $n$ | $I_1$ | $I_2$ | $I_3$ | $I_4$ |
|---|---|---|---|---|
| 13 | $9.2308 \times 10^{-1}$ | 1.0285 | 1.0022 | $9.9781 \times 10^{-1}$ |
| 21 | $9.4029 \times 10^{-1}$ | 1.0131 | 1.0023 | 1.0004 |
| 34 | $9.7059 \times 10^{-1}$ | 1.0060 | 1.0006 | 1.00002 |
| 55 | $9.7709 \times 10^{-1}$ | 1.0027 | 1.0001 | 1.000014 |

2. Denote

$$J_1 = \int_0^1 \int_0^1 \frac{x_1^2}{1 + x_2^2} dx_1 dx_2 = 2.6179\cdots \times 10^{-1},$$

$$J_2 = \int_0^1 \int_0^1 \int_0^1 (1 + 3x_1x_2x_3 + x_1^2x_2^2x_3^2) e^{x_1x_2x_3} dx_1 dx_2 dx_3$$
$$= 1.7182\cdots,$$

$$J_3 = \int_{-1}^1 \int_{-1}^1 \int_{-1}^1 \exp(\sin x_1 \sin x_2 \sin x_3) dx_1 dx_2 dx_3$$
$$= 8.0817\cdots,$$

$$J_4 = \int_0^2 \int_{x_1^2}^{2x_1} x_1 x_2^2 dx_1 dx_2 = 6.4$$

and

$$J_5 = \int_0^1 \int_{x_1}^{2x_1} \int_{x_1x_2}^{x_1^2x_2} x_1^3 x_2^3 x_3 dx_1 dx_2 dx_3 = -4.3356\cdots \times 10^{-2}.$$

Let $J_i'(1 \leqslant i \leqslant 5)$ denote the approximate values of $J_i(1 \leqslant i \leqslant 5)$ given by the quadrature formula constructed by $\mathscr{R}_3$ set (Cf. example 2 of §8.4) respectively. Then we have

**Table 2**

| $n$ | $J'_1$ | $J'_2$ | $J'_3$ | $J'_4$ | $J'_5$ |
|---|---|---|---|---|---|
| 20 | $3.5573 \times 10^{-1}$ | 1.4992 | 8.9476 | 8.0148 | $-6.7248 \times 10^{-2}$ |
| 83 | $2.9544 \times 10^{-1}$ | 1.6932 | 7.8366 | 6.9748 | $-5.2348 \times 10^{-2}$ |
| 418 | $2.6103 \times 10^{-1}$ | 1.7202 | 7.9947 | 6.3749 | $-4.2538 \times 10^{-2}$ |
| 1,692 | $2.6180 \times 10^{-1}$ | 1.7183 | 8.0844 | 6.3999 | $-4.3290 \times 10^{-2}$ |
| 3,802 | $2.6180 \times 10^{-1}$ | 1.7183 | 8.0817 | 6.4000 | $-4.3357 \times 10^{-2}$ |

3. Let

$$K = \int_{-1}^{1}\int_{-\sqrt{1-x_1^2}}^{\sqrt{1-x_1^2}}\int_{-\sqrt{1-x_1^2-x_2^2}}^{\sqrt{1-x_1^2-x_2^2}} \frac{dx_1dx_2dx_3}{x_1^2 + x_2^2 + (x_3 + 0.5)^2}$$
$$= 1.1460\cdots \times 10.$$

Now we give a comparison of the precisions of the approximate values of $K$ given by the quadrature formula constructed by the $\mathscr{R}_3$ set (Cf. example

**Table 3**

| Methods | $n$ | Approximate values of $K$ | Times in Djs-6 |
|---|---|---|---|
| Gauss Method | $2^3$ | $1.2409 \times 10$ | 2.6 sec. |
| | $4^3$ | $1.1324 \times 10$ | 15 sec. |
| | $8^3$ | $1.1391 \times 10$ | 104 sec. |
| | $16^3$ | $1.1425 \times 10$ | 789 sec. |
| Number Theoretic method | 20 | $1.3063 \times 10$ | 0.5 sec. |
| | 83 | $1.0928 \times 10$ | 1 sec. |
| | 418 | $1.1494 \times 10$ | 9 sec. |
| | 1,692 | $1.1460 \times 10$ | 41 sec. |

2 of §8.4) and the Cartesion product formula of Gaussian quadrature formula as follows.

*Remarks.* 1. The precision of numerical integration is often higher, if the integrand is replaced by a periodic function (Cf. §6.4—§6.5). But the integrand becomes complicated, if it has changed its variables. Hence careful analysis of the integrand is needed in practical use (Cf. D. Maisoneuve [1]).

2. If $s$ is large, the integral over $G_s$ may be divided into several parts and then the quadrature formula of lower dimensions can be applied to each part respectively. For example, suppose that $s = s' + s''$ and the good lattice points of $s'$ and $s''$ dimensions are $\mathbf{h}' (\bmod n')$ and $\mathbf{h}''$ $(\bmod n'')$ respectively. Then

$$\frac{1}{n'n''}\sum_{k=1}^{n'}\sum_{l=1}^{n''} f\left(\frac{k\mathbf{h}'}{n'}, \frac{l\mathbf{h}''}{n''}\right)$$

$$= \frac{1}{n'n''}\sum_{k=1}^{n'}\sum_{l=1}^{n''}\sum C(\mathbf{m}) e^{2\pi i\left(\frac{(\mathbf{h}',\mathbf{m}')k}{n'} + \frac{(\mathbf{h}'',\mathbf{m}'')l}{n''}\right)}$$

$$= \sum_{\substack{(\mathbf{h}',\mathbf{m}')\equiv 0 (\bmod n') \\ (\mathbf{h}'',\mathbf{m}'')\equiv 0 (\bmod n'')}} C(\mathbf{m}),$$

where $\mathbf{m} = (\mathbf{m}', \mathbf{m}'')$ in which $\mathbf{m}'$ and $\mathbf{m}''$ denote the integral vectors of $s'$ and $s''$ dimensions respectively. Hence

$$\sup_{f \in E_s^\alpha(C)} \left| \int_{G_s} f(\mathbf{x}) d\mathbf{x} - \frac{1}{n'n''}\sum_{k=1}^{n'}\sum_{l=1}^{n''} f\left(\frac{k\mathbf{h}'}{n'}, \frac{l\mathbf{h}''}{n''}\right)\right|$$

$$\leqslant C \sum_{\substack{(\mathbf{h}'\mathbf{m}')\equiv 0 (\bmod n') \\ (\mathbf{h}'',\mathbf{m}'')\equiv 0 (\bmod n'')}}' \frac{1}{(\|\mathbf{m}'\|\|\mathbf{m}''\|)^\alpha}.$$

It follows by the well-known method that the right hand side is dominated by

$$O((\min(n', n''))^{-\alpha} (\ln n'n'')^{\alpha(s-1)})$$

(Cf. §7.6). In general,the error is comparatively large. But some advantages may be obtained for some particular values of $n = n'n''$ (Cf. S. K, Zaramba [3]).

## Notes

Theorems 8.2 and 8.3: Cf. Zhang Roug Xiao [1], S. Haber [2] and M. Maisoneuve [1],

§ 2—§ 3: Cf. Hua Loo Keng and Wang Yuan [6,7].

§ 4—§ 5: Cf. Hua Loo Keng and Wang Yuan [1,4,5,6,7]. S. Haber, [2], Y. S. Moon [1] and Wang Yuan, Xu Guang Shan and Zhang Rong Xiao [1].

Theorems 6.4 and 6.5 were proved by N. M. Korobov [5,7] and Theorem 6.6 was obtained by Wang Yuan, Zhu Yao Cheng and Jian Yun Cui [1] and P. Keast [1,2].

Concerning the numerical integration over a domain $\mathscr{D} \fallingdotseq G_s$, we may refer also V. M. Colodov [2].

The examples of § 2 and § 9 were given by Xu Fong by the use of Djs—6.

*Chapter 9*

# Interpolation

## § 9.1. Introduction

Let $1 < n_1 < n_2 < \cdots$ be a sequence of integers and let

$$P_{n_l}(k) = (x_1^{(n_l)}(k), \cdots, x_s^{(n_l)}(k)), \quad 1 \leqslant k \leqslant n_l, l = 1, 2, \cdots$$

be a sequence of uniformly distributed sets in $G_s$. For any given function $f(\mathbf{x})$ on $G_s$, let

$$P_f(\mathbf{x}) = \sum_{k=1}^{n_l} f(P_{n_l}(k))\phi_{n_l,k}(\mathbf{x}), \tag{9.1}$$

where $\phi_{n_l,k}(\mathbf{x})(1 \leqslant k \leqslant n_l)$ are given functions.

If $P_f(\mathbf{x})$ converges to $f(\mathbf{x})$ according to a certain measure as $n_l \to \infty$, then $P_f(\mathbf{x})$ is called the approximate polynomial of $f(\mathbf{x})$. For simplicity, we use $P(\mathbf{x})$ or $P$ instead of $P_f(\mathbf{x})$.

The measures that are often used are as follows.

1. Absolute error

$$\sup_{\mathbf{x} \in G_s} |P(\mathbf{x}) - f(\mathbf{x})|.$$

2. Mean square error

$$\|P - f\|_2 = \left(\int_{G_s} |P - f|^2 d\mathbf{x}\right)^{1/2}.$$

Suppose that $f$ has an absolutely convergent Fourier expansion

$$f(\mathbf{x}) = \sum C(\mathbf{m}) e^{2\pi i(\mathbf{m},\mathbf{x})},$$

where

$$C(\mathbf{m}) = \int_{G_s} f(\mathbf{x}) e^{-2\pi i(\mathbf{m},\mathbf{x})} d\mathbf{x}.$$

Then by the method of numerical integration, we may use

$$\sum_{k=1}^{n_l} a_{n_l,k} f(P_{n_l}(k)) e^{-2\pi i(\mathbf{m},P_{n_l}(k))}$$

to approximate the Fourier coefficient $C(\mathbf{m})$ and so we may expect to use

$$P(\mathbf{x}) = \sum_{\|\mathbf{m}\| < N(n_l)} \sum_{k=1}^{n_l} \alpha_{n_l,k} f(P_{n_l}(k)) e^{2\pi i(\mathbf{m}, \mathbf{x}-P_{n_l}(k))}$$

to approximate $f(\mathbf{x})$. This is the simplest method to construct the approximate polynomial. We may also use other methods to construct the approximate polynomials. The results of this chapter are generalization and application of the multiple quadrature stated in previous chapters.

## § 9.2. The set of equi-distribution and interpolation

Let $m$ be an integer $\geqslant 2$, $n = m^s$, $\mathbf{l} = (l_1, \cdots, l_s)$

and
$$P(\mathbf{x}) = \frac{1}{n} \sum_{\substack{0 \leqslant l_i < m \\ 1 \leqslant i \leqslant s}} f\left(\frac{\mathbf{l}}{m}\right) \sum_{\|\mathbf{k}\| < N} e^{2\pi i\left(\mathbf{k}, \mathbf{x} - \frac{\mathbf{l}}{m}\right)}.$$

**Theorem 9.1.** *Suppose that* $\alpha > 1$ *and* $N = \left[\frac{m}{2}\right]$. *Then*

$$\sup_{f \in E_s^\alpha(C)} \|P - f\|_2 \leqslant C c(\alpha, s) n^{-\frac{2\alpha-1}{2s}} (\ln n)^{\frac{s-1}{2}}.$$

*Proof.* Suppose that $f \in E_s^\alpha(C)$ and

$$f(\mathbf{x}) = \sum C(\mathbf{m}) e^{2\pi i(\mathbf{m}, \mathbf{x})}.$$

Then

$$C(\mathbf{m}) - \frac{1}{n} \sum_{\substack{0 \leqslant l_i < m \\ 1 \leqslant i \leqslant s}} f\left(\frac{\mathbf{l}}{m}\right) e^{-2\pi i\left(\mathbf{m}, \frac{\mathbf{l}}{m}\right)}$$

$$= C(\mathbf{m}) - \frac{1}{n} \sum_{\substack{0 \leqslant l_i < m \\ 1 \leqslant i \leqslant s}} \sum C(\mathbf{k}) e^{2\pi i\left(\mathbf{k}-\mathbf{m}, \frac{\mathbf{l}}{m}\right)}$$

$$= C(\mathbf{m}) - \sum_{\substack{m \mid (k_i - m_i) \\ 1 \leqslant i \leqslant s}} C(\mathbf{k})$$

$$= - \sum_{\substack{m \mid r_i \\ 1 \leqslant i \leqslant s}}{}' C(\mathbf{r} + \mathbf{m}).$$

Since for integer $t$,

$$\int_0^1 e^{2\pi i t x} dx = \begin{cases} 1, & \text{if } t = 0, \\ 0, & \text{if } t \neq 0, \end{cases}$$

therefore

$$\begin{aligned} \|P - f\|_2^2 &= \int_{G_s} |P - f|^2 d\mathbf{x} \\ &= \int_{G_s} (P - f)(\overline{P - f}) d\mathbf{x} \\ &= \sum_{\|\mathbf{m}\| < N} \Big| \sum_{\substack{m | r_i \\ 1 \leqslant i \leqslant s}}{}' C(\mathbf{r} + \mathbf{m}) \Big|^2 + \sum_{\|\mathbf{m}\| \geqslant N} |C(\mathbf{m})|^2. \quad (9.2) \end{aligned}$$

Hence

$$\sup_{f \in E_s^\alpha(C)} \|P - f\|_2 \leqslant C^2(\Sigma_1 + \Sigma_2),$$

where

$$\Sigma_1 = \sum_{\|\mathbf{m}\| < N} \Big( \sum_{\substack{m | r_i \\ 1 \leqslant i \leqslant s}}{}' \frac{1}{\|\mathbf{r} + \mathbf{m}\|^\alpha} \Big)^2$$

and

$$\Sigma_2 = \sum_{\|\mathbf{m}\| \geqslant N} \frac{1}{\|\mathbf{m}\|^{2\alpha}}.$$

By Lemma 7.7, we have

$$\begin{aligned} \Sigma_1 &\leqslant s \sum_{\|\mathbf{m}\| < N} \left( 2 \sum_{r=1}^{\infty} \frac{1}{m^\alpha \left(r - \frac{1}{2}\right)^\alpha} \right)^2 \left( 1 + 2 \sum_{r=1}^{\infty} \frac{1}{m^\alpha \left(r - \frac{1}{2}\right)^\alpha} \right)^{2(s-1)} \\ &\leqslant c(\alpha, s) m^{-2\alpha} \sum_{\|\mathbf{m}\| < N} 1 \leqslant c(\alpha, s) m^{-2\alpha+1} (\ln n)^{s-1} \end{aligned}$$

and

$$\Sigma_2 \leqslant c(\alpha, s) m^{-2\alpha+1} (\ln n)^{s-1}.$$

The theorem follows.

**Theorem 9.2.** *Suppose that* $\alpha > 1$. *Then*

$$\sup_{f \in E_s^\alpha(C)} \|P - f\|_2 \geqslant \begin{cases} C, & \text{if } N > m, \\ \dfrac{C}{\sqrt{2\alpha - 1}} n^{-\frac{2\alpha - 1}{2s}}, & \text{if } N \leqslant m. \end{cases}$$

*Proof.* Take

$$f(\mathbf{x}) = \sum_{k=-\infty}^{\infty} \frac{e^{2\pi i k x_1}}{\bar{k}^\alpha}.$$

Then by (9.2),

$$\|P-f\|_2^2 = C^2 \sum_{\bar{k}<N} \left(\sum_{m|r}' \frac{1}{(r+k)^\alpha}\right)^2 + C^2 \sum_{\bar{k}>N} \frac{1}{\bar{k}^{2\alpha}}.$$

Hence

$$\|P-f\|_2^2 \geqslant C^2 \sum_{\bar{k}<N} \frac{1}{(m+k)^{2\alpha}} \geqslant C^2$$

for $N > m$ and

$$\|P-f\|_2^2 \geqslant C^2 \sum_{k\geqslant N} \frac{1}{k^{2\alpha}} \geqslant C^2 \int_N^\infty \frac{dt}{t^{2\alpha}}$$

$$= \frac{C^2}{2\alpha-1} N^{-2\alpha+1} \geqslant \frac{C^2}{2\alpha-1} m^{-2\alpha+1}$$

for $N \leqslant m$. The theorem follows.

It follows from Theorem 9.2 that the principal order $n^{-\frac{2\alpha-1}{2s}}$ of the error term in Theorem 9.1 can not be improved further. In the following, we shall study the lower estimate of the error term between the function and any of its approximate polynomial constructed by the set of equi-distribution.

**Lemma 9.1.** *Suppose that*

$$\phi(x) = \begin{cases} \left(\dfrac{\sin 2\pi m x}{2m}\right)^{\alpha-1}, & \text{if } 0 \leqslant x \leqslant \dfrac{1}{2m}, \\ 0, & \text{if } \dfrac{1}{2m} \leqslant x \leqslant 1 \end{cases}$$

*and* $\phi(x+1) = \phi(x)$, *where* $\alpha$ *is an integer* $> 1$. *Then*

$$f(\mathbf{x}) = \phi(x_1)\cdots\phi(x_s) \in E_s^\alpha(c(\alpha)^s).$$

*Proof.* Since

$$\phi(x) = \left(\frac{e^{2\pi i m x} - e^{-2\pi i m x}}{4im}\right)^{\alpha-1}$$

$$= \frac{1}{(4im)^{\alpha-1}} \sum_{\beta=0}^{\alpha-1} C_\beta^{\alpha-1} e^{2\pi i m(2\beta-\alpha+1)x},$$

therefore

$$\sup_{x\in G_1} |\phi^{(\alpha-1)}(x)| \leqslant \frac{|2\pi i m(\alpha-1)|^{\alpha-1} 2^{\alpha-1}}{|4im|^{\alpha-1}} = \pi^{\alpha-1}(\alpha-1)^{\alpha-1}$$

and

$$\sup_{x\in G_1} |\phi^{(\alpha)}(x)| \leqslant 2m\pi^\alpha(\alpha-1)^\alpha.$$

Since

$$\psi^{(\nu)}(0) = \psi^{(\nu)}\left(\frac{1}{2m}\right) = 0, \quad \nu = 0, 1, \cdots, \alpha - 2,$$

so

$$C(k) = \int_0^1 \psi(x) e^{-2\pi ikx} dx$$

$$= \frac{1}{(2\pi ik)^{\alpha-1}} \int_0^{\frac{1}{2m}} \psi^{(\alpha-1)}(x) e^{-2\pi ikx} dx$$

$$= \frac{-1}{(2\pi ik)^{\alpha}} \psi^{(\alpha-1)}(x) e^{-2\pi ikx} \Big|_0^{\frac{1}{2m}} + \frac{1}{(2\pi ik)^{\alpha}} \int_0^{\frac{1}{2m}} \psi^{(\alpha)}(x) e^{-2\pi ikx} dx$$

and

$$|C(k)| \leqslant c(\alpha) \bar{k}^{-\alpha}.$$

The lemma is proved.

**Theorem 9.3.** *Suppose that* $\alpha$ *is an integer* $>1$ *and* $P(\mathbf{x})$ *is an approximate polynomial of* $f$ *of the type* (9.1) *defined by the set of equi-distribution* $\mathbf{l}/m$. *Then*

$$\sup_{f \in E_s^\alpha(C)} \sup_{\mathbf{x} \in G_s} |f - P| \geqslant C c(\alpha, s) n^{-\frac{\alpha-1}{s}}.$$

*Proof.* Take

$$f(\mathbf{x}) = Cc(\alpha, s)\psi\left(x_1 + \frac{1}{2m}\right),$$

where $\psi(x)$ is defined in Lemma 9.1. Then we may choose suitable $c(\alpha, s)$ such that $f \in E_s^\alpha(C)$. Since

$$f\left(\frac{\mathbf{l}}{m}\right) = Cc(\alpha, s)\psi\left(\frac{l_1}{m} + \frac{1}{2m}\right) = 0, \quad 0 \leqslant l_i < m, 1 \leqslant i \leqslant s,$$

so $P\left(\frac{\mathbf{l}}{m}\right) = 0$. On the other hand

$$f\left(1 - \frac{1}{4m}, 0, \cdots, 0\right) = C c(\alpha, s) \psi\left(\frac{1}{4m}\right)$$

$$= C c(\alpha, s) m^{-\alpha+1} = C c(\alpha, s) n^{-\frac{\alpha-1}{s}}.$$

The theorem is proved.

It follows from Theorems 9.2 and 9.3 that the error terms between function and its approximate polynomials constructed by using the set of equidistribution are comparatively large. In the following, we shall use the $\mathscr{R}_s$ set and *glp* set to construct the approximate polynomial $P$ of the function $f$ such that the principal order of the error term between $P$ and $f$ is independent of $s$.

## § 9.3. Several lemmas

**Lemma 9.2.** *Let* $\mathbf{l}=(l_1,\cdots,l_s)$ *be an integral vector satisfying* $\|\mathbf{l}\|\geqslant 3^s$ *and let* $N$ *be a number satisfying* $1\leqslant N\leqslant \|\mathbf{l}\|/3^s$. *Then*

$$\sum_{\|\mathbf{m}\|\leqslant N}\frac{1}{\|\mathbf{l}+\mathbf{m}\|^\alpha}<\begin{cases} s!\,c(\alpha,\varepsilon)^s N^{1+\varepsilon}\|\mathbf{l}\|^{-\alpha}, & \text{if } 1\geqslant\alpha>0,\\ s!\,c(\alpha)^s N^{\alpha}\|\mathbf{l}\|^{-\alpha}, & \text{if } \alpha>1.\end{cases}$$

*Proof*. Suppose that $1\geqslant\alpha>0$. Since $N\leqslant \bar{l}_1/3$ and

$$\sum_{\bar{m}_1\leqslant N}\frac{1}{(\bar{l}_1+m_1)^\alpha}\leqslant\left(\frac{3}{2}\right)^\alpha\frac{3N}{\bar{l}_1^\alpha}$$

for $s=1$, the lemma is true for $s=1$. Suppose that $k\geqslant 1$ and the lemma holds for $1\leqslant s\leqslant k$. Now we proceed to prove that the lemma is also true for $s=k+1$. Obviously, it follows from $\bar{m}_1\cdots\bar{m}_{k+1}\leqslant\bar{l}_1\cdots\bar{l}_{k+1}/3^{k+1}$ that there exists at least an $m_i$ such that $\bar{m}_i<\bar{l}_i/2$, where $1\leqslant i\leqslant k+1$. Hence

$$\sum_{\bar{m}_1\cdots\bar{m}_{k+1}\leqslant N}\frac{1}{((\overline{l_1+m_1})\cdots(\overline{l_{k+1}+m_{k+1}}))^\alpha}\leqslant\sum\nolimits_1+\cdots+\sum\nolimits_{k+1},$$

where

$$\sum\nolimits_i=\sum_{\substack{\bar{m}_1\cdots\bar{m}_{k+1}\leqslant N\\ \bar{m}_i\leqslant\bar{l}_i/2}}\frac{1}{((\overline{l_1+m_1})\cdots(\overline{l_{k+1}+m_{k+1}}))^\alpha},\qquad 1\leqslant i\leqslant k+1.$$

Suppose that $N\leqslant\bar{l}_2\cdots\bar{l}_{k+1}/3^k$. Then by the induction hypothesis, we have

$$\begin{aligned}\sum\nolimits_1&\leqslant\frac{2^\alpha}{\bar{l}_1^\alpha}\sum_{\bar{m}_1\leqslant N}\ \sum_{\bar{m}_2\cdots\bar{m}_{k+1}\leqslant N/\bar{m}_1}\frac{1}{((\overline{l_2+m_2})\cdots(\overline{l_{k+1}+m_{k+1}}))^\alpha}\\ &\leqslant\frac{k!\,c(\alpha,\varepsilon)^kN^{1+\varepsilon}}{(\bar{l}_1\cdots\bar{l}_{k+1})^\alpha}\sum_{\bar{m}_1\leqslant N}\frac{1}{\bar{m}_1^{1+\varepsilon}}\leqslant\frac{k!\,c(\alpha,\varepsilon)^{k+1}N^{1+\varepsilon}}{(\bar{l}_1\cdots\bar{l}_{k+1})^\alpha}.\end{aligned}$$

Suppose that $N > \bar{l}_2 \cdots \bar{l}_{k+1}/3^k$. Then

$$\Sigma_1 \leqslant \sigma_1 + \sigma_2,$$

where

$$\sigma_1 = \frac{2^\alpha}{\bar{l}_1^\alpha} \sum_{\bar{m}_1 \leqslant 3^k N/\bar{l}_2 \cdots \bar{l}_{k+1}} \sum_{\bar{m}_2 \cdots \bar{m}_{k+1} \leqslant N/\bar{m}_1} \frac{1}{((\overline{l_2 + m_2}) \cdots (\overline{l_{k+1} + m_{k+1}}))^\alpha}$$

and

$$\sigma_2 = \frac{2^\alpha}{\bar{l}_1^\alpha} \sum_{3^k N/\bar{l}_2 \cdots \bar{l}_{k+1} < \bar{m}_1 \leqslant N} \sum_{\bar{m}_2 \cdots \bar{m}_{k+1} \leqslant N/\bar{m}_1} \frac{1}{((\overline{l_2 + m_2}) \cdots (\overline{l_{k+1} + m_{k+1}}))^\alpha}.$$

Evidently

$$\begin{aligned}\sigma_1 &\leqslant \frac{2^\alpha}{\bar{l}_1^\alpha} \sum_{\bar{m}_1 \leqslant 3^k N/\bar{l}_2 \cdots \bar{l}_{k+1}} \sum_{\bar{m}_2 \cdots \bar{m}_{k+1} \leqslant N/\bar{m}_1} \frac{(\bar{m}_2 \cdots \bar{m}_{k+1})^{1-\alpha+\varepsilon}}{(\bar{m}_2 \cdots \bar{m}_{k+1})^{1+\varepsilon}} \\ &\leqslant \frac{c(\alpha, \varepsilon)^k N^{1-\alpha+\varepsilon}}{\bar{l}_1^\alpha} \sum_{\bar{m}_1 \leqslant 3^k N/\bar{l}_2 \cdots \bar{l}_{k+1}} \frac{\bar{m}_1^\alpha}{\bar{m}_1^{1+\varepsilon}} \\ &\leqslant \frac{c(\alpha, \varepsilon)^k N^{1+\varepsilon}}{(\bar{l}_1 \cdots \bar{l}_{k+1})^\alpha} \sum \frac{1}{\bar{m}_1^{1+\varepsilon}} \leqslant \frac{c(\alpha, \varepsilon)^{k+1} N^{1+\varepsilon}}{(\bar{l}_1 \cdots \bar{l}_{k+1})^\alpha}\end{aligned}$$

and by the induction hypothesis, we have

$$\sigma_2 \leqslant \frac{k!\, c(\alpha, \varepsilon)^k N^{1+\varepsilon}}{(\bar{l}_1 \cdots \bar{l}_{k+1})^\alpha} \sum \frac{1}{\bar{m}_1^{1+\varepsilon}} \leqslant \frac{k!\, c(\alpha, \varepsilon)^{k+1} N^{1+\varepsilon}}{(\bar{l}_1 \cdots \bar{l}_{k+1})^\alpha}.$$

Hence it follows that

$$\Sigma_1 \leqslant \frac{k!\, c(\alpha, \varepsilon)^{k+1} N^{1+\varepsilon}}{(\bar{l}_1 \cdots \bar{l}_{k+1})^\alpha}.$$

Since $\Sigma_i$ satisfies also the above inequality for $2 \leqslant i \leqslant k+1$, the lemma follows by mathematical induction. For the case $\alpha > 1$, the lemma may be proved similarly.

**Lemma 9.3.** *Suppose that* $\alpha$ *and* $Q$ *are numbers satisfying* $1 \geqslant \alpha > 0$ *and* $Q \geqslant 1$. *If the congruence*

$$(\mathbf{a}, \mathbf{m}) = \sum_{k=1}^{s} a_k m_k \equiv 0 \pmod{n} \tag{9.3}$$

*has no solution in the domain*

$$\|\mathbf{m}\| \leqslant M, \quad \mathbf{m} \neq \mathbf{0}, \tag{9.4}$$

*then*

$$\sum_{\substack{(\mathbf{a},\mathbf{m})\equiv 0 \pmod n \\ \|\mathbf{m}\|\leqslant Q}}' \frac{1}{\|\mathbf{m}\|^{\alpha}} \leqslant c(\varepsilon)^s Q^{1-\alpha+\varepsilon} M^{-1}.$$

*Proof.* By Theorem 7.9, we have

$$\sum_{(\mathbf{a},\mathbf{m})\equiv 0 \pmod n}' \frac{1}{\|\mathbf{m}\|^{1+\varepsilon}} \leqslant c(\varepsilon)^s M^{-1}.$$

Hence

$$\sum_{\substack{(\mathbf{a},\mathbf{m})\equiv 0 \pmod n \\ \|\mathbf{m}\|\leqslant Q}}' \frac{1}{\|\mathbf{m}\|^{\alpha}} \leqslant Q^{1-\alpha+\varepsilon} \sum_{(\mathbf{a},\mathbf{m})\equiv 0 \pmod n}' \frac{1}{\|\mathbf{m}\|^{1+\varepsilon}}$$
$$\leqslant c(\varepsilon)^s Q^{1-\alpha+\varepsilon} M^{-1}.$$

The lemma is proved.

**Lemma 9.4.** *Suppose that* $n = p$ *and* $M = (4\zeta(1+\varepsilon)+2)^{-s} p^{1-\varepsilon}$. *Then there exists an integral vector* $\mathbf{a} = (1, a, \cdots a^{s-1})$ *such that the congruence* (9.3) *has no solution in the domain* (9.4).

*Proof.* Let $\mathbf{a} = (1, a, \cdots, a^{s-1})$. If $\mathbf{m} \neq 0$ and $\|\mathbf{m}\| \leqslant M$, then the congruence (9.3) has at most $s-1$ solutions in the range $1 \leqslant a \leqslant p$ (Cf. Lemma 4.5). Hence the total number of solutions of the congruence (9.3) in the domain $\|\mathbf{m}\| \leqslant M, \mathbf{m} \neq \mathbf{0}$ and $1 \leqslant a \leqslant p$ does not exceed

$$\sum_{\|\mathbf{m}\|\leqslant M}' \sum_{\substack{(\mathbf{a},\mathbf{m})\equiv 0 \pmod p \\ 1\leqslant a\leqslant p}} 1 \leqslant (s-1) \sum_{\|\mathbf{m}\|\leqslant M}' \frac{\|\mathbf{m}\|^{1+\varepsilon}}{\|\mathbf{m}\|^{1+\varepsilon}}$$
$$\leqslant (s-1)(2\zeta(1+\varepsilon)+1)^s M^{1+\varepsilon} < p/2.$$

Consequently, there exists an integer $a$ satisfying $1 \leqslant a \leqslant p$ such that the congruence (9.3) has no solution in the domain (9.4). The lemma is proved.

## § 9.4. The approximate formula of the function of $E_s^{\alpha}(C)$

In this section, we suppose that $\alpha > 1$ and use the notations

$$\Delta_1 = \sup_{f\in E_s^{\alpha}(C)} \left\| \frac{1}{n} \sum_{k=1}^{n} f\left(\frac{k\mathbf{a}}{n}\right) \sum_{\|\mathbf{m}\|<N} e^{2\pi i\left(\mathbf{m},\, \mathbf{x}-\frac{k\mathbf{a}}{n}\right)} - f \right\|_2$$

and

$$\Delta_2 = \sup_{f\in E_s^\alpha(C)} \sup_{\mathbf{x}\in G_s} \left| \frac{1}{n}\sum_{k=1}^{n} f\left(\frac{k\mathbf{a}}{n}\right) \sum_{\|\mathbf{m}\|<N} e^{2\pi i\left(\mathbf{m},\, \mathbf{x}-\frac{k\mathbf{a}}{n}\right)} - f \right|.$$

**Theorem 9.4.** *Suppose that* $N = [M^{\frac{2\alpha}{4\alpha-1}}]$. *If the congruence* (9.3) *has no solution in the domain* (9.4), *then*

$$\Delta_1 \leqslant Cs!^{1/2} c(\alpha, \varepsilon)^s M^{-\frac{\alpha(2\alpha-1)}{4\alpha-1}+\varepsilon}.$$

*Proof*. Clearly, we have

$$\Delta_1^2 \leqslant \Sigma_1 + \Sigma_2, \tag{9.5}$$

where

$$\Sigma_1 = \sup_{f\in E_s^\alpha(C)} \sum_{\|\mathbf{m}\|<N} \left| C(\mathbf{m}) - \frac{1}{n}\sum_{k=1}^{n} f\left(\frac{k\mathbf{a}}{n}\right) e^{-2\pi i\left(\mathbf{m},\frac{k\mathbf{a}}{n}\right)} \right|^2$$

and

$$\Sigma_2 = \sup_{f\in E_s^\alpha(C)} \sum_{\|\mathbf{m}\|\geqslant N} |C(\mathbf{m})|^2.$$

It follows by Lemma 3.6 that

$$\frac{1}{n}\sum_{k=1}^{n} f\left(\frac{k\mathbf{a}}{n}\right) e^{-2\pi i\left(\mathbf{m},\frac{k\mathbf{a}}{n}\right)} = \frac{1}{n}\sum_{k=1}^{n}\sum C(\mathbf{l}) e^{2\pi i(\mathbf{l}-\mathbf{m},\mathbf{a})k/n}$$

$$= \sum_{(\mathbf{a},\mathbf{l}-\mathbf{m})\equiv 0 \pmod n} C(\mathbf{l}) = \sum_{(\mathbf{a},\mathbf{r})\equiv 0 \pmod n} C(\mathbf{r}+\mathbf{m}),$$

$$\left| C(\mathbf{m}) - \frac{1}{n}\sum_{k=1}^{n} f\left(\frac{k\mathbf{a}}{n}\right) e^{-2\pi i\left(\mathbf{m},\frac{k\mathbf{a}}{n}\right)} \right| \leqslant C \sum_{(\mathbf{a},\mathbf{r})\equiv 0 \pmod n}{}' \frac{1}{\|\mathbf{r}+\mathbf{m}\|^\alpha}.$$

We may suppose that $N \leqslant M/3^s$. Since

$$\|\mathbf{r}\| \leqslant 2^s \|\mathbf{m}\|\,\|\mathbf{r}+\mathbf{m}\|,$$

therefore by Theorem 7.9 and Lemma 9.2, we have

$$\Sigma_1 \leqslant C^2 \sum_{\|\mathbf{m}\|<N} \left( \sum_{(\mathbf{a},\mathbf{r})\equiv 0 \pmod n}{}' \frac{1}{\|\mathbf{r}+\mathbf{m}\|^\alpha} \right)^2$$

$$= C^2 \sum_{\|\mathbf{m}\|<N} \sum_{(\mathbf{a},\mathbf{r})\equiv 0 \pmod n}{}' \frac{1}{\|\mathbf{r}+\mathbf{m}\|^\alpha} \sum_{(\mathbf{a},\mathbf{r}')\equiv 0 \pmod n}{}' \frac{\|\mathbf{m}\|^\alpha}{\|\mathbf{r}'\|^\alpha} \left( \frac{\|\mathbf{r}'\|}{\|\mathbf{m}\|\,\|\mathbf{r}'+\mathbf{m}\|} \right)^\alpha$$

$$\leqslant C^2 2^{\alpha s} N^\alpha \sum_{(\mathbf{a},\mathbf{r}')\equiv 0 \pmod n}{}' \frac{1}{\|\mathbf{r}'\|^2} \sum_{(\mathbf{a},\mathbf{r})\equiv 0 \pmod n}{}' \sum_{\|\mathbf{m}\|<N} \frac{1}{\|\mathbf{r}+\mathbf{m}\|^\alpha}$$

$$\leqslant C^2 s!\, c(\alpha)^s N^{2\alpha}\Bigg(\sum_{(\mathbf{a},\mathbf{r})\equiv 0 \pmod n}{}' \frac{1}{\|\mathbf{r}\|^{\alpha}}\Bigg)^2$$

$$\leqslant C^2 s!\, c(\alpha, \varepsilon)^s N^{2\alpha} M^{-2\alpha+\varepsilon} \leqslant C^2 s!\, c(\alpha, \varepsilon)^s M^{-\frac{2\alpha(2\alpha-1)}{4\alpha-1}+\varepsilon}. \tag{9.6}$$

The $\sum_2$ may be estimated as follows.

$$\sum\nolimits_2 \leqslant C^2 \sum_{\|\mathbf{m}\|\geqslant N} \frac{1}{\|\mathbf{m}\|^{2\alpha}} = C^2 \sum_{\|\mathbf{m}\|\geqslant N} \frac{\|\mathbf{m}\|^{-2\alpha+1+\varepsilon}}{\|\mathbf{m}\|^{1+\varepsilon}}$$

$$= C^2 N^{-2\alpha+1+\varepsilon} \sum \frac{1}{\|\mathbf{m}\|^{1+\varepsilon}} = C^2 c(\varepsilon)^s N^{-2\alpha+1+\varepsilon}$$

$$= C^2 c(\varepsilon)^s M^{-\frac{2\alpha(2\alpha-1)}{4\alpha-1}+\varepsilon}. \tag{9.7}$$

The theorem follows by (9.5), (9.6) and (9.7).

**Theorem 9.5.** *Suppose that* $N = [M^{\frac{\alpha}{2\alpha-1}}]$. *If the congruence* (9.3) *has no solution in* (9.4), *then*

$$\Delta_2 \leqslant Cs!\, c(\alpha, \varepsilon)^s M^{-\frac{\alpha(\alpha-1)}{2\alpha-1}+\varepsilon}.$$

*Proof.* Evidently

$$\Delta_2 \leqslant \sum\nolimits_1 + \sum\nolimits_2,$$

where

$$\sum\nolimits_1 = \sup_{f\in E_s^\alpha(C)} \sum_{\|\mathbf{m}\|<N} \left| C(\mathbf{m}) - \frac{1}{n}\sum_{k=1}^{n} f\left(\frac{k\mathbf{a}}{n}\right) e^{-2\pi i\left(\mathbf{m},\frac{k\mathbf{a}}{n}\right)}\right|$$

and

$$\sum\nolimits_2 = \sup_{f\in E_s^\alpha(C)} \sum_{\|\mathbf{m}\|\geqslant N} |C(\mathbf{m})|.$$

Similar to (9.6) and (9.7), we have

$$\sum\nolimits_1 \leqslant Cs!\, c(\alpha, \varepsilon)^s N^{\alpha} M^{-\alpha+\varepsilon} = Cs!\, c(\alpha, \varepsilon)^s M^{-\frac{\alpha(\alpha-1)}{2\alpha-1}+\varepsilon}$$

and

$$\sum\nolimits_2 \leqslant C \sum_{\|\mathbf{m}\|\geqslant N} \frac{1}{\|\mathbf{m}\|^{\alpha}} \leqslant Cc(\varepsilon)^s N^{-\alpha+1+\varepsilon}$$

$$\leqslant Cc(\varepsilon)^s M^{-\frac{\alpha(\alpha-1)}{2\alpha-1}+\varepsilon}.$$

The theorem follows.

By Theorem 9.4 and Lemma 7.5 (with the notation of §4.6), we have

**Theorem 9.6.** *Suppose that* $s = \dfrac{\varphi(m)}{2}$ *and* $\mathbf{a} = (c_1, \cdots, c_s)$. *Then*

$$\Delta_1 \leqslant Cc(\mathscr{R}_s, \alpha, \varepsilon) n^{-\left(\frac{1}{2}+\frac{1}{2(s-1)}\right)\frac{\alpha(2\alpha-1)}{4\alpha-1}+\varepsilon}.$$

From Theorem 9.4 and Lemma 9.4, we have:

**Theorem 9.7.** *Suppose that* $n = p$. *Then there exists an integral vector* $\mathbf{a}(=\mathbf{a}(p))$ *such that*

$$\Delta_1 \leqslant Cs!^{1/2} c(\alpha, \varepsilon)^s p^{-\frac{\alpha(2\alpha-1)}{4\alpha-1}+\varepsilon}.$$

## § 9.5. The approximate formula of the function of $Q_s^\alpha(C)$

Introduce the notations

$$\mu(\alpha) = \begin{cases} \dfrac{\alpha}{2}, & \text{if } \alpha > 1, \\[2ex] \dfrac{2\alpha^2}{1+4\alpha-\alpha^2}, & \text{if } 1 \geqslant \alpha > 0 \end{cases}$$

and

$$\Delta = \sup_{f \in Q_s^\alpha(C)} \left\| \frac{1}{n} \sum_{k=1}^{n} f\left(\frac{k\mathbf{a}}{n}\right) \sum_{\|\mathbf{m}\| < N} e^{2\pi i\left(\mathbf{m}, \mathbf{x} - \frac{k\mathbf{a}}{n}\right)} - f \right\|_2.$$

**Theorem 9.8.** *Suppose that* $N = [M^{\frac{\mu(\alpha)}{\alpha}}]$. *If the congruence* (9.3) *has no solution in* (9.1), *then*

$$\Delta \leqslant Cs!^{1/2} c(\alpha, \varepsilon)^s M^{-\mu(\alpha)+\varepsilon}.$$

*Proof.* Let

$$T = \begin{cases} [\log_2 M] + 1, & \text{if } \alpha > 1, \\ [\log_2 MN^{-1+\alpha/2}] + 1, & \text{if } 1 \geqslant \alpha > 0. \end{cases}$$

Then by Minkowski's inequality, we have

$$\Delta \leqslant \Sigma_1 + \Sigma_2 + \Sigma_3,$$

where

$$\Sigma_1 = \sup_{f \in Q_s^\alpha(C)} \left\| f(\mathbf{x}) - \sum_{t_0 \leqslant T}{}'' \varphi_{\mathbf{t}}(\mathbf{x}) \right\|_2,$$

$$\Sigma_2 = \sup_{f \in Q_s^\alpha(C)} \left\| \sum_{t_0 \leqslant T}{}'' \left( \varphi_{\mathbf{t}}(\mathbf{x}) - \frac{1}{n} \sum_{k=1}^{n} \varphi_{\mathbf{t}}\left(\frac{k\mathbf{a}}{n}\right) \sum_{\|\mathbf{m}\| < N} e^{2\pi i\left(\mathbf{m}, \mathbf{x} - \frac{k\mathbf{a}}{n}\right)} \right) \right\|_2$$

and

$$\Sigma_3 = \sup_{f \in Q_s^\alpha(C)} \left\| \frac{1}{n} \sum_{k=1}^{n} f\left(\frac{k\mathbf{a}}{n}\right) \sum_{\|\mathbf{m}\|<N} e^{2\pi i\left(\mathbf{m}, \mathbf{x} - \frac{k\mathbf{a}}{n}\right)} - \sum_{t_0 \leqslant T}'' \frac{1}{n} \sum_{k=1}^{n} \varphi_{\mathbf{t}}\left(\frac{k\mathbf{a}}{n}\right) \sum_{\|\mathbf{m}\|<N} e^{2\pi i\left(\mathbf{m}, \mathbf{x} - \frac{k\mathbf{a}}{n}\right)} \right\|_2.$$

It follows from Minkowski's inequality that

$$\Sigma_1 \leqslant \sup_{f \in Q_s^\alpha(C)} \sum_{t_0 > T}'' \|\varphi_{\mathbf{t}}\|_2 \leqslant C \sum_{t_0 > T}'' 2^{-(\alpha-\varepsilon)t_0 - \varepsilon t_0}$$
$$\leqslant C c(\alpha, \varepsilon)^s 2^{-(\alpha-\varepsilon)T} \leqslant C c(\alpha, \varepsilon)^s M^{-\mu(\alpha)+\varepsilon}.$$

Evidently

$$\Sigma_2 \leqslant \sigma_1 + \sigma_2,$$

where

$$\sigma_1 = \sup_{f \in Q_s^\alpha(C)} \left\| \sum_{t_0 \leqslant T}'' \sum_{\|\mathbf{m}\|<N} \left( C_{\mathbf{t}}(\mathbf{m}) - \frac{1}{n} \sum_{k=1}^{n} \varphi_{\mathbf{t}}\left(\frac{k\mathbf{a}}{n}\right) e^{-2\pi i\left(\mathbf{m}, \frac{k\mathbf{a}}{n}\right)} \right) e^{2\pi i(\mathbf{m}, \mathbf{x})} \right\|_2$$

and

$$\sigma_2 = \sup_{f \in Q_s^\alpha(C)} \left\| \sum_{t_0 \leqslant T}'' \sum_{\|\mathbf{m}\| \geqslant N} C_{\mathbf{t}}(\mathbf{m}) e^{2\pi i(\mathbf{m}, \mathbf{x})} \right\|_2.$$

Since

$$C_{\mathbf{t}}(\mathbf{m}) - \frac{1}{n} \sum_{k=1}^{n} \varphi_{\mathbf{t}}\left(\frac{k\mathbf{a}}{n}\right) e^{-2\pi i\left(\mathbf{m}, \frac{k\mathbf{a}}{n}\right)} = - \sum_{(\mathbf{a}, \mathbf{l}) \equiv 0 \pmod{n}}' C_{\mathbf{t}}(\mathbf{l} + \mathbf{m})$$

and

$$\|\mathbf{l}\| \leqslant 2^s \|\mathbf{m}\| \, \|\mathbf{l} + \mathbf{m}\|,$$

it follows by Theorem 6.2 that

$$\sigma_1^2 \leqslant \sup_{f \in Q_s^\alpha(C)} \left\| \sum_{t_0 \leqslant T}'' \sum_{\|\mathbf{m}\|<N} \sum_{(\mathbf{a}, \mathbf{l}) \equiv 0 \pmod{n}}' C_{\mathbf{t}}(\mathbf{l} + \mathbf{m}) e^{2\pi i(\mathbf{m}, \mathbf{x})} \right\|_2^2$$
$$\leqslant \sup_{f \in Q_s^\alpha(C)} \sum_{\|\mathbf{m}\|<N} \left( \sum_{t_0 \leqslant T}'' \sum_{(\mathbf{a}, \mathbf{l}) \equiv 0 \pmod{n}}' |C_{\mathbf{t}}(\mathbf{l} + \mathbf{m})| \right)^2$$
$$\leqslant C^2 c(\alpha)^s \sum_{\|\mathbf{m}\|<N} \left( \sum_{\substack{(\mathbf{a}, \mathbf{l}) \equiv 0 \pmod{n} \\ \|\mathbf{l}\| \leqslant 2^s + T_N}}' \frac{1}{\|\mathbf{l} + \mathbf{m}\|^\alpha} \right)^2$$
$$\leqslant C^2 c(\alpha)^s N^\alpha \sum_{\|\mathbf{m}\|<N} \sum_{\substack{(\mathbf{a}, \mathbf{l}) \equiv 0 \pmod{n} \\ \|\mathbf{l}\| \leqslant 2^s + T_N}}' \frac{1}{\|\mathbf{l} + \mathbf{m}\|^\alpha} \sum_{\substack{(\mathbf{a}, \mathbf{l}') \equiv 0 \pmod{n} \\ \|\mathbf{l}'\| \leqslant 2^s + T_N}}' \frac{1}{\|\mathbf{l}'\|^\alpha}.$$

We may suppose that $N \leqslant M/3^s$. Hence by Theorem 7.9, Lemmas 9.2 and 9.3, we have

$$\sigma_1^2 \leqslant \begin{cases} C^2 s!\, c(\alpha, \varepsilon)^s N^{2\alpha} M^{-2\alpha+\varepsilon}, & \text{if } \alpha > 1, \\ C^2 s!\, c(\alpha, \varepsilon)^s N^{3-\alpha+\varepsilon} M^{-2} 2^{2T(1-\alpha)+T\varepsilon}, & \text{if } 1 \geqslant \alpha > 0. \end{cases}$$

By Schwarz's inquality,

$$\begin{aligned}
\sigma_2^2 &\leqslant \sup_{f\in Q_s^\alpha(C)} \sum_{\|\mathbf{m}\|\geqslant N} \Big(\sum_{t_0\leqslant T}'' |C_{\mathbf{t}}(\mathbf{m})|\Big)^2 \\
&\leqslant \sup_{f\in Q_s^\alpha(C)} \sum_{\|\mathbf{m}\|\geqslant N} \sum_{t_0\leqslant T}'' 2^{-\varepsilon t_0/2} \sum_{t_0\leqslant T}'' 2^{\varepsilon t_0/2} |C_{\mathbf{t}}(\mathbf{m})|^2 \\
&\leqslant c(\varepsilon)^s \sup_{f\in Q_s^\alpha(C)} \sum_{t_0\leqslant T}'' \sum_{\|\mathbf{m}\|\geqslant N} 2^{\varepsilon t_0/2} |C_{\mathbf{t}}(\mathbf{m})|^2 \\
&\leqslant c(\varepsilon)^s \sup_{f\in Q_s^\alpha(C)} \sum_{t_0>\log_2 N}'' 2^{\varepsilon t_0/2} \|\varphi_{\mathbf{t}}\|_2^2 \\
&\leqslant C^2 c(\varepsilon)^s \sum_{t_0>\log_2 N}'' 2^{-2\alpha t_0+\varepsilon t_0/2} \\
&\leqslant C^2 c(\alpha, \varepsilon)^s N^{-2\alpha+\varepsilon}.
\end{aligned}$$

Hence

$$\Sigma_2 \leqslant C s!^{1/2} c(\alpha, \varepsilon)^s M^{-\mu(\alpha)+\varepsilon}.$$

$$\begin{aligned}
\Sigma_3 &\leqslant \sup_{f\in Q_s^\alpha(C)} \Big\| \sum_{t_0>T}'' \frac{1}{n} \sum_{k=1}^{n} \varphi_{\mathbf{t}}\Big(\frac{k\mathbf{a}}{n}\Big) \sum_{\|\mathbf{m}\|<N} e^{2\pi i\left(\mathbf{m}, \mathbf{x}-\frac{k\mathbf{a}}{n}\right)} \Big\|_2 \\
&\leqslant \sup_{f\in Q_s^\alpha(C)} \sum_{t_0>T}'' \|\varphi_{\mathbf{t}}\| \Big(\sum_{\|\mathbf{m}\|<N} 1\Big)^{1/2} \\
&\leqslant C \sum_{t_0>T}'' 2^{-\alpha t_0} \Big(\sum_{\|\mathbf{m}\|<N} \frac{N^{1+\varepsilon}}{\|\mathbf{m}\|^{1+\varepsilon}}\Big)^{1/2} \\
&\leqslant C c(\alpha, \varepsilon)^s N^{\frac{1}{2}+\frac{\varepsilon}{2}} 2^{-\alpha T+\frac{\varepsilon T}{2}} \\
&\leqslant C c(\alpha, \varepsilon)^s M^{-\mu(\alpha)+\varepsilon}.
\end{aligned}$$

The theorem follows.

From Theorem 9.8 and Lemma 7.5, we have

**Theorem 9.9.** *Suppose that* $s = \dfrac{\varphi(m)}{2}$ *and* $\mathbf{a} = (c_1, \cdots, c_s)$. *Then*

$$\Delta \leqslant C c(\mathscr{R}_s, \alpha, \varepsilon)\, n^{-\left(\frac{1}{2}+\frac{1}{2(s-1)}\right)\mu(\alpha)+\varepsilon}.$$

From Theorem 9.8 and Lemma 9.4, we have

**Theorem 9.10.** *Suppose that* $n=p$. *Then there exists an integral vector* $\mathbf{a}(=\mathbf{a}(p))$ *such that*

$$\Delta \leqslant Cs!^{1/2}c(\alpha, \varepsilon)^s p^{-\mu(\alpha)+\varepsilon}.$$

## § 9.6. The Bernoulli polynomial and the approximate polynomial

Suppose that $\alpha > 1$. We shall use the Bernoulli polynomials to express the function $f$ of $E_s^\alpha(C)$ as a definite integral over $G_s$ and then by the use of the results of numerical integration to obtain the approximate polynomial of $f$. The resuts so obtained are sharper than those given in §9.4 for certain values of $\alpha$.

**Lemma 9.5.** *If* $f_i \in E_s^\alpha(C_i)(i=1,2)$, *then* $f_1 f_2 \in E_s^\alpha(C_1 C_2 c(\alpha)^s)$.

*Proof.* Let $C_i(\mathbf{m})$ and $C(\mathbf{m})$ be the Fourier coefficients of $f_i(i=1,2)$ and $f$ respectively. Then

$$\begin{aligned} f(\mathbf{x}) &= \sum_{\mathbf{n}} \sum_{\mathbf{k}} C_1(\mathbf{n}) C_2(\mathbf{k}) e^{2\pi i(\mathbf{n}+\mathbf{k}, \mathbf{x})} \\ &= \sum_{\mathbf{m}} \left( \sum_{\mathbf{n}} C_1(\mathbf{n}) C_2(\mathbf{m}-\mathbf{n}) \right) e^{2\pi i(\mathbf{m}, \mathbf{x})} \\ &= \sum_{\mathbf{m}} C(\mathbf{m}) e^{2\pi i(\mathbf{m}, \mathbf{x})}, \end{aligned}$$

where

$$C(\mathbf{m}) = \sum_{\mathbf{n}} C_1(\mathbf{n}) C_2(\mathbf{m}-\mathbf{n}).$$

Since

$$\begin{aligned} \sum_{n=-\infty}^{\infty} \frac{1}{(\bar{n}(\overline{m-n}))^\alpha} &= \sum_{|n| \leqslant |m|/2} \frac{1}{(\bar{n}(\overline{m-n}))^\alpha} + \sum_{|n| > |m|/2} \frac{1}{(\bar{n}(\overline{m-n}))^\alpha} \\ &\leqslant \frac{2^\alpha}{\bar{m}^\alpha} \left( \sum_{|n| \leqslant |m|/2} \frac{1}{\bar{n}^\alpha} + \sum_{|n| > |m|/2} \frac{1}{(\overline{m-n})^\alpha} \right) \\ &\leqslant \frac{2^{\alpha+1}}{\bar{m}^\alpha} \sum \frac{1}{\bar{n}^\alpha} = c(\alpha) \bar{m}^{-\alpha}, \end{aligned}$$

therefore

$$
\begin{aligned}
|C(\mathbf{m})| &\leqslant \left|\sum_{\mathbf{n}} C_1(\mathbf{n})C_2(\mathbf{m}-\mathbf{n})\right| \\
&\leqslant C_1C_2\sum_{\mathbf{n}}\frac{1}{(\|\mathbf{n}\|\,\|\mathbf{m}-\mathbf{n}\|)^{\alpha}} \\
&= C_1C_2\prod_{i=1}^{s}\left(\sum_{n_i=-\infty}^{\infty}\frac{1}{(\bar{n}_i\,(\overline{m_i-n_i}))^{\alpha}}\right) \\
&= C_1C_2c(\alpha)^s\|\mathbf{m}\|^{-\alpha}.
\end{aligned}
$$

The lemma is proved.

**Lemma 9.6.** *Suppose that $f$ has $r$-th continuous derivatives and $f \in E_1^{\alpha}(C)$. Then*

$$f(x) = \int_0^1 \sum_{\tau=0}^{1} f^{(\nu\tau)}(y)\varphi_\nu^{\tau}(y-x)dy$$

*for $\nu = 1, \cdots, r$, where*

$$\varphi_\nu(x) = \frac{(-1)^{\nu-1}}{\nu!}B_\nu(\{x\})$$

*in which $B_\nu(x)$ denotes the $\nu$-th Bernoulli polynomial.*

*Proof.* Since

$$
\begin{aligned}
&\int_0^1 \sum_{\tau=0}^{1} f^{(\nu\tau)}(y)\varphi_\nu^{\tau}(y-x)dy \\
&\quad = \int_0^1 f(y)dy + \frac{(-1)^{\nu-1}}{\nu!}\int_0^1 f^{(\nu)}(y)B_\nu(\{y-x\})dy \\
&\quad = \int_0^1 f(y)dy + \frac{(-1)^{\nu-1}}{\nu!}\int_0^1 f^{(\nu)}(x+y)B_\nu(y)dy,
\end{aligned}
$$

the lemma is reduced to proving by induction the assertion that the right hand side is equal to $f(x)$. Since

$$
\begin{aligned}
\int_0^1 f'(x+y)B_1(y)dy &= f(x+y)B_1(y)\Big|_0^1 - \int_0^1 f(x+y)dy \\
&= f(x) - \int_0^1 f(y)dy,
\end{aligned}
$$

the assertion holds for $\nu = 1$. Suppose that $r \geqslant \nu \geqslant 2$ and the assertion holds for any positive integer less than $\nu$. Since

$$B_\nu(1) = B_\nu(0), \quad B_\nu'(y) = \nu B_{\nu-1}(y)$$

by Lemma 6.6, therefore

$$\begin{aligned}
&\frac{(-1)^{\nu-1}}{\nu!}\int_0^1 f^{(\nu)}(x+y)B_\nu(y)dy \\
&= \frac{(-1)^{\nu-1}}{\nu!} f^{(\nu-1)}(x+y)B_\nu(y)\Big|_0^1 \\
&\quad - \frac{(-1)^{\nu-1}}{\nu!}\int_0^1 f^{(\nu-1)}(x+y)B_\nu'(y)dy \\
&= \frac{(-1)^{\nu-2}}{(\nu-1)!}\int_0^1 f^{(\nu-1)}(x+y)B_{\nu-1}(y)dy \\
&= \cdots = f(x) - \int_0^1 f(y)dy.
\end{aligned}$$

Hence the assertion holds for $\nu$. The lemma follows.

**Lemma 9.7.** *Suppose that $f$ has continuous derivatives $f(\mathbf{x})^{(r_1,\cdots,r_s)}(0 \leqslant r_1, \cdots, r_s \leqslant r)$ and $f \in E_s^\alpha(C)$. Then*

$$f(\mathbf{x}) = \int_{G_s} \sum_{\tau_1,\cdots,\tau_s=0}^{1} f(\mathbf{y})^{(\tau_1\nu,\cdots,\tau_s\nu)} \prod_{i=1}^{s} \varphi_\nu^{\tau_i}(y_i - x_i)d\mathbf{y}$$

*holds for $\nu = 1, \cdots, r$.*

*Proof.* The lemma is true for $s = 1$ by Lemma 9.6. Now suppose that $s \geqslant 2$ and the lemma holds for any positive integer less than $s$. Obviously $f \in E_{s-1}^\alpha(Cc(\alpha))$. Hence it follows by the induction hypothesis that

$$\begin{aligned}
f(\mathbf{x}) = \int_{G_{s-1}} &\sum_{\tau_1,\cdots,\tau_{s-1}=0}^{1} f(y_1,\cdots,y_{s-1},x_s)^{(\tau_1\nu,\cdots,\tau_{s-1}\nu,0)} \\
&\cdot \prod_{i=1}^{s-1} \varphi_\nu^{\tau_\nu}(y_\nu - x_\nu)dy_1\cdots dy_{s-1}
\end{aligned} \tag{9.8}$$

holds for $\nu = 1, \cdots, r$. Since

$$\begin{aligned}
&f(y_1,\cdots,y_{s-1},x_s)^{(\tau_1\nu,\cdots,\tau_{s-1}\nu,0)} \\
&= \int_0^1 \sum_{\tau_s=0}^{1} f(\mathbf{y})^{(\tau_1\nu,\cdots,\tau_s\nu)}\varphi_\nu^{\tau_s}(y_s - x_s)dy_s
\end{aligned} \tag{9.9}$$

by Lemma 9.6, the lemma follows by substituting (9.9) into (9.8).

Introduce the notation

$$\Delta = \sup_{f \in E_s^\alpha(C)} \sup_{\mathbf{x} \in G_s} \left| \frac{1}{n} \sum_{k=1}^{n} \sum_{\tau_1, \cdots, \tau_s = 0}^{1} f\left(\frac{k\mathbf{a}}{n}\right)^{(\tau_1 r, \cdots, \tau_s r)} \cdot \prod_{i=1}^{s} \varphi_r^{\tau_i}\left(\frac{a_i k}{n} - x_i\right) - f \right|.$$

**Theorem 9.11.** *Suppose that* $\alpha > 3$. *If the congruence* (9.3) *has no solutions in* (9.4), *then*

$$\Delta \leqslant C c(\alpha, \varepsilon)^s M^{-\gamma+\varepsilon},$$

*where* $\gamma = \min(r, \alpha - r)$ *and* $r = \left[\frac{\alpha+1}{2}\right]$.

*Proof.* Suppose that $f \in E_s^\alpha(C)$. Then

$$f(\mathbf{x})^{(\tau_1 r, \cdots, \tau_s r)} = (2\pi i)^{(\tau_1 + \cdots + \tau_s) r} \sum \left(\prod_{i=1}^{s} m_i^{\tau_i r}\right) C(\mathbf{m}) e^{2\pi i (\mathbf{m}, \mathbf{x})},$$

where $\tau_i = 0$ or $1 (1 \leqslant i \leqslant s)$. Since

$$\left|\left(\prod_{i=1}^{s} m_i^{\tau_i r}\right) C(\mathbf{m})\right| \leqslant C \|\mathbf{m}\|^{-\alpha} \prod_{i=1}^{s} |m_i|^{\tau_i r} \leqslant C \|\mathbf{m}\|^{-\alpha+r},$$

we have $f(\mathbf{x})^{(\tau_1 r, \cdots, \tau_s r)} \in E_s^{\alpha-r}(C(2\pi)^{rs})$, where

$$\alpha - r \geqslant \alpha - \frac{\alpha+1}{2} = \frac{\alpha-1}{2} > 1.$$

And so it follows from Lemma 9.7 that

$$f(\mathbf{x}) = \sum_{\tau_1, \cdots, \tau_s = 0}^{1} \int_{G_s} f(\mathbf{y})^{(\tau_1 r, \cdots, \tau_s r)} \prod_{i=1}^{s} \varphi_r^{\tau_i}(y_i - x_i) d\mathbf{y}. \tag{9.10}$$

Let $C(k)$ be the Fourier coefficient of $B_r(\{x\})$. Then $B_r^{(t)}(0) = B_r^{(t)}(1)$ for $t \leqslant r-2$ by Lemma 6.7 and so

$$\begin{aligned} C(k) &= \int_0^1 B_r(x) e^{-2\pi i k x} dx \\ &= \frac{-1}{(2\pi i k)^r} B_r^{(r-1)}(x) e^{-2\pi i k x} \Big|_0^1 + \frac{1}{(2\pi i k)^r} \int_0^1 B_r^{(r)}(x) e^{-2\pi i k x} dx \\ &= c(r) |k|^{-r} \end{aligned}$$

for $k \neq 0$, i.e., $B_r(\{x\}) \in E_1^r(c(r))$. Since for any given $\mathbf{x}$,

$$\prod_{i=1}^{s} \varphi_r^{\tau_i}(y_i - x_i) = \frac{(-1)^{(\tau_1+\cdots+\tau_s)(r-1)}}{r!^{\tau_1+\cdots+\tau_s}} \prod_{i=1}^{s} B_r^{\tau_i}(\{y_i - x_i\}) \in E_s^r(c(r)^s),$$

we have

$$f(\mathbf{y})^{(\tau_1 r,\cdots,\tau_s r)} \prod_{i=1}^{s} \varphi_r^{\tau_i}(y_i - x_i) \in E_s^\gamma(Cc(\alpha)^s)$$

by Lemma 9.5. The theorem follows from (9.9) and Theorem 7.9.

For the case $\alpha = 2k (k = 2, 3, \cdots)$, we have $r = \alpha/2$ and $\gamma = \alpha/2$. Hence Theorem 9.11 is sharper than Theorems 9.4 and 9.5.

From Theorem 9.11 and Lemma 7.5, we have

**Theorem 9.12.** *Suppose that* $s = \dfrac{\varphi(m)}{2}$ *and* $\mathbf{a} = (c_1, \cdots, c_s)$. *Then*

$$\Delta \leqslant Cc(\mathscr{R}_s, \alpha, \varepsilon) n^{-\left(\frac{1}{2}+\frac{1}{2(s-1)}\right)\gamma+\varepsilon}.$$

By Theorem 9.11 and Lemma 9.4, we have

**Theorem 9.13.** *Suppose that* $n = p$. *Then there exists an integral vector* $\mathbf{a} = \mathbf{a}(p)$ *such that*

$$\Delta \leqslant Cc(\alpha, \varepsilon)^s p^{-\gamma+\varepsilon}.$$

## § 9.7. The $\Omega$ results

**Lemma 9.8.** *Suppose that* $\xi$ *is a real number and* $p_k/q_k$ *is the* $k$*-th convergent of* $\xi$. *Then*

$$\left|\frac{p_k}{q_k} - \xi\right| \leqslant \frac{1}{q_k q_{k+1}}.$$

(Cf. Hua Loo Keng [2], Chap. 10).

**Theorem 9.14.** *For any given integer* $N$ *satisfying* $n > N \geqslant 1$ *and any integral vector* $\mathbf{a}$, *we have*

$$\Delta = \sup_{f \in H_s^\alpha(C)} \left\| \frac{1}{n} \sum_{k=1}^{n} f\left(\frac{k\mathbf{a}}{n}\right) \sum_{\|\mathbf{m}\|<N} e^{-2\pi i\left(\mathbf{m}, \frac{k\mathbf{a}}{n} - \mathbf{x}\right)} - f \right\|_2$$

$$\geqslant \frac{C}{2(2\pi)^{2\alpha}} n^{-\alpha/2}.$$

*Proof*. Evidently, we may suppose that $N > 1$. Otherwise we have $\Delta \geqslant C$ for $f(\mathbf{x}) = C$. We may assume also that $(a_1, \cdots, a_s, n) = 1$ and $a_1 = -1$. Let $p_t/q_t$ be the $t$-th convergent of $\frac{p_h}{q_h} = \frac{a_2}{n}$ and $N$ satisfy

$$1 = q_0 < \cdots < q_k \leqslant N < q_{k+1} < \cdots < q_h \leqslant n.$$

Let

$$K = a_2 q_k - n p_k.$$

Then

$$|K| \leqslant \frac{n}{q_{k+1}} < \frac{n}{N}$$

by Lemma 9.8. Take

$$f(\mathbf{x}) = \frac{C}{4(2\pi)^{2\alpha}} \left( \frac{e^{2\pi i (K x_1 + x_2)}}{\bar{K}^\alpha} + \frac{e^{-2\pi i (K x_1 + x_2)}}{\bar{K}^\alpha} + \frac{e^{2\pi i N x_1}}{N^\alpha} + \frac{e^{-2\pi i N x_1}}{N^\alpha} \right).$$

Then

$$\Delta^2 \geqslant \sum_{\bar{m}_1 \bar{m}_2 < N} \left| \sum_{l_1 \equiv a_2 l_2 \, (\mathrm{mod}\, n)}' C(l_1 + m_1, l_2 + m_2, 0, \cdots, 0) \right|^2 + 2 \frac{C^2}{16(2\pi)^{4\alpha}} N^{-2\alpha}$$

$$\geqslant \sum_{\bar{m}_1 \bar{m}_2 < K} (C(K + m_1, q_k + m_2, 0, \cdots, 0)$$

$$+ C(-K + m_1, -q_k + m_2, 0, \cdots, 0))^2 + \frac{C^2}{8(2\pi)^{4\alpha}} N^{-2\alpha}$$

$$\geqslant C(K, 1, 0, \cdots, 0)^2 + C(-K, -1, 0, \cdots, 0)^2 + \frac{C^2}{8(2\pi)^{4\alpha}} N^{-2\alpha}$$

$$\geqslant \frac{C^2}{8(2\pi)^{4\alpha}} (n^{-2\alpha} N^{2\alpha} + N^{-2\alpha}) \geqslant \frac{C^2}{4(2\pi)^{4\alpha}} n^{-\alpha}.$$

The theorem is proved.

**Theorem 9.15.** *For any given functions $\phi_{n,k}(\mathbf{x}) (1 \leqslant k \leqslant n)$ and integral vector $\mathbf{a}$, we have*

$$\sup_{f \in H_s^\alpha(C)} \sup_{\mathbf{x} \in G_s} \left| \sum_{k=1}^{n} f\left(\frac{k\mathbf{a}}{n}\right) \phi_{n,k}(\mathbf{x}) - f \right| \geqslant \frac{C}{(2\pi)^\alpha} n^{-\alpha/2}.$$

To prove Theorem 9.15, we shall need

**Lemma 9.9.** *The congruence*

$$a_1n_1 + a_2n_2 \equiv 0 \pmod n \tag{9.11}$$

*has solution satisfying*

$$|n_1| \leqslant \sqrt{n}, \quad |n_2| \leqslant \sqrt{n}, \quad (n_1, n_2) \neq (0, 0). \tag{9.12}$$

*Proof.* Since

$$([\sqrt{n}\,] + 1)([\sqrt{n}\,] + 1) > n,$$

the number of integer pairs $(x_1, x_2)$ $(x_1, x_2 = 0, 1, \cdots, [\sqrt{n}\,])$ is greater than $n$. Hence there exist two different pairs $(x_1', x_2')$ and $(x_1'', x_2'')$ such that

$$a_1x_1' + a_2x_2' \equiv a_1x_1'' + a_2x_2'' \pmod n.$$

Take $n_1 = x_1' - x_1''$ and $n_2 = x_2' - x_2''$. Then $n_1$ and $n_2$ satisfy (9.11) and (9.12). The lemma is proved.

The proof of Theorem 9.15. Let $n_1, n_2$ satisfy (9.11) and (9.12). Without loss of generality, we may suppose that $n_2 \neq 0$. Then we have

$$a_1n_1k \equiv -a_2n_2k \pmod n \tag{9.13}$$

for any integer $k$. Take

$$f(\mathbf{x}) = C\,\frac{e^{2\pi i n_1 x_1} - e^{-2\pi i n_2 x_2}}{2(2\pi)^\alpha N^\alpha} \in H_s^\alpha(C),$$

where $N = \max(|n_1|, |n_2|)$. Then

$$f\left(\frac{k\mathbf{a}}{n}\right) = C\,\frac{e^{\frac{2\pi i a_1 n_1 k}{n}} - e^{-\frac{2\pi i a_2 n_2 k}{n}}}{2(2\pi)^\alpha N^\alpha} = 0$$

by (9.13) for $k = 1, \cdots, n$. Let $x_1 = 0$ and $x_2 = \dfrac{1}{2|n_2|}$. Then

$$f\left(0, \frac{1}{2|n_2|}, 0, \cdots, 0\right) = \frac{C}{(2\pi)^\alpha} N^{-\alpha} \geqslant \frac{C}{(2\pi)^\alpha} n^{-\alpha/2}.$$

Hence the theorem follows.

## Notes

Theorem 9.3: Cf. N. M. Korobov [7].

Lemma 9.2: Cf. Wang Yuan [1].

Theorems of the type of Theorem 9.4 were first proved by V. S. Rjabenkii [1] and S. A. Smoljak [1] and their results were obtained by Theorem 9.7 with $O(p^{-\frac{\alpha}{2}+\frac{1}{4}+\varepsilon})$ and $N = [p^{1/2}]$ instead of $O(p^{\frac{-\alpha(2\alpha-1)}{4\alpha-1}+\varepsilon})$ and $N = [p^{\frac{2\alpha}{4\alpha-1}}]$ respectively. Theorem 9.7 was given by Wang Yuan [1,2].

Theorem 9.8: Cf. Hua Loo Keng and Wang Yuan [6,7].

Theorem 9.11: Cf. N. M. Korobov [6,7].

Theorems 9.14 and 9.15 were proved by Hua Loo Keng and Wang Yuan [7] and N. M. Korobov [7] respectively.

*Chapter 10*

# Approximate Solution of Integral Equations and Differential Equations

## § 10.1. Several lemmas

**Lemma 10.1.** *If* $\sum_{i=1}^{s}\sum_{j=1}^{s} \alpha_{ij}x_i x_j (\alpha_{ij}=\alpha_{ji})$ *is a semi-positive definite quadratic form, then*

$$0 \leqslant \det(\alpha_{ij}) \leqslant \prod_{i=1}^{s} \alpha_{ii}.$$

*Proof.* Since the matrix $(\alpha_{ij})(1 \leqslant i, j \leqslant s)$ may be reduced to diagonal form

$$(\alpha_{ij}) = \Lambda(\gamma_{ij}\delta_{ij})\Lambda', \quad 1 \leqslant i, j \leqslant s,$$

where $\gamma_{ii} \geqslant 0 (1 \leqslant i \leqslant s)$ and

$$\Lambda = \begin{pmatrix} 1 & 0 & 0 \cdots 0 \\ \lambda_{21} & 1 & 0 \cdots 0 \\ & & \cdots \\ \lambda_{s1} & \lambda_{s2} & \lambda_{s3} \cdots 1 \end{pmatrix}$$

(Cf. Hua Loo Keng [2], Chap. 14), we have

$$\det(\alpha_{ij}) = \prod_{i=1}^{s} \gamma_{ii}$$

and

$$\alpha_{ii} = \sum_{j=1}^{s}\sum_{k=1}^{s} \lambda_{ij}\gamma_{jk}\delta_{jk}\lambda_{ik} = \gamma_{ii} + \sum_{j<i} \lambda_{ij}^2 \gamma_{jj} \geqslant \gamma_{ii}$$

The lemma follows.

From Lemma 10.1, Hadamard's inequality can be easily derived.

**Lemma 10.2.** *Let $\beta_{ij}(1 \leqslant i, j \leqslant s)$ be real numbers. Then*

$$|\det(\beta_{ij})| \leqslant \prod_{i=1}^{s} \left(\sum_{j=1}^{s} \beta_{ij}^2\right)^{1/2}.$$

*Proof.* Let

$$\sum_{k=1}^{s} \left(\sum_{i=1}^{s} \beta_{ik} x_i\right)^2 = \sum_{i=1}^{s} \sum_{j=1}^{s} \left(\sum_{k=1}^{s} \beta_{ik}\beta_{jk}\right) x_i x_j = \sum_{i=1}^{s} \sum_{j=1}^{s} \alpha_{ij} x_i x_j.$$

Then

$$\det(\alpha_{ij}) = (\det(\beta_{ij}))^2, \quad 1 \leqslant i, j \leqslant s$$

and

$$\alpha_{ii} = \sum_{k=1}^{s} \beta_{ik}^2, \quad 1 \leqslant i \leqslant s.$$

Hence the lemma follows by Lemma 10.1.

The geometrical meaning of Lemma 10.2 is that the volume of a parallelepiped does not exceed the product of the lengthes of its edges.

**Lemma 10.3.** *Let $a_{ij}(1 \leqslant i, j \leqslant s)$ be real numbers and $\tilde{A}_s(k) = \det(a'_{ij})$ $(1 \leqslant i, j \leqslant s)$, where $0 \leqslant k \leqslant s$ and*

$$a'_{ij} = \begin{cases} 1 + a_{ii}, & \text{if } j = i, 1 \leqslant i \leqslant k, \\ a_{ij}, & \text{otherwise.} \end{cases}$$

*Let $A_s(k)$ donoto tho upper bound of $|\tilde{A}_s(k)|$ under the condition $|a_{ij}| \leqslant \gamma/s$ $(1 \leqslant i, j \leqslant s)$, where $\gamma$ is a constant. Then there exist two constants $\gamma_1$ and $\gamma_2$ such that*

$$A_s(s-1) \leqslant \frac{\gamma_1}{s}, \quad A_s(s) \leqslant \gamma_2.$$

*Proof.* We express the determinant $\tilde{A}_s(k)$ as a sum of two determinants $A$ and $B$, where $(1, 0, \cdots, 0)$ and $(a_{11}, a_{12}, \cdots, a_{1s})$ are the first rows of $A$ and $B$ respectively and the other elements of $A$ and $B$ are the same as the corresponding elements of $\tilde{A}_s(k)$. Then

$$A_s(k) \leqslant A_{s-1}(k-1) + A_s(k-1).$$

If $k \geqslant 2$, then

$$A_{s-1}(k-1) \leqslant A_{s-2}(k-2) + A_{s-1}(k-2)$$

and

$$A_s(k-1) \leqslant A_{s-1}(k-2) + A_s(k-2).$$

Hence

$$A_s(k) \leqslant A_{s-2}(k-2) + 2A_{s-1}(k-2) + A_s(k-2).$$

Consequently

$$A_s(k) \leqslant A_{s-k}(0) + C_1^k A_{s-k+1}(0) + \cdots + A_s(0).$$

Take $k = s-1$. Then

$$A_s(s-1) \leqslant \sum_{\nu=1}^{s} C_{\nu-1}^{s-1} A_\nu(0).$$

Since $|a_{i_j}| \leqslant \gamma/s$, we have

$$A_\nu(0) \leqslant \nu^{\nu/2}\left(\frac{\gamma}{s}\right)^\nu$$

by Lemma 10.2. Hence

$$A_s(s-1) \leqslant \sum_{\nu=1}^{s} C_{\nu-1}^{s-1}\nu^{\nu/2}\left(\frac{\gamma}{s}\right)^\nu \leqslant \frac{1}{s}\sum_{\nu=1}^{\infty}\frac{\gamma^\nu\nu^{\nu/2}}{(\nu-1)!} = \frac{\gamma_1}{s}.$$

Using Lemma 10.2 again, we have

$$\begin{aligned} A_s(s) &\leqslant ((|1+a_{11}|^2 + \cdots + |a_{1s}|^2)\cdots(|a_{s1}|^2 + \cdots + |1+a_{ss}|^2))^{1/2} \\ &\leqslant \left(\left(1+\frac{\gamma}{s}\right)^2 + (s-1)\frac{\gamma^2}{s^2}\right)^{s/2} = \left(1 + \frac{2\gamma+\gamma^2}{s}\right)^{s/2} \\ &\leqslant e^{\gamma+\frac{\gamma^2}{2}} = \gamma_2. \end{aligned}$$

The lemma is proved.

For simplicity, we use capital Latin letters to denote the $s$-dimensional vector.

**Lemma 10.4.** *Suppose that $F(Q_1, \cdots, Q_r) \in H_{rs}^\alpha(C)(\alpha > 0)$. If the quadrature formula given by the set $M_k(1 \leqslant k \leqslant n)$*

$$\int_{G_{rs}} F(Q_1, \cdots, Q_r)dQ_1\cdots dQ_r$$

$$= \frac{1}{n^r}\sum_{k_1,\cdots,k_r=1}^{n} F(M_{k_1}, \cdots, M_{k_r}) + O(\varepsilon(n))$$

*holds for* $r=1$, *where* $\varepsilon(n)=o(1)$ (as $n\to\infty$) *and the constant implied by the symbol "O" depends on* $C, \alpha, r, s$ *only, then it holds for* $r>1$ *also.*

*Proof.* We shall prove the lemma by induction. The lemma is true for $r=1$ by the assumption. Suppose that $r\geqslant 2$ and the lemma is true for integers less than $r$. Since $F(Q_1,\cdots,Q_r)\in H^{\alpha}_{(r-1)s}(C)$ for given $Q_r$ and $F(Q_1,\cdots,Q_r)\in H^{\alpha}_{s}(C)$ for given $Q_1,\cdots,Q_{r-1}$, we have

$$\begin{aligned}
&\int_{G_{rs}} F(Q_1,\cdots,Q_r)dQ_1\cdots dQ_r \\
&\quad=\int_{G_s} dQ_r\int_{G_{(r-1)s}} F(Q_1,\cdots,Q_{r-1},Q_r)dQ_1\cdots dQ_{r-1} \\
&\quad=\frac{1}{n^{r-1}}\sum_{k_1,\cdots,k_{r-1}=1}^{n}\int_{G_s} F(M_{k_1},\cdots,M_{k_{r-1}},Q_r)dQ_r+O(\varepsilon(n)) \\
&\quad=\frac{1}{n^{r}}\sum_{k_1,\cdots,k_r=1}^{n} F(M_{k_1},\cdots,M_{k_r})+O(\varepsilon(n)).
\end{aligned}$$

and the lemma follows by induction.

## § 10.2. The approximate solution of the Fredholm integral equation of second type

In this section, we shall study the problem of approximate solution of the Fredholm integral equation of second type

$$\varphi(P)=\lambda\int_{G_s} K(P,Q)\varphi(Q)dQ+f(P), \tag{10.1}$$

where $f\in H^{\alpha}_{s}(C)$ and $K\in H^{\alpha}_{2s}(C)$.

Let

$$D(\lambda)=1+\sum_{\nu=1}^{\infty}\frac{(-1)^{\nu}}{\nu!}\lambda^{\nu}\int_{G_{\nu s}} K\begin{pmatrix}P_1,\cdots,P_\nu\\ P_1,\cdots,P_\nu\end{pmatrix}dP_1\cdots dP_\nu$$

denote the Fredholm kernel of the equation (10.1), where

$$K\begin{pmatrix}P_1,\cdots,P_\nu\\ Q_1,\cdots,Q_\nu\end{pmatrix}=\det\left(K(P_i,Q_j)\right),\quad 1\leqslant i,j\leqslant\nu.$$

Let

$$\Delta(\lambda)=\det\left(\delta_{ij}-\frac{\lambda}{n}K(M_i,M_j)\right),\quad 1\leqslant i,j\leqslant n.$$

We suppose that

$$D(\lambda) \neq 0.$$

**Theorem 10.1.** *Suppose that*

$$\sup_{f \in H_s^\alpha(C)} \left| \int_{G_s} F(P)dP - \frac{1}{n} \sum_{k=1}^{n} F(M_k) \right| \leqslant Cc(\alpha, s)\varepsilon(n),$$

*where* $\varepsilon(n) = o(1)$ (as $n \to \infty$). *Let* $\tilde{\varphi}(M_k)(1 \leqslant k \leqslant n)$ *denote the solution of the system of linear equations*

$$\tilde{\varphi}(M_j) = \frac{\lambda}{n} \sum_{k=1}^{n} K(M_j, M_k)\tilde{\varphi}(M_k) + f(M_j), \quad 1 \leqslant j \leqslant n. \quad (10.2)$$

*Then the solution of* (10.1) *may be expressed by*

$$\varphi(P) = f(P) + \frac{\lambda}{n} \sum_{k=1}^{n} K(P, M_k)\tilde{\varphi}(M_k) + O(\varepsilon(n)),$$

*where the constant implied by the symbol "O" depends on* $\lambda, K$ *and* $f$ *only.*

To prove Theorem 10.1, we shall need

**Lemma 10.5.** *Let*

$$D_r(\lambda) = 1 + \sum_{\nu=1}^{r} \frac{(-1)^\nu}{\nu!} \lambda^\nu \int_{G_{\nu s}} K\begin{pmatrix} P_1, \cdots, P_\nu \\ P_1, \cdots, P_\nu \end{pmatrix} dP_1 \cdots dP_\nu.$$

*Then*

$$|D_r(\lambda) - D(\lambda)| \leqslant \frac{1}{2^r},$$

*if* $r$ *is sufficiently large.*

*Proof.* By Lemma 10.2,

$$\left| K\begin{pmatrix} P_1, \cdots, P_\nu \\ P_1, \cdots, P_\nu \end{pmatrix} \right| \leqslant \nu^{\nu/2} C^\nu.$$

Take $r$ sufficiently large such that

$$r \geqslant (2e|\lambda|C)^2 \quad \text{and} \quad \frac{3}{2^r} \leqslant \frac{1}{2}|D(\lambda)|. \quad (10.3)$$

Then

$$|D(\lambda) - D_r(\lambda)| = \left| \sum_{\nu=r+1}^{\infty} \frac{(-1)^\nu}{\nu!} \lambda^\nu \int_{G_{\nu s}} K\begin{pmatrix} P_1, \cdots, P_\nu \\ P_1, \cdots, P_\nu \end{pmatrix} dP_1 \cdots dP_\nu \right|$$

$$\leqslant \sum_{\nu=r+1}^{\infty} \frac{(|\lambda| C)^\nu \nu^{\nu/2}}{\nu!}.$$

Since $r! > r^r e^{-r}$ and

$$\frac{\nu!(|\lambda|C)^{\nu+1}(\nu+1)^{\frac{\nu+1}{2}}}{(\nu+1)!(|\lambda|C)^\nu \nu^{\nu/2}} = \frac{|\lambda|C}{\sqrt{\nu+1}}\left(1+\frac{1}{\nu}\right)^{\nu/2}$$

$$\leqslant \frac{|\lambda| C\sqrt{e}}{\sqrt{r+2}} \leqslant \frac{1}{2\sqrt{e}} < \frac{1}{2}$$

for $\nu \geqslant r+1$, therefore

$$|D(\lambda) - D_r(\lambda)| \leqslant \frac{(|\lambda|C)^{r+1}(r+1)^{\frac{r+1}{2}}}{(r+1)!} \sum_{\nu=0}^{\infty} \frac{1}{2^\nu}$$

$$\leqslant \frac{2(e|\lambda|C)^{r+1}}{(r+1)^{\frac{r+1}{2}}} \leqslant \frac{1}{2^r},$$

The lemma is proved.

**Lemma 10.6.** *There exists constant $n_0 = n_0(\lambda, K, f)$ such that*

$$|\Delta(\lambda)| \geqslant \frac{1}{2}|D(\lambda)|$$

*for $n > n_0$.*

*Proof.* Choose $r$ satisfying (10.3). Let

$$\Delta_r(\lambda) = 1 + \sum_{\nu=1}^{r} \frac{(-1)^\nu}{\nu!} \lambda^\nu \frac{1}{n^\nu} \sum_{k_1,\cdots,k_\nu=1}^{n} K\begin{pmatrix} M_{k_1}, \cdots, M_{k_\nu} \\ M_{k_1}, \cdots, M_{k_\nu} \end{pmatrix}.$$

Then in a way similar to the proof of Lemma 10.5, we have

$$|\Delta(\lambda) - \Delta_r(\lambda)| \leqslant \left| \sum_{\nu=r+1}^{\infty} \frac{(-1)^\nu}{\nu!} \lambda^\nu \frac{1}{n^\nu} \sum_{k_1,\cdots,k_\nu=1}^{n} K\begin{pmatrix} M_{k_1}, \cdots, M_{k_\nu} \\ M_{k_1}, \cdots, M_{k_\nu} \end{pmatrix} \right|$$

$$\leqslant \sum_{\nu=r+1}^{\infty} \frac{(|\lambda|C)^\nu \nu^{\nu/2}}{\nu!} \leqslant \frac{1}{2^r}.$$

The quadrature formula

$$\int_{G_{\nu s}} K\begin{pmatrix} P_1, \cdots, P_\nu \\ P_1, \cdots, P_\nu \end{pmatrix} dP_1 \cdots dP_\nu$$

$$= \frac{1}{n^\nu} \sum_{k_1, \cdots, k_\nu = 1}^{n} K\begin{pmatrix} M_{k_1}, \cdots, M_{k_\nu} \\ M_{k_1}, \cdots, M_{k_\nu} \end{pmatrix} + O(\varepsilon(n))$$

holds for $\nu = 1$ by the assumption of the lemma. Since

$$K\begin{pmatrix} P_1, \cdots, P_\nu \\ P_1, \cdots, P_\nu \end{pmatrix} \in H^{\alpha}_{\nu s}(Cc(\nu, \alpha, s)) \text{ for } K(P, Q) \in H^{\alpha}_{2s}(C),$$

therefore it holds for $\nu > 1$ by Lemma 10.4. Hence

$$D_r(\lambda) - \Delta_r(\lambda) = \sum_{\nu=1}^{r} \frac{(-1)^\nu \lambda^\nu}{\nu!} \left( \int_{G_{\nu s}} K\begin{pmatrix} P_1, \cdots, P_\nu \\ P_1, \cdots, P_\nu \end{pmatrix} dP_1 \cdots dP_\nu \right.$$

$$\left. - \frac{1}{n^\nu} \sum_{k_1, \cdots, k_\nu = 1}^{n} K\begin{pmatrix} M_{k_1}, \cdots, M_{k_\nu} \\ M_{k_1}, \cdots, M_{k_\nu} \end{pmatrix} \right) = O(\varepsilon(n)),$$

where the constant implied by the symbol "$O$" depends on $\lambda, K, f$ only, i.e., there exists constant $n_0 = n_0(\lambda, K, f)$ such that

$$|D_r(\lambda) - \Delta_r(\lambda)| \leqslant \frac{1}{2^r}$$

for $n > n_0$. Then by Lemma 10.5,

$$|D(\lambda) - \Delta(\lambda)| \leqslant |D(\lambda) - D_r(\lambda)| + |D_r(\lambda) - \Delta_r(\lambda)|$$

$$+ |\Delta_r(\lambda) - \Delta(r)| \leqslant \frac{3}{2^r} \leqslant \frac{1}{2}|D(r)|.$$

Hence

$$|\Delta(\lambda)| \geqslant |D(\lambda)| - |D(\lambda) - \Delta(\lambda)| \geqslant \frac{1}{2}|D(\lambda)|.$$

The lemma is proved.

**Lemma 10.7.** *Let* $\varphi(P)$ *denote the solution of* (10.1). *Then*

$$\varphi(P) \in H^{\alpha}_{s}(c(\lambda, K, f)).$$

*Proof.* Since

$$\varphi(P) - f(P) = \lambda \int_{G_s} K(P, Q)\varphi(Q)dQ,$$

therefore

$$|(\varphi(P) - f(P))^{(\alpha,\cdots,\alpha)}| \leqslant \sup_{Q \in G_s} |K(P, Q)^{(\alpha,\cdots,\alpha)}|\,|\lambda| \int_{G_s} |\varphi(Q)|dQ$$
$$\leqslant c(\lambda, K, f).$$

Hence

$$\varphi(P) - f(P) \in H_s^\alpha(c(\lambda, K, f))$$

and so

$$\varphi(P) = f(P) + (\varphi(P) - f(P)) \in H_s^\alpha(c(\lambda, K, f)).$$

The lemma is proved.

The proof of Theorem 10.1. It follows by Lemma 10.7 that

$$K(P, Q)\varphi(Q) \in H_s^\alpha(c(\lambda, K, f))$$

for given $P$. Hence

$$\varphi(P) = \frac{\lambda}{n} \sum_{k=1}^{n} K(P, M_k)\varphi(M_k) + f(P) + O(\varepsilon(n)) \qquad (10.4)$$

and

$$\varphi(M_j) = \frac{\lambda}{n} \sum_{k=1}^{n} K(M_j, M_k)\varphi(M_k) + f(M_j) + O(\varepsilon(n)), \quad 1 \leqslant j \leqslant n.$$

From (10.2), we have the system of linear equations

$$z_j = \sum_{k=1}^{n} a_{jk} z_k + b_j, \quad 1 \leqslant j \leqslant n,$$

where

$$z_j = \varphi(M_j) - \tilde{\varphi}(M_j),$$

$$a_{jk} = \frac{\lambda}{n} K(M_j, M_k),$$

and

$$b_j = O(\varepsilon(n)).$$

Let $\Delta_k(\lambda)$ denote the determinant obtained from $\Delta(\lambda)$ by replacing its $k$-th column

$$\left(-\frac{\lambda}{n} K(M_1, M_k), \cdots, 1 - \frac{\lambda}{n} K(M_k, M_k), \cdots, -\frac{\lambda}{n} K(M_n, M_k)\right)'$$

by

$$(b_1, \cdots, b_n)'.$$

Then we have

$$z_j = \frac{\Delta_j(\lambda)}{\Delta(\lambda)}, \quad 1 \leqslant j \leqslant n.$$

When $n$ is sufficiently large, we have

$$|\Delta(\lambda)| > \frac{1}{2}|D(\lambda)| > 0$$

by Lemma 10.6. Further more since

$$\left|\frac{\lambda}{n} K(M_j, M_k)\right| \leqslant \frac{|\lambda| C}{n},$$

therefore by Lemma 10.3, we have

$$|\Delta_j(\lambda)| \leqslant |b_j B_j| + \sum_{\substack{1 \leqslant k \leqslant n \\ k \neq j}} |b_k B_k|$$

$$\leqslant \gamma_2 |b_j| + \frac{\gamma_1}{n} \sum_{k=1}^{n} |b_k| = O(\varepsilon(n)),$$

where $B_k$ denotes the cofactor of $b_k$ in $\Delta_k(\lambda)$. Hence

$$z_j = O(\varepsilon(n)), \quad 1 \leqslant j \leqslant n.$$

Substituting into (10.4), the theorem follows.

Especially, let $M_k (1 \leqslant k \leqslant n)$ be the sets introduced in Chap. 4. We obtain various approximate formulas for the solutions of the equation (10.1).

## § 10.3. The approximate solution of the Volterra integral equation of second type

In this section, we shall study the problem of approximate solution of the Volterra equation of second type

$$\varphi(x) = \int_0^x K(x, y)\varphi(y)dy + f(x), \tag{10.5}$$

where $f \in H_1^\alpha(C)$ and $K(x, y) \in H_2^\alpha(C)$.

Introduce the notations

$$\mu(\alpha)=\begin{cases}\frac{\alpha}{2}, & \text{if } \alpha>1,\\ \frac{2\alpha^2}{1+4\alpha-\alpha^2}, & \text{if } 1\geqslant\alpha>0,\end{cases}$$

$$q=[p^{\frac{\mu(\alpha)}{\alpha}}],$$

$$Q=\left[\mu(\alpha)\frac{\log_2 p}{\log_2\log_2 3p}\right]$$

and

$$B_{k,\nu,p}(a)=\sum_{\overline{m}_1\cdots\overline{m}_\nu<q} e^{-2\pi i(m_1+m_2a+\cdots+m_\nu a^{\nu-1})k/p}$$

$$\cdot\int_0^x\int_0^{x_1}\cdots\int_0^{x_{\nu-1}} e^{2\pi i(m_1x_1+\cdots+m_\nu x_\nu)}dx_1\cdots dx_\nu.$$

**Theorem 10.2.** *There exists an integer $a(=a(p))$ such that the solution of (10.5) may be represented as*

$$\varphi(x)=f(x)+\frac{1}{p}\sum_{k=1}^{p}\sum_{\nu=1}^{Q}B_{k,\nu,p}(a)K\left(x,\frac{k}{p}\right)K\left(\frac{k}{p},\frac{ak}{p}\right)\cdots$$

$$\cdots K\left(\frac{a^{\nu-2}k}{p},\frac{a^{\nu-1}k}{p}\right)f\left(\frac{a^{\nu-1}k}{p}\right)+O(p^{-\mu(\alpha)+\varepsilon}),$$

*where the constant implied by the symbol "O" depends only on $K, f, \varepsilon$.*

*Proof*. The solution of the equation (10.5) is given by the Neumann series

$$\varphi(x)=f(x)+\sum_{\nu=1}^{\infty}\varphi_\nu(x),$$

where

$$\varphi_\nu(x)=\int_0^x\int_0^{x_1}\cdots\int_0^{x_{\nu-1}}R_\nu dx_1\cdots dx_\nu$$

and

$$R_\nu=R_\nu(x,x_1,\cdots,x_\nu)=K(x,x_1)K(x_1,x_2)\cdots K(x_{\nu-1},x_\nu)f(x_\nu).$$

Since $R_\nu\in H^\alpha_{\nu+1}(2^{(\alpha+1)(\nu+1)}C^{\nu+1})$, we have

$$|\varphi_\nu(x)|\leqslant 2^{(\alpha+1)(\nu+1)}C^{\nu+1}\int_0^x\int_0^{x_1}\cdots\int_0^{x_{\nu-1}}dx_1\cdots dx_\nu$$

$$\leqslant\frac{2^{(\alpha+1)(\nu+1)}C^{\nu+1}}{\nu!}$$

and

$$\left|\sum_{\nu=Q+1}^{\infty}\varphi_\nu(x)\right| \leqslant \sum_{\nu=Q+1}^{\infty}\frac{2^{(\alpha+1)(\nu+1)}C^{\nu+1}}{\nu!} \leqslant \frac{c(C,\alpha)^Q}{Q!} \leqslant c(K,f,\varepsilon)p^{-\mu(\alpha)+\varepsilon}.$$

Let

$$S_\nu = S_\nu(x, x_1, \cdots, x_\nu)$$

$$= \frac{1}{p}\sum_{k=1}^{p}K\left(x,\frac{k}{p}\right)K\left(\frac{k}{p},\frac{ak}{p}\right)\cdots K\left(\frac{a^{\nu-2}k}{p},\frac{a^{\nu-1}k}{p}\right)f\left(\frac{a^{\nu-1}k}{p}\right)$$

$$\cdot \sum_{\overline{m}_1\cdots\overline{m}_\nu<q} e^{-2\pi i(m_1+m_2a+\cdots+m_\nu a^{\nu-1})k/p+2\pi i(m_1x_1+\cdots+m_\nu x_\nu)}.$$

Then it follows from Theorem 9.10 that there exists an integer $a(=a(p))$ such that

$$\|R_\nu - S_\nu\|_2 \leqslant C^{\nu+1}c(\alpha,\varepsilon)^{\nu+1}\nu!^{1/2}p^{-\mu(\alpha)+\varepsilon/2}$$

and

$$\left|\varphi_\nu(x) - \int_0^x\cdots\int_0^{x_{\nu-1}} S_\nu dx_1\cdots dx_\nu\right|$$

$$\leqslant \int_0^x\cdots\int_0^{x_{\nu-1}}|R_\nu - S_\nu|dx_1\cdots dx_\nu$$

$$\leqslant \left(\int_0^x\cdots\int_0^{x_{\nu-1}}dx_1\cdots dx_\nu\right)^{1/2}\|R_\nu - S_\nu\|_2$$

$$\leqslant C^{\nu+1}c(\alpha,\varepsilon)^\nu p^{-\mu(\alpha)+\varepsilon/2}.$$

The theorem follows.

## § 10.4. The eigenvalue and eigenfunction of the Fredholm equation

For $f(x) = 0$, the Fredholm equation of second type

$$\varphi(x) = \lambda\int_0^1 K(x,y)\varphi(y)dy \tag{10.6}$$

is called the homogeneous equation. If there is a $\lambda$ such that (10.6) has a non-zero solution $\varphi(x)$, then $\lambda$ is called the eigenvalue of the kernel $K(x,y)$ and $\varphi(x)$ the eigenfunction of $K(x,y)$ corresponding to $\lambda$. The maximum number of linear independent eigenfunctions over the complex number field corresponding to an eigenvalue $\lambda$ is called the multiplicity of $\lambda$.

In this section, we suppose that $K(x,y) > 0$ and $K(x,y) = K(y,x)$

for $(x, y) \in G_2$ and that $K(x, y) \in H_2^a(C)$. We shall study the problem of the approximate solution of the least eigenvalue and its corresponding eigenfunction of the equation (10.6). First, We shall mention some well known results for the integral equation (Cf. V. S. Vladimirov [1]).

**Lemma 10.8.** *The equation* (10.6) *has eigenvalues. The number of its eigenvalues is denumerable. The eigenvalues are all real and have no finite limit point. Moreover the multiplicity of every eigenvalue is finite.*

Now we arrange the eigenvalues of the equation (10.6) according to their absolute values

$$|\lambda_1| \leqslant |\lambda_2| \leqslant \cdots, \tag{10.7}$$

where if $\lambda$ has the multiplicity $k$, then it will appear $k$ times in (10.7). The corresponding eigenfunctions are denoted by

$$\varphi_1, \varphi_2, \cdots.$$

Without loss of generality, we may suppose that

$$\|\varphi_i\|_2 = 1, \quad i = 1, 2, \cdots.$$

**Lemma 10.9.** $\lambda_1$ *is positive and simple and* $\varphi_1(x) > 0 (x \in G_1)$.

By Lemmas 10.8 and 10.9, we have

$$0 < \lambda_1 < |\lambda_2| \leqslant \cdots.$$

Suppose that $f \in H_1^a(C)$ and $f$ is a non-negative real function satisfying $\|f\|_2 = 1$, for example $f(x) = 1$. Then it follows by Schwarz's inequality that

$$0 < c_1 = \int_0^1 f(x)\varphi_1(x)dx \leqslant 1.$$

Denote

$$\Phi_s(x) = \frac{\int_{G_s} R_s d\mathbf{x}}{\|R_s\|_2}, \quad s \geqslant 1$$

and

$$\Lambda_s = \frac{\|R_{s-1}\|_2}{\|R_s\|_2}, \quad s \geqslant 2,$$

where

$$\begin{aligned} R_s &= R_s(x, x_1, \cdots, x_s) \\ &= K(x, x_1)K(x_1, x_2)\cdots K(x_{s-1}, x_s)f(x_s). \end{aligned}$$

**Lemma 10.10.**

$$0 \leqslant \Lambda_s - \lambda_1 \leqslant \frac{1-c_1^2}{c_1^2} \cdot \frac{\lambda_1}{2}\left(\frac{\lambda_1}{|\lambda_2|}\right)^{2s-2}, \qquad s \geqslant 2$$

and

$$\|\Phi_s - \varphi_1\|_2 \leqslant \frac{\sqrt{1-c_1^2}}{c_1}\left(\frac{\lambda_1}{|\lambda_2|}\right)^s, \qquad s \geqslant 1.$$

Introduce the notations

$$R_s^* = R_s^*(x) = \frac{1}{n}\sum_{k=1}^{n} R_s\left(x, \frac{a_1 k}{n}, \cdots, \frac{a_s k}{n}\right)$$

and

$$\tilde{R}_s = \left(\frac{1}{n}\sum_{k=1}^{n} R_s\left(\frac{a_1 k}{n}, \cdots, \frac{a_{s+1} k}{n}\right)^2\right)^{1/2}.$$

**Theorem 10.3.** *Suppose that the congruence*

$$\sum_{i=1}^{s+1} a_i m_i \equiv 0 \pmod{n}$$

*has no solution in the domain*

$$\overline{m}_1 \cdots \overline{m}_{s+1} \leqslant M, \qquad (m_1, \cdots, m_{s+1}) \neq (0, \cdots, 0).$$

*Then*

$$\left|\frac{\tilde{R}_{s-1}}{\tilde{R}_s} - \lambda_1\right| \leqslant \frac{1-c_1^2}{c_1^2} \cdot \frac{\lambda_1}{2}\left(\frac{\lambda_1}{|\lambda_2|}\right)^{2s-2} + c(K, f, \varepsilon) M^{-\alpha+\varepsilon}, \qquad s \geqslant 2 \tag{10.8}$$

*and*

$$\left\|\frac{R_s^*(x)}{\tilde{R}_s} - \varphi_1(x)\right\|_2 \leqslant \frac{\sqrt{1-c_1^2}}{c_1}\left(\frac{\lambda_1}{|\lambda_2|}\right)^s + c(K, f, \varepsilon) M^{-\alpha+\varepsilon}, \qquad s \geqslant 1. \tag{10.9}$$

*Proof.* Since

$$R_s \in H_{s+1}^\alpha(2^{(\alpha+1)(s+1)} C^{s+1}), \qquad R_s^2 \in H_{s+1}^\alpha(2^{2(\alpha+1)(s+1)} C^{2(s+1)}),$$

we have

$$\left| \|R_s\|_2^2 - \tilde{R}_s^2 \right| \leqslant c(K, f, \varepsilon) M^{-\alpha+\varepsilon}$$

by Theorem 7.9 and so

$$\left| \|R_s\|_2 - \tilde{R}_s \right| \leqslant c(K, f, \varepsilon) M^{-\alpha+\varepsilon}.$$

Since $\|R_s\|_2 = c(K, f) > 0$, we may suppose that $\tilde{R}_s = c(K, f) > 0$. Hence

$$\left|\frac{\tilde{R}_{s-1}}{\tilde{R}_s}-\Lambda_s\right| = \frac{|\tilde{R}_s\|R_{s-1}\|_2-\tilde{R}_{s-1}\|R_s\|_2|}{\tilde{R}_s\|R_s\|_2}$$

$$\leqslant \frac{|\,\|R_s\|_2(\|R_{s-1}\|_2-\tilde{R}_{s-1})+\|R_{s-1}\|_2(\tilde{R}_s-\|R_s\|_2)\,|}{c(K,f)}$$

$$\leqslant c(K,f,\varepsilon)M^{-\alpha+\varepsilon}.$$

Consequently, (10.8) follows by Lemma 10.10 and

$$\left|\frac{\tilde{R}_{s-1}}{\tilde{R}_s}-\lambda_1\right| \leqslant \left|\frac{\tilde{R}_{s-1}}{\tilde{R}_s}-\Lambda_s\right| + |\Lambda_s-\lambda_1|.$$

By Theorem 7.9

$$\left\|\int_{G_s} R_s d\mathbf{x} - R_s^*\right\|_2 \leqslant \sup_{x\in G_1}\left|\int_{G_s} R_s d\mathbf{x} - R_s^*\right| \leqslant c(K,f,\varepsilon)M^{-\alpha+\varepsilon}$$

and so by Minkowski's inequality, we have

$$\left\|\frac{R_s^*}{\tilde{R}_s}-\Phi_s\right\|_2 \leqslant c(K,f)\left\|\tilde{R}_s\int_{G_s} R_s d\mathbf{x} - \|R_s\|_2 R_s^*\right\|_2$$

$$= c(K,f)\left\|\tilde{R}_s\left(\int_{G_s} R_s d\mathbf{x} - R_s^*\right) + R_s^*(\tilde{R}_s-\|R_s\|_2)\right\|_2$$

$$\leqslant c(K,f)\left(\left\|\tilde{R}_s\left(\int_{G_s} R_s d\mathbf{x} - R_s^*\right)\right\|_2 + \left\|R_s^*(\tilde{R}_s-\|R_s\|_2)\right\|_2\right)$$

$$\leqslant c(K,f)M^{-\alpha+\varepsilon}.$$

Hence (10.9) follows by Lemma 10.10 and

$$\left\|\frac{R_s^*}{\tilde{R}_s}-\varphi_1\right\|_2 \leqslant \left\|\frac{R_s^*}{\tilde{R}_s}-\Phi_s\right\| + \|\Phi_s-\varphi_1\|_2.$$

The theorem is proved.

## § 10.5. The Cauchy problem of the partial differential equation of the parabolic type

In this section, we shall study the approximate solution of the parabolic equation

$$\frac{\partial u}{\partial t} = \left(\frac{\partial^2}{\partial x_1^2}+\cdots+\frac{\partial^2}{\partial x_s^2}\right)u,$$

$$0\leqslant t\leqslant T,\quad -\infty<x_\nu<\infty (1\leqslant\nu\leqslant s).$$

Suppose that the initial condition is

$$u(0, \mathbf{x}) = f(\mathbf{x}) \in E_s^\alpha(C).$$

where $\alpha > 1$.

**Theorem 10.4.** *If the congruence*

$$(\mathbf{a}, \mathbf{m}) = \sum_{i=1}^{s} a_i m_i \equiv 0 \pmod{n} \tag{10.10}$$

*has no solution in the domain*

$$\|\mathbf{m}\| \leqslant M, \quad \mathbf{m} \neq \mathbf{0}, \tag{10.11}$$

*then*

$$\sup_{\mathbf{x} \in G_s} \left| u(t, \mathbf{x}) - \sum_{\|\mathbf{m}\| < N} \left( \frac{1}{n} \sum_{k=1}^{n} f\left(\frac{k\mathbf{a}}{n}\right) e^{-\frac{2\pi i (\mathbf{a}, \mathbf{m}) k}{n}} \right) e^{-4\pi^2 (\mathbf{m}, \mathbf{m}) t + 2\pi i (\mathbf{m}, \mathbf{x})} \right|$$

$$\leqslant Cc(\alpha, s, \varepsilon) M^{-\frac{\alpha(\alpha-1)}{2\alpha-1} + \varepsilon}.$$

*where* $N = [M^{\frac{\alpha}{2\alpha-1}}]$.

*Proof.* Let

$$u(t, \mathbf{x}) = \sum C(t, \mathbf{m}) e^{2\pi i (\mathbf{m}, \mathbf{x})}.$$

Then

$$\sum \frac{d}{dt} C(t, \mathbf{m}) e^{2\pi i (\mathbf{m}, \mathbf{x})} = \frac{\partial u}{\partial t} = \left( \frac{\partial^2}{\partial x_1^2} + \cdots + \frac{\partial^2}{\partial x_s^2} \right) u$$

$$= - \sum C(t, \mathbf{m}) 4\pi^2 (\mathbf{m}, \mathbf{m}) e^{2\pi i (\mathbf{m}, \mathbf{x})}$$

and so by the comparison of the coefficients of $e^{2\pi i (\mathbf{m}, \mathbf{x})}$, we have

$$\frac{d}{dt} C(t, \mathbf{m}) = - 4\pi^2 (\mathbf{m}, \mathbf{m}) C(t, \mathbf{m}),$$

$$\int \frac{dC(t, \mathbf{m})}{C(t, \mathbf{m})} = - 4\pi^2 (\mathbf{m}, \mathbf{m}) \int dt,$$

$$C(t, \mathbf{m}) = c e^{-4\pi^2 (\mathbf{m}, \mathbf{m}) t}.$$

Since

$$C(0, \mathbf{m}) = C(\mathbf{m}),$$

where $C(\mathbf{m})$ is the Fourier coefficient of $f(\mathbf{x})$, therefore $c = C(\mathbf{m})$ and

$$C(t, \mathbf{m}) = C(\mathbf{m}) e^{-4\pi^2 (\mathbf{m}, \mathbf{m}) t}.$$

Hence

$$u(t, \mathbf{x}) = \sum C(\mathbf{m}) e^{-4\pi^2(\mathbf{m},\mathbf{m})t+2\pi i(\mathbf{m},\mathbf{x})}$$

and

$$\sup_{\mathbf{x}\in \bar{G}_s} \left| u(t, \mathbf{x}) - \sum_{\|\mathbf{m}\|<N} \left( \frac{1}{n} \sum_{k=1}^{n} f\left(\frac{k\mathbf{a}}{n}\right) e^{-\frac{2\pi i(\mathbf{a},\mathbf{m})k}{n}} \right) e^{-4\pi(\mathbf{m},\mathbf{m})t+2\pi i(\mathbf{m},\mathbf{x})} \right|$$

$$\leqslant \Sigma_1 + \Sigma_2,$$

where

$$\Sigma_1 = \sup_{f\in E_s(C)} \sum_{\|\mathbf{m}\|<N} \left| C(\mathbf{m}) - \frac{1}{n} \sum_{k=1}^{n} f\left(\frac{k\mathbf{a}}{n}\right) e^{\frac{-2\pi i(\mathbf{a},\mathbf{m})k}{n}} \right|$$

and

$$\Sigma_2 = \sup_{f\in E_s^{\alpha}(C)} \sum_{\|\mathbf{m}\|\geqslant N} |C(\mathbf{m})|.$$

By the argument of the proof of Theorem 9.5, the theorem follows.

## § 10.6. The Dirichlet problem of the partial differential equation of the elliptic type

In this section, we shall study the approximate solution of the elliptic equation

$$\left(\frac{\partial^2}{\partial x_1^2} + \cdots + \frac{\partial^2}{\partial x_s^2}\right) u = f. \tag{10.12}$$

Suppose that $f \in E_s^{\alpha}(C)(\alpha > 1)$ and $f$ is an odd function with respect to each variable and that $u(\mathbf{x}) = 0$ if $\mathbf{x}$ belongs to the boundary of $G_s$.

Introduce the notations

$$0 \leqslant \omega \leqslant 1,$$

$$\nu(\alpha, \omega) = \frac{\alpha(\alpha+\omega-1)}{2\alpha-1}$$

and

$$g(\mathbf{x}) = \sum B(\mathbf{m}) C(\mathbf{m}) e^{2\pi i(\mathbf{m},\mathbf{x})}, \tag{10.13}$$

where $C(\mathbf{m})$ is the Fourier coefficient of $f$ and $B(\mathbf{m})$ satisfies

$$|B(\mathbf{m})| \leqslant \frac{1}{\|\mathbf{m}\|^{\omega}}.$$

**Theorem 10.5.** *Suppose that* $s \geqslant 2$. *If the congruence* (10.10) *has no solution in* (10.11), *then*

$$\sup_{\mathbf{x} \in G_s} \left| u(\mathbf{x}) - \sum_{k=1}^{n} f\left(\frac{k\mathbf{a}}{n}\right) \psi_k(\mathbf{x}) \right| \leqslant Cc(\alpha, s, \varepsilon) M^{\nu\left(\alpha, \frac{2}{s}\right)+\varepsilon},$$

*where*

$$\psi_k(\mathbf{x}) = -\frac{1}{4\pi^2 n} \sum_{\|\mathbf{m}\| < N}{}' \frac{e^{2\pi i\left(\mathbf{m}, \mathbf{x} - \frac{k\mathbf{a}}{n}\right)}}{(\mathbf{m}, \mathbf{m})}, \quad 1 \leqslant k \leqslant n$$

*and* $N = [M^{\frac{\alpha}{2\alpha-1}}]$.

To prove the theorem, we shall need

**Lemma 10.11.** *If* $a_i \geqslant 0 (1 \leqslant i \leqslant s)$, *then*

$$(a_1 \cdots a_s)^{1/s} \leqslant \frac{a_1 + \cdots + a_s}{s}$$

(Cf. Hua Loo Keng [2], Chap. 20).

**Lemma 10.12.** *If the congruence* (10.10) *has no solution in* (10.11), *then*

$$\mathfrak{S} = \sup_{\mathbf{x} \in G_s} \left| g(\mathbf{x}) - \sum_{k=1}^{n} f\left(\frac{k\mathbf{a}}{n}\right) \chi_k(\mathbf{x}) \right| \leqslant Cc(\alpha, \omega, s, \varepsilon) M^{-\nu(\alpha, \omega)+\varepsilon},$$

*where*

$$\chi_k(\mathbf{x}) = \frac{1}{n} \sum_{\|\mathbf{m}\| < N} B(\mathbf{m}) e^{2\pi i\left(\mathbf{m}, \mathbf{x} - \frac{k\mathbf{a}}{n}\right)}, \quad 1 \leqslant k \leqslant n.$$

*Proof.* By (10.13), we have

$$\mathfrak{S} \leqslant \Sigma_1 + \Sigma_2,$$

where

$$\Sigma_1 = \sum_{\|\mathbf{m}\| < N} |B(\mathbf{m})| \left| C(\mathbf{m}) - \frac{1}{n} \sum_{k=1}^{n} f\left(\frac{k\mathbf{a}}{n}\right) e^{\frac{-2\pi i(\mathbf{a}, \mathbf{m})k}{n}} \right|$$

and

$$\Sigma_2 = \sum_{\|\mathbf{m}\| \geqslant N} |B(\mathbf{m})| |C(\mathbf{m})|.$$

Since

$$\left| C(\mathbf{m}) - \frac{1}{n} \sum_{k=1}^{n} f\left(\frac{k\mathbf{a}}{n}\right) e^{\frac{-2\pi i(\mathbf{a},\mathbf{m})k}{n}} \right| \leqslant C \sum_{(\mathbf{a},\mathbf{l})\equiv 0 \pmod n}{}' \frac{1}{\|\mathbf{l}+\mathbf{m}\|^{\alpha}}$$

(Cf. §9.4), we have

$$\begin{aligned}
\Sigma_1 &\leqslant C \sum_{k=0}^{[\log_2 N]} \sum_{2^{-k-1} \leqslant \|\mathbf{m}\| < 2^{-k}N} \frac{1}{\|\mathbf{m}\|^{\omega}} \sum_{(\mathbf{a},\mathbf{l})\equiv 0 \pmod n}{}' \frac{1}{\|\mathbf{l}+\mathbf{m}\|^{\alpha}} \\
&\leqslant C \sum_{k=0}^{[\log_2 N]} (2^{-k-1}N)^{-\omega} \sum_{\|\mathbf{m}\| < 2^{-k}N} \sum_{(\mathbf{a},\mathbf{l})\equiv 0 \pmod n}{}' \frac{1}{\|\mathbf{l}+\mathbf{m}\|^{\alpha}} \\
&\leqslant Cc(\alpha, s) N^{\alpha-\omega} \sum_{k=0}^{[\log_2 N]} 2^{-(\alpha-\omega)k} \sum_{(\mathbf{a},\mathbf{l})\equiv 0 \pmod n}{}' \frac{1}{\|\mathbf{l}\|^{\alpha}} \\
&\leqslant Cc(\alpha, s, \varepsilon) N^{\alpha-\omega} M^{-\alpha+\varepsilon} \sum_{k=0}^{\infty} 2^{-(\alpha-\omega)k} \\
&\leqslant Cc(\alpha, \omega, s, \varepsilon) M^{-\nu(\alpha,\omega)+\varepsilon}
\end{aligned}$$

by Theorem 7.9 and Lemma 9.2. Take $\varepsilon < \alpha - 1$. Then

$$\begin{aligned}
\Sigma_2 &\leqslant C \sum_{\|\mathbf{m}\| \geqslant N} \frac{1}{\|\mathbf{m}\|^{\alpha+\omega}} \leqslant C \sum_{\|\mathbf{m}\| \geqslant N} \frac{1}{\|\mathbf{m}\|^{\alpha+\omega-1-\varepsilon} \|\mathbf{m}\|^{1+\varepsilon}} \\
&\leqslant CN^{-\alpha-\omega+1+\varepsilon} \sum \frac{1}{\|\mathbf{m}\|^{1+\varepsilon}} \leqslant Cc(s, \varepsilon) M^{-\nu(\alpha,\omega)+\varepsilon}.
\end{aligned}$$

The lemma follows.

The proof is of Theorem 10.5. Let

$$g(\mathbf{x}) = \sum B(\mathbf{m}) C(\mathbf{m}) e^{2\pi i(\mathbf{m},\mathbf{x})},$$

where $C(\mathbf{m})$ is the Fourier coefficient of $f$ and

$$B(\mathbf{m}) = \begin{cases} 0, & \text{if } \mathbf{m} = \mathbf{0}, \\ -\dfrac{1}{4\pi^2(\mathbf{m},\mathbf{m})}, & \text{if } \mathbf{m} \neq \mathbf{0}. \end{cases} \tag{10.14}$$

We shall prove that $g(\mathbf{x})$ is the solution of (10.12). Since $f(\mathbf{x})$ is an odd function with respect to each variable, therefore

$$C(\mathbf{0}) = 0$$

and

$$C(m_1, \cdots, m_\nu, \cdots, m_s) = -C(m_1, \cdots, -m_\nu, \cdots, m_s).$$

It follows that

$$\left(\frac{\partial^2}{\partial x_1^2} + \cdots + \frac{\partial^2}{\partial x_s^2}\right) g = \sum{}' C(\mathbf{m}) e^{2\pi i(\mathbf{m},\mathbf{x})} = f(\mathbf{x})$$

and

$$g(x_1, \cdots, -x_\nu, \cdots, x_s) = -\frac{1}{4\pi^2} \sum{}' \frac{C(\mathbf{m})}{(\mathbf{m}, \mathbf{m})} e^{2\pi i(\mathbf{m},\mathbf{x}) - 4\pi i m_\nu x_\nu}$$

$$= \frac{1}{4\pi^2} \sum{}' \frac{C(\mathbf{m})}{(\mathbf{m}, \mathbf{m})} e^{2\pi i(\mathbf{m},\mathbf{x})} = -g(\mathbf{x}).$$

Hence $g(\mathbf{x})$ is a solution of (10.12) and vanishes on the boundary of $G_s$, i.e.,

$$u(\mathbf{x}) = g(\mathbf{x}).$$

For $\mathbf{m} \neq \mathbf{0}$, we have

$$(\mathbf{m}, \mathbf{m}) \geqslant s \|\mathbf{m}\|^{2/s}$$

by Lemma 10.11 and so

$$|B(\mathbf{m})| \leqslant \frac{1}{4\pi^2 s \|\mathbf{m}\|^{2/s}}.$$

by (10.14). Hence the theorem follows by Lemma 10.12.

## § 10.7. Several remarks

1. We may also use the method of § 10.2 to treat the problem of the approximate solution of the Fredholm integral equation of second type, if $K \in B_{2s}$ and $f \in B_s$.

2. The method of § 10.3 may be used to treat the problem of the approximate solution of integral equation of the type

$$\varphi(x_1, \cdots, x_{s+l}) = \int_0^1 \cdots \int_0^1 \int_0^{x_{s+1}} \cdots \int_0^{x_{s+l}} K(x_1, \cdots, x_{s+l}, y_1, \cdots, y_{s+l})$$

$$\cdot \varphi(y_1, \cdots, y_{s+l}) dy_1 \cdots dy_{s+l} + f(x_1, \cdots, x_{s+l}),$$

where $s \geqslant 0$ and $l \geqslant 1$ and where $f \in H_{s+l}^\alpha(C)$ and $K \in H_{2(s+l)}^\alpha(C)$ (or $f \in B_{s+l}$ and $K \in B_{2(s+l)}$).

3. We may also use the method of § 10.3 to treat the problem of the approximate solution of the Fredholm equation of second type, if $\lambda$ is suf-

ficiently small.

4. The method of § 10.3 may be used also to treat the problem of the approximate solution of the linear parabolic equation

$$\frac{\partial u(t,\mathbf{x})}{\partial t} = \left(\sum_{i=1}^{s} \frac{\partial^2}{\partial x_i^2}\right) u(t,\mathbf{x}) + \sum_{i=1}^{s} a_i(t,\mathbf{x}) \frac{\partial u(t,\mathbf{x})}{\partial x_i}$$
$$+ a(t,\mathbf{x})u(t,\mathbf{x}) + f(t,\mathbf{x}),$$
$$0 \leqslant t \leqslant T, \quad -\infty < x_i < \infty (1 \leqslant i \leqslant s),$$

where $u\ (0,\mathbf{x}) = 0$ and $a_i, a$ and $f$ belong to $H_{s+1}^{\alpha}(C)$.

Since the solution of the partial differential equation may be represented as the solution of the Volterra integral equation, it is given by the well-known Neumann series and so it may be represented approximately by a finite sum.

5. The method of § 10.4 may be used to treat the problem of the approximate solution of the eigenvalue and eigenfunction of the homogeneous Fredholm integral equation of second type

$$\varphi(P) = \lambda \int_{G_s} K(P, Q)\varphi(Q)dQ,$$

where $K(P,Q) \in H_{2s}^{\alpha}(C)$ or $K(P,Q) \in B_{2s}$.

6. The number theoretic method may be used also to arrange the experimental design and to find the optimal points of a function in a bounded reqion.

## Notes

Theorem 10.1 was first proved by N. M. Korobov [3,7] for the functions $f \in E_s^{\alpha}(C)$ and $K \in E_{2s}^{\alpha}(C)$, where $\alpha > 1$ (Cf. also I. F. Sarygin [1], Wang Yuan [2] and Hua Loo Keng and Wang Yuan [3,6,7]).

Theorem 10.2 is an improvement of an earlier theorem of Yu. N. Sahov [1,3] (Cf. Wang Yuan [1,2] and Hua Loo Keng and Wang Yuan [3,6,7]).

Concerning the eigenvalues and eigenfunctions of the Fredholm equation, we refer also Yu. N. Sahov [2].

Theorem 10.5 is an improvement of an earlier theorem of N. M. Korobov [7] (Cf. Wang Yuan [3,4]).

Concerning the problems stated in § 10.7, we refer N. M. Korobov [7], Hua Loo Keng and Wang Yuan [3,6,7], E. Hlawka [4,5], V. S. Rjabenkii [2], Wang Yuan, Zhu Yao Cheng and Jian Yun Cui [1], V. T. Stojancev [1], Xu Guang Shan [1] and Wang Yuan and Fang Kai Tai [1].

# *Appendix* Tables

We use the notations $W_l(n, \mathbf{h})$ (Cf. § 8.1) and $\rho(n, \mathbf{h}) = \min_{\mathbf{m}} \|\mathbf{m}\|$, where $\mathbf{m}$ runs over the integral vectors with $\mathbf{m} \neq \mathbf{0}$ and $(\mathbf{h},\mathbf{m}) \not\equiv 0 \pmod{n}$.

Table (1) is given by the Fibonacci sequence $F_m (= F_{2,m})$, i.e.

$$n = F_m, \quad \mathbf{h} = (1, F_{m-1}).$$

Tables (2)—(12) are given by the methods of Chap. 8 (Cf. § 8.8). In table (11), the $\mathbf{h}(n)$'s coresponding to $s = 12$ and 13 are obtained by neglecting the components $h_{13}, h_{14}$ and $h_{14}$ respectively and we use the notation $W_2\ (s,n,\mathbf{h})$ instead of $W_2(n,\mathbf{h})$.

**1** $(s = 2,\ h_1 = 1,\ h_2 = F_{m-1},\ n = F_m)$*

| $n$ | 13 | 21 | 34 | 55 | 89 |
|---|---|---|---|---|---|
| $W_2(n,\mathbf{h})$ | $4.7586 \times 10^{-1}$ | $2.0909 \times 10^{-1}$ | $8.9745 \times 10^{-2}$ | $3.8148 \times 10^{-2}$ | $1.6033 \times 10^{-2}$ |
| $n$ | 144 | 233 | 377 | 610 | 987 |
| $W_2(n,\mathbf{h})$ | $6.6851 \times 10^{-3}$ | $2.7673 \times 10^{-3}$ | $1.1388 \times 10^{-3}$ | $4.6619 \times 10^{-4}$ | $1.8900 \times 10^{-4}$ |
| $n$ | 1,597 | 2,584 | 4,181 | 6,765 | 10,946 |
| $W_2(n,\mathbf{h})$ | $7.7127 \times 10^{-5}$ | $3.1200 \times 10^{-5}$ | $1.2581 \times 10^{-5}$ | $5.0595 \times 10^{-6}$ | $2.0293 \times 10^{-6}$ |
| $n$ | 17,711 | 28,657 | 46,368 | 75,025 | |
| $W_2(n,\mathbf{h})$ | $8.1206 \times 10^{-7}$ | $3.2376 \times 10^{-7}$ | $1.2819 \times 10^{-7}$ | $5.1270 \times 10^{-8}$ | |

2 ($s=3,\ h_1=1$)

| $n$ | $h_2$ | $h_3$ | $\rho(n,\mathbf{h})$ | $W_2(n,\mathbf{h})$ | $W_4(n,\mathbf{h})$ |
|---|---|---|---|---|---|
| 21 | 3 | 8 | 3 | 2.3320 | $1.1700 \times 10^{-1}$ |
| 35 | 11 | 16 | 5 | 1.1074 | $2.1437 \times 10^{-2}$ |
| 66 | 10 | 24 | 8 | $3.9332 \times 10^{-1}$ | $2.8304 \times 10^{-3}$ |
| 86 | 30 | 40 | 10 | $2.6836 \times 10^{-1}$ | $1.3069 \times 10^{-3}$ |
| 135 | 29 | 42 | 13 | $1.4577 \times 10^{-1}$ | $3.4114 \times 10^{-4}$ |
| 185 | 26 | 64 | 20 | $8.5667 \times 10^{-2}$ | $1.0860 \times 10^{-4}$ |
| 266 | 27 | 69 | 27 | $5.0586 \times 10^{-2}$ | $3.5702 \times 10^{-5}$ |
| 418 | 90 | 130 | 40 | $2.1688 \times 10^{-2}$ | $6.3870 \times 10^{-6}$ |
| 597 | 63 | 169 | 55 | $1.3007 \times 10^{-2}$ | $2.0000 \times 10^{-6}$ |
| 828 | 285 | 358 | 72 | $7.7157 \times 10^{-3}$ | $7.1265 \times 10^{-7}$ |
| 1,010 | 140 | 237 | 86 | $5.2751 \times 10^{-3}$ | $2.9203 \times 10^{-7}$ |
| 1,220 | 319 | 510 | 108 | $3.6308 \times 10^{-3}$ | $1.1552 \times 10^{-7}$ |
| 1,459 | 256 | 373 | 114 | $2.9263 \times 10^{-3}$ | $9.7463 \times 10^{-8}$ |
| 1,626 | 572 | 712 | 140 | $2.1506 \times 10^{-3}$ | $4.4293 \times 10^{-8}$ |
| 1,958 | 202 | 696 | 162 | $1.5620 \times 10^{-3}$ | $2.2093 \times 10^{-8}$ |
| 2,440 | 638 | 1,002 | 216 | $1.0313 \times 10^{-3}$ | $8.5161 \times 10^{-9}$ |
| 3,237 | 456 | 1,107 | 252 | $7.0670 \times 10^{-4}$ | $4.9006 \times 10^{-9}$ |
| 4,044 | 400 | 1,054 | 308 | $4.5620 \times 10^{-4}$ | $1.8752 \times 10^{-9}$ |
| 5,037 | 580 | 1,997 | 390 | $3.3527 \times 10^{-4}$ | $9.8872 \times 10^{-10}$ |
| 6,066 | 600 | 1,581 | 460 | $2.3416 \times 10^{-4}$ | $4.6664 \times 10^{-10}$ |
| 8,191 | 739 | 5,515 | 364 | $1.7 \times 10^{-4}$ | $4.0 \times 10^{-10}$ |
| 10,007 | 544 | 5,733 | 400 | $1.3 \times 10^{-4}$ | $2.5 \times 10^{-10}$ |
| 20,039 | 5,704 | 12,319 | 396 | $6.4 \times 10^{-5}$ | |
| 28,117 | 19,449 | 5,600 | 585 | $3.0 \times 10^{-5}$ | |
| 39,029 | 10,607 | 26,871 | 570 | $2.1 \times 10^{-5}$ | |
| 57,091 | 48,188 | 21,101 | 1,084 | $9.8 \times 10^{-6}$ | |
| 82,001 | 21,252 | 67,997 | 1,978 | $4.1 \times 10^{-6}$ | |
| *140,052 | 34,590 | 112,313 | | $3.33 \times 10^{-6}$ | |
| *314,694 | 77,723 | 252,365 | | $1.23 \times 10^{-6}$ | |

**3** ($s=4,\ h_1=1$)

| $n$ | $h_2$ | $h_3$ | $h_4$ | $\rho(n, \mathbf{h})$ | $W_2(n, \mathbf{h})$ | $W_4(n, \mathbf{h})$ |
|---|---|---|---|---|---|---|
| 60 | 8 | 18 | 22 | 4 | 3.3875 | $8.5025 \times 10^{-2}$ |
| 118 | 18 | 40 | 52 | 6 | 1.4214 | $1.5513 \times 10^{-2}$ |
| 180 | 8 | 46 | 74 | 8 | $8.1807 \times 10^{-1}$ | $5.2230 \times 10^{-3}$ |
| 286 | 16 | 94 | 138 | 12 | $4.4143 \times 10^{-1}$ | $1.5466 \times 10^{-3}$ |
| 440 | 21 | 136 | 216 | 15 | $2.5001 \times 10^{-1}$ | $4.3550 \times 10^{-4}$ |
| 562 | 53 | 89 | 221 | 20 | $1.8208 \times 10^{-1}$ | $2.0716 \times 10^{-4}$ |
| 732 | 248 | 294 | 324 | 24 | $1.1232 \times 10^{-1}$ | $8.5499 \times 10^{-5}$ |
| 932 | 116 | 288 | 314 | 26 | $8.0987 \times 10^{-2}$ | $4.7288 \times 10^{-5}$ |
| 1,142 | 150 | 187 | 274 | 32 | $6.7770 \times 10^{-2}$ | $2.9213 \times 10^{-5}$ |
| 1,354 | 492 | 550 | 658 | 40 | $4.5581 \times 10^{-2}$ | $1.2280 \times 10^{-5}$ |
| 2,129 | 766 | 1,281 | 1,906 | 32 | $2.7 \times 10^{-2}$ | |
| 3,001 | 174 | 266 | 1,269 | 46 | $1.7 \times 10^{-2}$ | |
| 4,001 | 113 | 766 | 2,537 | 51 | $1.1 \times 10^{-2}$ | |
| 5,003 | 792 | 1,889 | 191 | 32 | $9.2 \times 10^{-3}$ | |
| 6,007 | 1,351 | 5,080 | 3,086 | 80 | $5.9 \times 10^{-3}$ | |
| 8,191 | 2,488 | 5,939 | 7,859 | 72 | $3.8 \times 10^{-3}$ | |
| 10,007 | 1,206 | 3,421 | 2,842 | 84 | $3.0 \times 10^{-3}$ | |
| 20,039 | 19,668 | 17,407 | 14,600 | 60 | $1.6 \times 10^{-3}$ | |
| 28,117 | 17,549 | 1,900 | 24,455 | 144 | $6.5 \times 10^{-4}$ | |
| 39,029 | 30,699 | 34,367 | 605 | 135 | $4.9 \times 10^{-4}$ | |
| 57,091 | 52,590 | 48,787 | 38,790 | 268 | $2.8 \times 10^{-4}$ | |
| 82,001 | 57,270 | 58,903 | 17,672 | 260 | $1.7 \times 10^{-4}$ | |
| 100,063 | 92,313 | 24,700 | 95,582 | 352 | $1.1 \times 10^{-4}$ | |
| *147,312 | 136,641 | 116,072 | 76,424 | | $8.5376 \times 10^{-5}$ | |

**4** ($s=5,\ h_1=1$)

| $n$ | $h_2$ | $h_3$ | $h_4$ | $h_5$ | $\rho(n,\mathbf{h})$ | $W_2(n,\mathbf{h})$ | $W_4(n,\mathbf{h})$ |
|---|---|---|---|---|---|---|---|
| 1,069 | 63 | 762 | 970 | 177 | 6 | $7.4\times10^{-1}$ | $4.7\times10^{-3}$ |
| 1,543 | 58 | 278 | 694 | 134 | 8 | $4.2\times10^{-1}$ | $1.5\times10^{-3}$ |
| 2,129 | 618 | 833 | 1,705 | 1,964 | 9 | $3.1\times10^{-1}$ | $1.1\times10^{-3}$ |
| 3,001 | 408 | 1,409 | 1,681 | 1,620 | 18 | $1.7\times10^{-1}$ | $1.3\times10^{-4}$ |
| 4,001 | 1,534 | 568 | 3,095 | 2,544 | 17 | $1.2\times10^{-1}$ | $1.1\times10^{-4}$ |
| 5,003 | 840 | 177 | 3,593 | 1,311 | 16 | $9.2\times10^{-2}$ | $7.6\times10^{-5}$ |
| 6,007 | 509 | 780 | 558 | 1,693 | 22 | $7.0\times10^{-2}$ | $3.5\times10^{-5}$ |
| 8,191 | 1,386 | 4,302 | 7,715 | 3,735 | 30 | $4.3\times10^{-2}$ | $1.0\times10^{-5}$ |
| 10,007 | 198 | 9,183 | 6,967 | 8,507 | 36 | $3.4\times10^{-2}$ | $7.2\times10^{-6}$ |
| 15,019 | 10,641 | 2,640 | 6,710 | 784 | 18 | $2.9\times10^{-2}$ | $2.5\times10^{-5}$ |
| 20,039 | 11,327 | 11,251 | 12,076 | 18,677 | 21 | $1.8\times10^{-2}$ | $1.2\times10^{-5}$ |
| 33,139 | 32,133 | 17,866 | 21,281 | 32,247 | 60 | $8.5\times10^{-3}$ | $8.9\times10^{-7}$ |
| 51,097 | 44,672 | 45,346 | 7,044 | 14,242 | 35 | $5.4\times10^{-3}$ | $1.5\times10^{-6}$ |
| 71,053 | 33,755 | 65,170 | 12,740 | 6,878 | 80 | $2.8\times10^{-3}$ | $1.1\times10^{-5}$ |
| 100,063 | 90,036 | 77,477 | 27,253 | 6,222 | 96 | $1.7\times10^{-3}$ | $5.8\times10^{-8}$ |
| *374,181 | 343,867 | 255,381 | 310,881 | 115,892 | | $1.01\times10^{-3}$ | |

**5** ($s=6,\ h_1=1$)

| $n$ | $h_2$ | $h_3$ | $h_4$ | $h_5$ | $h_6$ | $\rho(n,\mathbf{h})$ | $W_2(n,\mathbf{h})$ | $W_4(n,\mathbf{h})$ |
|---|---|---|---|---|---|---|---|---|
| 2,129 | 41 | 1,681 | 793 | 578 | 279 | 4 | 2.0 | $1.9\times10^{-2}$ |
| 3,001 | 233 | 271 | 122 | 1,417 | 51 | 8 | 1.3 | $5.9\times10^{-3}$ |
| 4,001 | 1,751 | 1,235 | 1,945 | 844 | 1,475 | 6 | $9.5\times10^{-1}$ | $4.4\times10^{-3}$ |
| 5,003 | 2,037 | 1,882 | 1,336 | 4,803 | 2,846 | 8 | $6.8\times10^{-1}$ | $1.6\times10^{-3}$ |
| 6,007 | 312 | 1,232 | 5,943 | 4,060 | 5,250 | 9 | $5.6\times10^{-1}$ | $1.1\times10^{-3}$ |
| 8,191 | 1,632 | 1,349 | 6,380 | 1,399 | 6,070 | 12 | $3.7\times10^{-1}$ | $5.0\times10^{-4}$ |
| 10,007 | 2,240 | 4,093 | 1,908 | 931 | 3,984 | 12 | $2.9\times10^{-1}$ | $3.8\times10^{-4}$ |
| 15,019 | 8,743 | 8,358 | 6,559 | 2,795 | 772 | 8 | $2.0\times10^{-1}$ | $6.9\times10^{-4}$ |
| 20,039 | 5,557 | 150 | 11,951 | 2,461 | 9,179 | 12 | $1.3\times10^{-1}$ | $1.7\times10^{-4}$ |
| 33,139 | 18,236 | 1,831 | 19,143 | 5,522 | 22,910 | 18 | $6.8\times10^{-2}$ | $3.5\times10^{-5}$ |
| 51,097 | 9,931 | 7,551 | 29,682 | 44,446 | 17,340 | 24 | $4.2\times10^{-2}$ | $1.8\times10^{-5}$ |
| 71,053 | 18,010 | 3,155 | 50,203 | 6,605 | 13,328 | 18 | $3.3\times10^{-2}$ | $2.5\times10^{-5}$ |
| 100,063 | 43,307 | 15,440 | 39,114 | 43,534 | 39,955 | 30 | $1.8\times10^{-2}$ | $4.5\times10^{-6}$ |
| *114,174 | 107,538 | 88,018 | 15,543 | 80,974 | 56,747 | | $1.47\times10^{-2}$ | |
| *302,686 | 285,095 | 233,344 | 41,204 | 214,668 | 150,441 | | $4.06\times10^{-3}$ | |

**6** $(s = 7,\ h_1 = 1)$

| $n$ | $h_2$ | $h_3$ | $h_4$ | $h_5$ | $h_6$ | $h_7$ | $\rho(n,\mathbf{h})$ | $W_2(n,\mathbf{h})$ | $W_4(n,\mathbf{h})$ |
|---|---|---|---|---|---|---|---|---|---|
| *3,997 | 3,888 | 3,564 | 3,034 | 2,311 | 1,417 | 375 | | 5.8 | $1.1 \times 10^{-1}$ |
| *11,215 | 10,909 | 10,000 | 8,512 | 6,485 | 3,976 | 1,053 | | 1.9 | $2.2 \times 10^{-2}$ |
| 15,019 | 12,439 | 2,983 | 8,607 | 7,041 | 7,210 | 6,741 | 6 | 1.2 | |
| 24,041 | 1,833 | 18,190 | 21,444 | 23,858 | 1,135 | 12,929 | 6 | $6.9 \times 10^{-1}$ | $2.5 \times 10^{-3}$ |
| 33,139 | 7,642 | 9,246 | 5 ,584 | 23,035 | 32,241 | 30,396 | 6 | $5.0 \times 10^{-1}$ | $1.9 \times 10^{-3}$ |
| 46,213 | 37,900 | 17,534 | 41,873 | 32,280 | 15,251 | 26,909 | 12 | $3.3 \times 10^{-1}$ | $3.5 \times 10^{-4}$ |
| 57,091 | 35,571 | 45,299 | 51,436 | 34,679 | 1,472 | 8,065 | 12 | $2.5 \times 10^{-1}$ | $3.0 \times 10^{-4}$ |
| 71,053 | 31,874 | 36,082 | 13,810 | 6,605 | 68,784 | 9,848 | 10 | $2.1 \times 10^{-1}$ | $4.5 \times 10^{-4}$ |
| *84,523 | 82,217 | 75,364 | 64,149 | 48,878 | 29,969 | 7,936 | | $2.0 \times 10^{-1}$ | $6.2 \times 10^{-4}$ |
| 100,063 | 39,040 | 62,047 | 89,839 | 6,347 | 30,892 | 64,404 | 16 | $1.4 \times 10^{-1}$ | $1.4 \times 10^{-4}$ |
| *172,155 | 167,459 | 153,499 | 130,657 | 99,554 | 61,040 | 18,165 | | $7.3 \times 10^{-2}$ | $5.4 \times 10^{-5}$ |
| *234,646 | 228,245 | 209,218 | 178,084 | 135,691 | 83,197 | 22,032 | | $8.0 \times 10^{-2}$ | $4.6 \times 10^{-5}$ |
| *462,891 | 450,265 | 412,730 | 351,310 | 267,681 | 164,124 | 43,464 | | $1.9 \times 10^{-2}$ | $3.4 \times 10^{-6}$ |
| *769,518 | 748,528 | 686,129 | 584,024 | 444,998 | 272,843 | 72,255 | | $1.2 \times 10^{-2}$ | |
| *957,838 | 931,711 | 854,041 | 726,949 | 553,900 | 339,614 | 89,937 | | $8.0 \times 10^{-3}$ | |

7 ($s = 8, h_1 = 1$)

| $n$ | $h_2$ | $h_3$ | $h_4$ | $h_5$ | $h_6$ | $h_7$ | $h_8$ | $\rho(n,\mathbf{h})$ | $W_2(n,\mathbf{h})$ | $W_4(n,\mathbf{h})$ |
|---|---|---|---|---|---|---|---|---|---|---|
| *3,997 | 3,888 | 3,564 | 3,034 | 2,311 | 1,417 | 375 | 3,211 | | $2.8 \times 10$ | 2.7 |
| *11,215 | 10,909 | 10,000 | 8,512 | 6,485 | 3,976 | 1,053 | 9,010 | | 9.6 | $3.5 \times 10^{-1}$ |
| 24,041 | 17,441 | 21,749 | 5,411 | 12,326 | 3,144 | 21,024 | 6,252 | 3 | 3.9 | $4.4 \times 10^{-2}$ |
| *28,832 | 27,850 | 24,938 | 20,195 | 13,782 | 5,918 | 25,703 | 15,781 | | 3.5 | $4.2 \times 10^{-2}$ |
| 33,139 | 3,520 | 29,553 | 3,239 | 1,464 | 16,735 | 19,197 | 3,019 | 6 | 2.7 | $1.2 \times 10^{-2}$ |
| 46,213 | 5,347 | 30,775 | 35,645 | 11,403 | 16,894 | 32,016 | 16,600 | 4 | 1.9 | $1.4 \times 10^{-2}$ |
| 57,091 | 17,411 | 46,802 | 9,779 | 16,807 | 35,302 | 1,416 | 47,755 | 6 | 1.5 | $4.6 \times 10^{-3}$ |
| 71,053 | 60,759 | 26,413 | 24,409 | 48,215 | 51,048 | 19,876 | 29,096 | 6 | 1.2 | $4.2 \times 10^{-3}$ |
| *84,523 | 82,217 | 75,364 | 64,149 | 48,878 | 29,969 | 7,936 | 67,905 | | $9.9 \times 10^{-1}$ | $2.2 \times 10^{-3}$ |
| 100,063 | 4,344 | 58,492 | 29,291 | 60,031 | 10,486 | 22,519 | 60,985 | 9 | $7.6 \times 10^{-1}$ | $1.0 \times 10^{-3}$ |
| *172,155 | 167,459 | 153,499 | 130,657 | 99,554 | 61,040 | 18,165 | 138,308 | | $4.6 \times 10^{-1}$ | $1.2 \times 10^{-3}$ |
| *234,646 | 228,245 | 209,218 | 178,084 | 135,691 | 83,197 | 22,032 | 188,512 | | $4.1 \times 10^{-1}$ | $1.3 \times 10^{-3}$ |
| *462,891 | 450,265 | 412,730 | 351,310 | 267,681 | 164,124 | 43,464 | 371,882 | | $1.6 \times 10^{-1}$ | $2.2 \times 10^{-4}$ |
| *769,518 | 748,528 | 686,129 | 584,024 | 444,998 | 272,843 | 72,255 | 618,224 | | $1.2 \times 10^{-1}$ | $5.3 \times 10^{-4}$ |
| *957,838 | 931,711 | 854,041 | 726,949 | 553,900 | 339,614 | 89,937 | 769,518 | | $8.0 \times 10^{-2}$ | $1.4 \times 10^{-4}$ |

**8** ($s=9,\ h_1=1$)

| $n$ | $h_2$ | $h_3$ | $h_4$ | $h_5$ | $h_6$ | $h_7$ | $h_8$ | $h_9$ | $\rho(n,\mathbf{h})$ | $W_2(n,\mathbf{h})$ | $W_4(n,\mathbf{h})$ |
|---|---|---|---|---|---|---|---|---|---|---|---|
| *3,997 | 3,888 | 3,564 | 3,034 | 2,311 | 1,417 | 375 | 3,211 | 1,962 | | $1.2\times10^2$ | 6.7 |
| *11,215 | 10,909 | 10,000 | 8,512 | 6,485 | 3,976 | 1,053 | 9,010 | 5,506 | | $4.2\times10$ | $9.2\times10^{-1}$ |
| 33,139 | 68 | 4,624 | 16,181 | 6,721 | 26,221 | 26,661 | 23,442 | 3,384 | 3 | $1.4\times10$ | $1.7\times10^{-1}$ |
| *42,570 | 41,409 | 37,957 | 32,308 | 24,617 | 15,094 | 3,997 | 34,200 | 20,901 | | $1.0\times10$ | $1.2\times10^{-1}$ |
| 46,213 | 8,871 | 40,115 | 20,065 | 30,352 | 15,654 | 42,782 | 17,966 | 33,962 | 3 | 9.5 | $2.1\times10^{-1}$ |
| 57,091 | 20,176 | 12,146 | 23,124 | 2,172 | 33,475 | 5,070 | 42,339 | 36,122 | 4 | 7.5 | $4.6\times10^{-2}$ |
| 71,053 | 26,454 | 13,119 | 27,174 | 17,795 | 22,805 | 43,500 | 45,665 | 49,857 | 4 | 6.0 | $4.3\times10^{-2}$ |
| 100,063 | 70,893 | 53,211 | 12,386 | 27,873 | 56,528 | 16,417 | 17,628 | 14,997 | 6 | 4.1 | $1.3\times10^{-2}$ |
| 159,053 | 60,128 | 101,694 | 23,300 | 43,576 | 57,659 | 42,111 | 85,501 | 93,062 | 8 | 2.5 | $6.6\times10^{-3}$ |
| *172,155 | 167,459 | 153,499 | 130,657 | 99,554 | 61,040 | 18,165 | 138,308 | 84,523 | | 2.4 | $7.6\times10^{-3}$ |
| *234,646 | 228,245 | 209,218 | 178,084 | 135,691 | 83,197 | 22,032 | 188,512 | 115,204 | | 1.9 | $2.0\times10^{-2}$ |
| *462,891 | 450,265 | 412,730 | 351,310 | 267,681 | 164,124 | 43,464 | 371,882 | 227,266 | | $9.8\times10^{-1}$ | $1.1\times10^{-2}$ |
| *769,518 | 748,528 | 686,129 | 584,024 | 444,998 | 272,843 | 72,255 | 618,224 | 377,811 | | $5.3\times10^{-1}$ | $1.4\times10^{-3}$ |
| *957,838 | 931,711 | 854,041 | 726,949 | 553,900 | 339,614 | 89,937 | 769,518 | 470,271 | | $4.1\times10^{-1}$ | $1.9\times10^{-3}$ |

**9** ($s=10,\ h_1=1$)

| $n$ | $h_2$ | $h_3$ | $h_4$ | $h_5$ | $h_6$ | $h_7$ | $h_8$ | $h_9$ | $h_{10}$ | $\rho(n,\mathbf{h})$ | $W_2(n,\mathbf{h})$ | $W_4(n,\mathbf{h})$ |
|---|---|---|---|---|---|---|---|---|---|---|---|---|
| *4,661 | 4,574 | 4,315 | 3,889 | 3,304 | 2,570 | 1,702 | 715 | 4,289 | 3,122 | | $4.5\times10^2$ | $2.2\times10$ |
| *13,587 | 13,334 | 12,579 | 11,337 | 9,631 | 7,492 | 4,961 | 2,084 | 12,502 | 9,100 | | $1.6\times10^2$ | 8.7 |
| *24,076 | 23,628 | 22,290 | 20,090 | 17,066 | 13,276 | 8,790 | 3,692 | 22,153 | 16,125 | | $8.8\times10$ | 5.1 |
| *58,358 | 57,271 | 54,030 | 48,695 | 41,366 | 32,180 | 21,307 | 8,950 | 53,697 | 39,086 | | $3.6\times10$ | 2.4 |
| 85,633 | 37,677 | 35,345 | 3,864 | 54,821 | 74,078 | 30,354 | 57,935 | 51,906 | 56,279 | 2 | $2.4\times10$ | 2.3 |
| 103,661 | 45,681 | 57,831 | 80,987 | 9,718 | 51,556 | 55,377 | 37,354 | 4,353 | 27,595 | 2 | $2.1\times10$ | $4.2\times10^{-1}$ |
| 115,069 | 65,470 | 650 | 95,039 | 77,293 | 98,366 | 70,366 | 74,605 | 55,507 | 49,201 | 2 | $1.7\times10$ | $3.0\times10^{-1}$ |
| 130,703 | 64,709 | 53,373 | 17,385 | 5,244 | 29,008 | 52,889 | 66,949 | 51,906 | 110,363 | 4 | $1.4\times10$ | $8.9\times10^{-2}$ |
| 155,093 | 90,485 | 20,662 | 110,048 | 102,303 | 148,396 | 125,399 | 124,635 | 10,480 | 44,198 | 4 | $1.2\times10$ | $6.9\times10^{-2}$ |
| *805,098 | 790,101 | 745,388 | 671,792 | 570,685 | 443,949 | 293,946 | 123,470 | 740,795 | 539,222 | | 2.3 | $2.8\times10^{-2}$ |

**10** $(s = 11,\ h_1 = 1)$*

| $n$ | 4,661 | 13,587 | 24,076 | 58,358 | 297,974 | 698,047 | 1,243,423 | 2,226,963 | 7,494,007 |
|---|---|---|---|---|---|---|---|---|---|
| $h_2$ | 4,574 | 13,334 | 23,628 | 57,271 | 294,481 | 685,041 | 1,228,845 | 2,200,854 | 7,354,408 |
| $h_3$ | 4,315 | 12,579 | 22,290 | 54,030 | 284,041 | 646,274 | 1,185,282 | 2,122,833 | 6,938,211 |
| $h_4$ | 3,889 | 11,337 | 20,090 | 48,695 | 266,778 | 582,461 | 1,113,244 | 1,993,814 | 6,253,169 |
| $h_5$ | 3,304 | 9,631 | 17,066 | 41,366 | 242,894 | 494,796 | 1,013,577 | 1,815,311 | 5,312,043 |
| $h_6$ | 2,570 | 7,492 | 13,276 | 32,180 | 212,668 | 384,914 | 887,449 | 1,589,415 | 4,132,365 |
| $h_7$ | 1,702 | 4,961 | 8,790 | 21,307 | 176,456 | 254,860 | 736,338 | 1,318,777 | 2,736,109 |
| $h_8$ | 715 | 2,084 | 3,692 | 8,950 | 134,682 | 107,051 | 562,016 | 1,006,567 | 1,149,286 |
| $h_9$ | 4,289 | 12,502 | 22,153 | 53,697 | 87,835 | 642,292 | 366,527 | 656,448 | 6,895,461 |
| $h_{10}$ | 3,122 | 9,100 | 16,125 | 39,086 | 36,464 | 467,527 | 152,163 | 272,523 | 5,019,180 |
| $h_{11}$ | 1,897 | 5,529 | 9,797 | 23,747 | 279,147 | 284,044 | 1,164,860 | 2,086,257 | 3,049,402 |
| $W_2(n,\mathbf{h})$ | $1.9 \times 10^3$ | $6.6 \times 10^2$ | $3.7 \times 10^2$ | $1.5 \times 10^2$ | $3.1 \times 10$ | $1.2 \times 10$ | 8.1 | 3.6 | $6.4 \times 10^{-1}$ |
| $W_4(n,\mathbf{h})$ | $7.7 \times 10$ | $2.0 \times 10$ | $1.1 \times 10$ | 4.2 | $7.0 \times 10^{-1}$ | $7.6 \times 10^{-2}$ | | | |

11 ($s = 12, 13, 14, h_1 = 1$)*

| $n$ | 18,984 | 53,328 | 77,431 | 297,974 | 1,243,423 | 2,428,705 | 14,753,436 | 19,984,698 | 34,248,063 |
|---|---|---|---|---|---|---|---|---|---|
| $h_2$ | 18,761 | 52,703 | 76,523 | 294,481 | 1,228,845 | 2,400,231 | 14,580,465 | 19,750,396 | 33,846,536 |
| $h_3$ | 18,096 | 50,834 | 73,810 | 284,041 | 1,185,282 | 2,315,141 | 14,063,582 | 19,050,236 | 32,646,662 |
| $h_4$ | 16,996 | 47,745 | 69,324 | 266,778 | 1,113,244 | 2,174,435 | 13,208,845 | 17,892,427 | 30,662,508 |
| $h_5$ | 15,475 | 43,470 | 63,118 | 242,894 | 1,013,577 | 1,979,761 | 12,026,276 | 16,290,543 | 27,917,337 |
| $h_6$ | 13,549 | 38,061 | 55,264 | 212,668 | 887,449 | 1,733,402 | 10,529,739 | 14,263,366 | 24,443,334 |
| $h_7$ | 11,242 | 31,580 | 45,854 | 176,456 | 736,338 | 1,438,245 | 8,736,780 | 11,834,661 | 20,281,228 |
| $h_8$ | 8,581 | 24,104 | 34,998 | 134,682 | 562,016 | 1,097,753 | 6,668,420 | 9,032,903 | 15,479,816 |
| $h_9$ | 5,596 | 15,720 | 22,825 | 87,835 | 366,527 | 715,916 | 4,348,908 | 5,890,941 | 10,095,390 |
| $h_{10}$ | 2,323 | 6,526 | 9,476 | 36,464 | 152,163 | 297,211 | 1,805,439 | 2,445,610 | 4,191,077 |
| $h_{11}$ | 17,785 | 49,959 | 72,539 | 279,147 | 1,164,860 | 2,275,252 | 13,821,268 | 18,722,002 | 32,084,164 |
| $h_{12}$ | 14,053 | 39,477 | 57,320 | 220,583 | 920,477 | 1,797,913 | 10,921,619 | 14,794,199 | 25,353,030 |
| $h_{13}$ | 10,158 | 28,534 | 41,430 | 159,433 | 665,302 | 1,299,495 | 7,893,924 | 10,692,946 | 18,324,655 |
| $h_{14}$ | 6,143 | 17,255 | 25,054 | 96,414 | 402,327 | 785,841 | 4,773,681 | 6,466,329 | 11,081,440 |
| $W_2(12,n,\mathbf{h})$ | $1.3 \times 10^3$ | $7.2 \times 10^2$ | $5.1 \times 10^2$ | $1.3 \times 10^2$ | $3.1 \times 10$ | $1.6 \times 10$ | | | |
| $W_4(12,n,\mathbf{h})$ | $6.9 \times 10$ | $1.5 \times 10$ | $1.3 \times 10$ | 1.9 | $4.2 \times 10^{-1}$ | | | | |
| $W_2(13,n,\mathbf{h})$ | $5.7 \times 10^3$ | $3.1 \times 10^3$ | $2.1 \times 10^3$ | $5.6 \times 10^2$ | $1.3 \times 10^2$ | $6.9 \times 10$ | $1.0 \times 10$ | 8.8 | 4.0 |
| $W_4(13,n,\mathbf{h})$ | $1.5 \times 10^2$ | $3.7 \times 10$ | $2.6 \times 10$ | 5.9 | 1.8 | $3.6 \times 10^{-1}$ | | | |
| $W_2(14,n,\mathbf{h})$ | $3.8 \times 10^4$ | $1.3 \times 10^4$ | $9.2 \times 10^3$ | $2.4 \times 10^3$ | $5.8 \times 10^2$ | $3.0 \times 10^2$ | $4.7 \times 10$ | $3.5 \times 10$ | $2.0 \times 10$ |
| $W_4(14,n,\mathbf{h})$ | $5.1 \times 10^2$ | $1.8 \times 10^2$ | $1.4 \times 10^2$ | $3.6 \times 10$ | $1.1 \times 10$ | 5.4 | | | |

**12** ($s = 15, 16, 17, 18, h_1 = 1$)*

| $n$ | 70,864 | 139,489 | 1,139,691 | 2,422,957 | 4,395,774 | 14,271,038 | 55,879,244 |
|---|---|---|---|---|---|---|---|
| $h_2$ | 70,353 | 138,484 | 1,131,480 | 2,398,094 | 4,364,102 | 14,168,215 | 55,476,633 |
| $h_3$ | 68,825 | 135,476 | 1,106,904 | 2,323,761 | 4,269,316 | 13,860,486 | 54,271,700 |
| $h_4$ | 66,291 | 130,487 | 1,066,142 | 2,200,720 | 4,112,097 | 13,350,069 | 52,273,127 |
| $h_5$ | 62,768 | 123,553 | 1,009,487 | 2,030,234 | 3,893,578 | 12,640,642 | 49,495,314 |
| $h_6$ | 58,283 | 114,724 | 937,347 | 1,814,052 | 3,615,335 | 11,737,315 | 45,958,274 |
| $h_7$ | 52,867 | 104,063 | 850,242 | 1,554,392 | 3,279,371 | 10,646,597 | 41,687,493 |
| $h_8$ | 46,559 | 91,647 | 748,799 | 1,253,920 | 2,888,108 | 9,376,347 | 36,713,742 |
| $h_9$ | 39,405 | 77,566 | 633,750 | 915,717 | 2,444,365 | 7,935,718 | 31,072,856 |
| $h_{10}$ | 31,457 | 61,921 | 505,923 | 543,256 | 1,951,338 | 6,335,088 | 24,805,477 |
| $h_{11}$ | 22,772 | 44,825 | 366,239 | 140,357 | 1,412,580 | 4,585,990 | 17,956,764 |
| $h_{12}$ | 13,412 | 26,401 | 215,705 | 2,134,112 | 831,972 | 2,701,027 | 10,576,061 |
| $h_{13}$ | 3,445 | 6,781 | 55,406 | 1,683,011 | 213,699 | 693,780 | 50,314,090 |
| $h_{14}$ | 63,806 | 125,597 | 1,026,186 | 1,214,641 | 3,957,988 | 12,849,750 | 41,669,876 |
| $h_{15}$ | 52,844 | 104,019 | 849,882 | 733,806 | 3,277,986 | 10,642,098 | 32,725,430 |
| $h_{16}$ | 41,501 | 81,691 | 667,455 | | 2,574,365 | 8,357,770 | 23,545,197 |
| $h_{17}$ | 29,859 | 58,775 | 480,219 | | 1,852,197 | 6,013,224 | 14,195,319 |
| $h_{18}$ | 18,002 | 35,435 | 289,522 | | 1,116,683 | 3,625,352 | 2,716,545 |
| $W_2(15,n,\mathbf{h})$ | $4.3 \times 10^4$ | $2.2 \times 10^4$ | $2.7 \times 10^3$ | $1.3 \times 10^3$ | $7.0 \times 10^2$ | $2.2 \times 10^2$ | $5.5 \times 10$ |
| $W_4(15,n,\mathbf{h})$ | $4.6 \times 10^2$ | $2.4 \times 10^2$ | $2.4 \times 10$ | $1.2 \times 10$ | 6.8 | 2.8 | |
| $W_2(16,n,\mathbf{h})$ | $1.9 \times 10^5$ | $9.4 \times 10^4$ | $1.2 \times 10^4$ | | $3.0 \times 10^3$ | $9.2 \times 10^2$ | $2.4 \times 10^2$ |
| $W_4(16,n,\mathbf{h})$ | $1.5 \times 10^3$ | $7.5 \times 10^2$ | $8.1 \times 10$ | | $2.3 \times 10$ | 7.2 | $7.2 \times 10^{-1}$ |
| $W_2(17,n,\mathbf{h})$ | $8.0 \times 10^5$ | $4.0 \times 10^5$ | $5.0 \times 10^4$ | | $1.3 \times 10^4$ | $4.0 \times 10^3$ | $1.0 \times 10^3$ |
| $W_4(17,n,\mathbf{h})$ | $4.5 \times 10^3$ | $2.3 \times 10^3$ | $2.7 \times 10^2$ | | $7.0 \times 10$ | $2.3 \times 10$ | 4.1 |
| $W_2(18,n,\mathbf{h})$ | $3.4 \times 10^6$ | $1.7 \times 10^6$ | $2.1 \times 10^5$ | | $5.5 \times 10^4$ | $1.7 \times 10^4$ | $4.3 \times 10^3$ |
| $W_4(18,n,\mathbf{h})$ | $1.4 \times 10^4$ | $7.3 \times 10^3$ | $8.7 \times 10^2$ | | $2.3 \times 10^2$ | $7.1 \times 10$ | |

# Bibliography


C. R. Adams, and J. A. Clarkson,

[1] On definitions of bounded variation for function of two variables, *Trans. Amer. Math. Soc.*, **35**, 1933, 824—854.

[2] Properties of functions $f(x, y)$ of bounded variation, *Trans. Amer. Math. Soc.*, **36**, 1934, 711—730.

N. S. Bahvalov,

[1] Approximate computation of multiple integrals, *Vestnik Moskow Univ., Ser. Mat. Meh. Astr. Fiz. Him.*, **4**, 1959, 3—18.

[2] An estimate of the main remainder term in quadrature formula, *Z. Vycisl. Mat. i Mat. Fiz.*, **1**, 1961, 64—77.

[3] On embedding theorems for class of functions with bounded derivatives, *Vestnik Moskow Univ., Ser. Mat. Meh. Astr. Fiz. Him.*, **3**, 1963, 7—16.

[4] Optimal convergence bounds for quadrature processes and integration methods of Monte Carlo type for classes of functions, *Z. Vycisl. Mat. i Mat. Fiz.*, **4**, suppl., 1964, 5—63.

A. Baker,

[1] On some Diophantine inequalities involving the exponential function, *Canad. J. Math.*, **17**, 1965, 616—626.

L. Bernstein,

[1] The Jacobi-Perron algorism, its theory and applications, Lec. Not. in Math., Springer Verlag, 207, 1971.

J. W. S. Cassels,

[1] An introduction to Diophantine approximation, Camb. Univ. Press, 1957.

H. Conroy,

[1] Molecular Schrödinger equation, VIII: A new method for the evaluation of multidimensional integrals, *J. Chemical Phys.*, **47**, 1967, 5307—5318.

R. Cranley and T. N. L. Patterson,

[1] Randomization of number theoretic methods for multiple integration, *SIAM J. Numer. Anal.*, **13**, 1976, 904 —914.

P. J. Davis and P. Rabinowitz,

[1] Some Monte Carlo experiments in computing multiple integrals, *Math. Tables Aids Comput.*; **10**, 1956, 1—8.

P. Erdös and P. Turán,

[1] On a problem in the theory of uniform distribution, I, *Indag. Math.*, **10**, 1948, 370—378.

I. M. Gelfand, A. S. Frolov and N. N. Cencov,

[1] The computation of continuous integrals by the Monte Carlo method, *Izv. Vyss.*

*Ucebn., Zaved. Mat.*, **5**, 1958, 32—45.

S. Haber,

[1] Numerical evaluation of multiple integrals, *SIAM Rev.*; **12**, 1970, 481—526.

[2] Experiments on optimal coefficients, Applications of number theory to numerical analysis (S. K. Zaremba, ed.), Academic Press, New York, 1972, 11—37.

S. Haber and C. F. Osgood,

[1] On the sum $\sum\langle n\alpha\rangle^{-1}$ and numerical integration, *Pacific J. Math.*, **31**, 1969, 383—394.

J. H. Halton,

[1] On the efficiency of certain quasi-random sequences of points in evaluating multi-dimensional integrals, *Numer. Math.*, **2**, 1960, 84—90.

J. M. Hammersley,

[1] Monte Carlo methods for solving multivariable problems, *Ann. New York Acad. Sci.*, **86**, 1960, 844—874.

G. H. Hardy,

[1] On double Fourier series and especially those which represent the double zeta function with real and incommensurable parameters, *Quart. J. Math.*, Oxford, **37**, 1906, 53—79.

C. B. Haselgrove,

[1] A method for numerical integration, *Math. Comp.*, **15**, 1961, 323—337.

E. Hlawka,

[1] Funktionen von beschränkter Variation in der Theorie Gleichverteilung, *Ann. Mat. pure Appl.*, **54**, 1961, 325—333.

[2] Über die Diskrepanz mehrdimensionaler Folgen mod 1, *Math. Z.*, **77**, 1961, 273—284.

[3] Zur angenäherten Berechnung mehrfacher Integrale, *Monatsh, Math.*, **66**, 1962, 140—151.

[4] Uniform distribution modulo 1 and numerical analysis, *Compositio Math.*, **16**, 1964, 92—105.

[5] Trigonometrische Interpolation bei Funktionen Von mehreren Variablen, *Acta Arith.*, **9**, 1964, 305—320.

Hsu Li Zhi and Zhou Yun Shi,

[1] Numerical evaluation of multiple integrals, Science Press, Beijing, 1980.

Hua Loo Keng,

[1] Corrigendum on a paper of Su Jia Ju concerning the 5-th algebraic equation, *Science*, 2, **15**, 1930, 307.

[2] Introduction to number theory, Science Press, Beijing, 1956.

[3] Starting from "Yang Hui triangle", Qing Nian Press, Beijing, 1956.

[4] Additive prime number theory, Science Press, Beijing, 1957.

[5] The estimation of trigonometrial sum and its application in number theory, Science Press, Beijing, 1963.

Hua Loo Keng and Wang Yuan,

[1] Remarks concerning numerical integration, *Sci. Record*, (N. S.), **4**, 1960, 8—11.

[2] Numerical evaluation of integrals, Science Press, Beijing, 1961.

[3] Numerical integration and its applications, Science Press, Beijing, 1963.

[4] On Diophantine approximations and numerical integrations, (I) *Sci. Sin.*; 6, **13**, 1964,

1007—1008 (II) *Sci. Sin.*, 6, **13**, 1964, 1009—1010.

[5] On numerical integration of periodic functions of several variables, *Sci. Sin.*, 7, **14**, 1965, 964—978.

[6] On uniform distribution and numerical analysis (Number theoretic method), (I) *Kexue Tongbao*, **3**, 1973, 112—114, (II) *Kexue Tongbao*, **4**, 1973, 165—166, (III) *Kexue Tongbao*, **12**, 1974, 559—560.

[7] On uniform distribution and numerical analysis (Number theoretic method), (I) *Sci. Sin.*, 4, **16**, 1973, 483—505, (II) *Sci. Sin.*, 3, **17**, 1974, 331—348, (III) *Sci. Sin.*, 2, **18**, 1975, 184—198.

[8] A note on simultaneous Diophantine approximations to algebraic integers, *Sci. Sin.*, 5, **20**, 1977, 563—567.

Hua Loo Keng, Wang Yuan and Pei Ding Yi,

[1] On a set of independent units of cyclotomic field, *Ziran Zazhi*, 5, 1978, 6.

P. Keast,

[1] Multi-dimensional quadrature formula, *Tech. Rep.*, **40**, Dept. of Computer Science, Toronto Univ., 1972.

[2] Optimal parameters for multi-dimensional integrals, *SIAM J. on Numer. Anal.*, **10**, 1973, 831—838.

G. Kedem and S. K. Zaremba,

[1] A table of good lattice point in three dimensions, *Numer. Math.*, **23**, 1974, 175—180.

A. Khintchine,

[1] Metrical problems of irrational numbers, *Uspehi Mat. Nauk SSSR*, 1, 1936. 7—37.

I. F. Koksma,

[1] Een algemeene stelling uit de theorie der gelijkmatige Verdeeling modulo 1, *Math.* B (Zutphen), **11**, 1942—1943, 7—11.

[2] Some theorems on Diophantine inequalities, *Math. Cent. Amer.*, Scriptum, **5**, 1950, 1—51.

N. M. Korobov,

[1] Approximate calculation of multiple integrals with the aid of methods in the theory of numbers, *Dokl. Akad. Nauk SSSR*, **115**, 1957, 1062—1065.

[2] The approximate computation of multiple integrals, *Dokl. Akad. Nauk SSSR*, **124**, 1959, 1207—1210.

[3] On the approximate solultion of integral equations, *Dokl. Akad. Nauk SSSR*, **128**, 1959, 233—238.

[4] Computation of multiple integrals by the method of optimal coefficients, *Vestnik Moskow Univ. ser. Mat. Meh. Astr. Fiz. Him.*, **4**, 1959, 19—25.

[5] Properties and calculation of optimal coefficients, *Dokl. Akad. Nauk SSSR*, **132**, 1960, 1009—1012.

[6] Application of number-theoretic nets to integral equations and interpolation formulas, Trudy Mat. Inst. Steklov, **60**, 1961, 195—210.

[7] Number theoretic methods in approximate analysis, Fizmatigiz, Moscow, 1963.

[8] Some problems in the theory of Diophantine approximation, *Uspehi Mat. Nauk SSSR*, 3, **22**, 1967, 73—118.

J. M. Krause,

[1] Fouriersche Reihen mit zwei veranderlichen Grössen, *Ber. Verh. sächs. Akad. Wiss. Leipzig. Math-naturw. Kl.;* **55**,1903, 164—197.

L. Kuipers and H, Niederreiter,
[1] Uniform distribution of sequences, Wiley, New York, 1974.

E. Landau,
[1] Vorlesungen über Zahlentheorie, III Chelsea pub. Co., New York, 1947.

J. J. Liang,
[1] On the integral basis of the maximal real subfield of a cyclotomcic field, *J. Reine angew, Math.*, **286/287**, 1976, 223—226.

K. Mahler,
[1] On a paper by A. Baker on the approximation of rational powers of *e*, *Acta Arith.*, **27**, 1975, 61—87.

D. Maisonneuve,
[1] Recherche et utilisation des "bons treillis". Programmation et résultats numériques, Applications of Number Theory to Numerical Analysis (S. K. Zaremba, ed.), Academic Press, New York, 1972, 121—201.

J. M. Masley and H. L. Montgomery,
[1] Cyclotomic fields with unique facterization, *J. Reine angew. Math.*, **286/287**, 1976, 248—256,

H. Minkowski,
[1] Über periodische Approximationen algebraischer Zahlen, *Acta Math.*, **26**, 1902, 333—351.

Y. S. Moon,
[1] Some numerical experiments on number-theoretic methods in the approximation of multi-dimensional integrals, *Tech.Rep.*, **72**, Dept. of Computer Science, Toronto Univ., 1974.

H. Niederreiter,
[1] Methods for estimating discrepancy, Applications of Number Theory to Numerical Analysis (S. K. Zaremba, ed.), Academic Press, New York, 1972, 203—236.
[2] Application of Diophantine approximations to numerical integration, Diophantine Approximation and Its Applications (C. F. Osgood, ed.), Academic Press, New York, 1973, 129—199.
[3] Pseudo-random numbers and optimal coefficients, *Advances in Math.*, **26**, 1977, 99—181.
[4] Existence of good lattice points in the sense of Hlawka, *Monatsh. Math.*, **86**, 1978, 203—219.
[5] Quasi-Monte Carlo methods and pseudo-random numbers, *Bull. Amer. Math. Soc.*, 6, **84**, 1978, 957—1041.

L. G. Peck,
[1] On uniform distribution of algebraic numbers, *Proc. Amer. Math. Soc.*, **4**, 1953, 440—443.

O. Perron,
[1] Grundlagen fuer eine Theorie des Jacobische Kettenhruchalgorithmus, *Math. Ann.*, **64**, 1907, 1—76.

C. Pisot,
[1] La répartition modulo 1 et les nombres algebriques, *Ann. Sc. Norm. Sup. Pisa.*, **2**, 1938, 205—248.

K. Ramachandra,
[1] On the units of cyclotomic fields, *Acta Arith.*, **12**, 1966, 165—173.

G. N. Raney,

[ 1 ] Generalization of Fibonacci sequence to $n$ dimensions, *Canad. J. Math.*, **18**, 1966, 332—349.

R. D. Richtmyer,

[ 1 ] The evaluation of definite integrals and a quasi-Monte Carlo method based on the properties of algebraic numbers, *Report LA*-1342, *Los Alamos Sci. Lab.*, Los Alamos, N. M., 1951.

V. S. Rjabenkii,

[ 1 ] Tables and interpolation of a certain class of functions, *Dokl. Akad. Nauk SSSR*, **131**, 1960, 1025—1027.

[ 2 ] A way of obtaining difference schemes and the use of number theoretic nets for the solution of the Caucly problem by the method of finite differences, Trudy Mat. Inst. Steklov, **60**, 1961, 232—237.

K. F. Roth,

[ 1 ] On irreqularities distribution, *Mathematika*, 1, 1954, 73—79.

Yu. N. Sahov,

[ 1 ] On the approximate solution of Volterra equation of second type by the method of iteration, *Dokl. Akad. Nauk SSSR*, **128**, 1959, 1136—1139.

[ 2 ] On calculating the eigenvalues of a multi-dimensional symmetric kernel using number-theoretic nets, *Z, Vycisl, Mat. i Mat. Fiz.*, **3**, 1963, 988—997.

[ 3 ] On the approximate solution of multi-dimensional linear Volterra equation of second type by the method of interation, *Z. Vycisl. Mat. i Mat. Fiz.*, **4**, Suppl., 1964, 75—100.

[ 4 ] The calculation of integrals of increasing multiplicity, *Z. Vycisl. Mat. i Mat. Fiz.*, **5**, 1965, 911—916.

A. I. Saltykov,

[ 1 ] Tables for computing multiple integrals by the method of optimal coefficients, *Z. Vycisl. Mat. i Mat. Fiz.*, **3**, 1963, 181—186.

I. F. Sarygin,

[ 1 ] The use of number-theoretic methods of integration in the case of non-periodic functions, *Dokl. Akad. Nauk. SSSR*, **132**, 1960, 71—74.

[ 2 ] A lower estimate for the error of quadrature formulas for certain classes of functions, *Z. Vycisl. Mat. i Mat. Fiz.*, **3**, 1963, 370—376.

W. M. Schmidt,

[ 1 ] Metrical theorems on fractional parts of sequences, *Trans. Amer. Math. Soc.*; **110**, 1964, 493—518.

[ 2 ] Simultaneous approximation to algebraic numbers by rationals, *Acta Math.*, **125**, 1970, 189—201.

[ 3 ] Inregularities of distribution, VII, *Acta Arich.*, **21**, 1972, 45—50.

[ 4 ] Diophantine approximation, Lec. Not. in Math. Springer Verlag, 785, 1980.

Shih Shu Chung,

[ 1 ] Une generalization des «bons trellis», *C. R. Acad. Sc. Paris*, 290, 1980, 527—530.

W. Sinnott,

[ 1 ] On the stickelberger ideal and the circular units of a cyclotomic field (to appear).

S. A. Smoljak,

[ 1 ] Interpolation and quadrature formulas for the classes $W_s^a$ and $E_s^a$, *Dokl. Akad. Nauk SSSR*, **131**, 1960, 1028—1031.

I. M. Sobol,

[1] An exact estimate of the error in multidimensional quadrature formulas for functions of the classes $\widetilde{W}_1$ and $\widetilde{H}_1$, *Z. Vycisl. Mat. i Mat. Fiz.*, **1**, 1961, 208—216.

V. M. Solodov,

[1] On the calculation of multiple integrals, *Dokl. Akad. Nauk SSSR*, **127**, 1959, 753—756.

[2] Integration over regions different from the unit cube, *Z. Vycisl. Mat. i Mat. Fiz.*, **8**, 1968, 1334—1341.

V. T. Stojancev,

[1] Solution of the Cauchy problem for a parabolic equation by a quasi-Monte-Carlo method, *Z. Vycisl. Mat. i Mat. Fiz.*, **13**, 1973, 1153—1160.

A. H. Stroud,

[1] Approximate calculation of multiple integrals, Prentice-Hall, 1971.

T. Van Aardenne Ehrenfest,

[1] On the imposibility of a just distribution, *Indag. Math.*, **11**, 1949, 264—269.

J. G. Van der Corput,

[1] Verteilungs funktionen, I, *Proc. Akad. Amsterdam*, **38**, 1935, 813—821.

T. Vijayaraghavan,

[1] On the fractional parts of the powers of a number, II, *Proc. Cambridge Phil. Soc.*, **37**, 1941, 349—357.

I. M. Vinogradov,

[1] The method of trigonometrical sums in the theory of numbers, Fizmatgiz, Moscow, 1971.

V. S. Vladimirov,

[1] Equations of mathematical physics, Marcel Dekkel, New York, 1971.

Wang Yuan,

[1] A note on interpolation of a certain class of functions, *Sci. Sin.*, 6, **10**, 1960, 632—636.

[2] On numerical integration and its applications (number-theoretic method), *Shuxue Jinzhan*, 1, **5**, 1962, 1—44.

[3] Remarks on the interpolation of a certain class of functions, *Sci. Sin.*, 4, **14**, 1965, 629—631.

[4] On interpolation of a certain class of functions, *Kexue Tongbao*, 9, 1966, 387—389.

[5] On Diophantine approxination and approximate analysis (to appear)

Wang Yuan and Fang Kai Tai,

[1] A note on uniform distribution and experimental design, *Kexue Tongbao* (to appear).

Wang Yuan, Xu Guang Shan and Zhang Rong Xiao,

[1] On number-theoretic method of numerical integration in multi-dimensional space, I, *Acta Math. Appl. Sin.*, 2, 1978, 106—114, II (to appear).

Wang Yuan, Zhu Yao Cheng and Jian Yun Cui,

[1] Several remarks on number-theoretic method in numerical analysis, *Journal of the Univ. of sci. and tech. of China*, 1965, 213—218.

A. Weil,

[1] On some exponential sums, *Proc. Nat. Acad. Sci.*, USA, **34**, 1948, 204—207.

H. Weyl,

[1] Über die Gleichverteilung der Zahlen mod. Eins, *Math. Ann.*, **77**, 1916, 313—352.

Xie Ting Fan and Pei Ding Yi,

[1] On the irreducibility of polynomials, *Kexue Tongbao*, 9, 1975, 414—415.

Xu Guang Shan,

[1] On the approximate solution of the Cauchy problem of parabolic equation, *Kexue Tongbao*, 8, 1975, 361—364.

S. K. Zaremba,

[1] Good lattice points, discrepancy and numerical integration, *Ann. Mat. Pure Appl.*, **73**, 1966, 293—317.

[2] La méthode des "bons treillis" pour le calcul numérique des intégrales multiples, Applications of number theory to numerical analysis (S. K. Zaremba, ed.), Academic Press, New York, 1972, 39—119.

[3] On Cartesian products of good lattices, *Math. Comp.*, **30**, 1976, 546—552.

Zhang Rong Xiao,

[1] On approximate calculation of multiple integrals (number-theoretic method), *Journal of the Univ. of sci. and tech. of China*, 1964, 76—87.

H. Hasse

# Number Theory

English Translation Edited and Prepared for Publication by H. G. Zimmer

1980. 49 figures. XVII, 638 pages
(Grundlehren der mathematischen Wissenschaften, Band 229)
ISBN 3-540-08275-1
Distribution rights for all countries except the Socialist Countries:
Springer-Verlag Berlin-Heidelberg-New York

**Contents:** The Foundations of Arithmetic in the Rational Number Field. - The Theory of Valued Fields. - The Foundations of Arithmethic in Algebraic Number Fields. - Index of Names. - Subject Index.

Hasse's classic work, originally published in 1949 and then in a second, thoroughly revised edition in 1962, is now available in English, revised once more. The main topic of the book is the foundations of number theory in algebraic number fields and algebraic function fields in one indeterminate. Hasse's approach derives from the works of Kronecker and Kummer, and of his own teacher Hensel, whose valuation theory plays a major role in the book. Traditionally this treatment stands in contrast to the ideal-theoretic approach historically associated with Dedekind, Hilbert, and Emmy Noether. The publication of this English edition coincides with the growing interest in the divisor-theoretic methods of Kronecker and Kummer, exemplified by the recent publication (also by Springer-Verlag) of Kummer's collected works. Hasse's book is still up-to-date and remains the only comprehensive presentation of this approach. Only the most important results are presented in Theorem-Proof style, and much space is devoted to giving insight into the structure of the subjects discussed by expounding on them from many points of view.

Springer-Verlag
Berlin
Heidelberg
New York